T0250143

Mobility Management and Quality-of-Service for Heterogeneous Networks

RIVER PUBLISHERS SERIES IN INFORMATION SCIENCE AND TECHNOLOGY

Volume 3

Consulting Series Editor

KWANG-CHENG CHEN
National Taiwan University
Taiwan

Other books in this series:

Volume 1
Traffic and Performance Engineering for Heterogeneous Networks
Demetres D. Kouvatsos
February 2009
ISBN 978-87-92329-16-5

Volume 2
Performance Modelling and Analysis of Heterogeneous Networks
Demetres D. Kouvatsos
February 2009
ISBN 978-87-92329-18-9

Mobility Management and Quality-of-Service for Heterogeneous Networks

Editor

Demetres D. Kouvatsos

*PERFORM – Networks & Performance Engineering
Research Unit, University of Bradford, U.K.*

River Publishers

Aalborg

ISBN 978-87-92329-20-2 (hardback)

Published, sold and distributed by:
River Publishers
P.O. Box 1657
Algade 42
9000 Aalborg
Denmark

Tel.: +45369953197
www.riverpublishers.com

To Diana

Table of Contents

Preface

Considerable attention is currently devoted towards quality-of-service issues and mobility challenges such as those arising from the integration of the next and future generation Internets over terrestrial digital TV, mobile user location management and multi-service heterogeneous networks. These problems have to be addressed and resolved before the establishment of a global and wide-scale integrated broadband network infrastructure for the efficient support of multimedia applications with QoS guarantees. Of particular importance is the wireless Internet with its many applications based on the convergence of heterogeneous wireless networks supported by internetworking and the evolution of wireless access and switching technologies.

Mobility Management and Quality-of-Service for Heterogeneous Networks presents recent advances in networks of diverse technology and the Internet reflecting state-of-the-art research achievements in mobility and access network management, performance enhancement, optimal admission control and methodologies for QoS. The book contains 19 extended/revised research papers, which have their roots in the series of the HET-NETs International Working Conferences on the 'Performance Modelling and Evaluation of Heterogeneous Networks'. These events were staged under the auspices of the EU Networks of Excellence (NoE) Euro-NGI and Euro-FGI and are associated with the NoE Work-packages WP.SEA.6.1 and WP.SEA.6.3, respectively.

The research papers are classified into five technical parts dealing with current research themes in Mobility Management, Optimal Admission Control, Performance Modelling Studies, Access Network Coverage and Quality of Service (QoS).

In *Part One* 'Mobility Management', Negru et al. focus on the mobility aspects of next generation IP networks technology, IPv6, over Digital Video Broadcasting (DVB) mechanisms and present mobility issues and challenges arising from the integration of IPv6 over the terrestrial digital TV standard, DVB-T. Stamoulakatos et al. propose a methodology that combines cellular signaling measurements with pattern recognition techniques, performed by a hidden Markov model and macroscopic traffic filtering supporting location based services for mobile users, subject to mobile terminal location. Martínez-Arrúe et al. devise a novel random-directional model, based on a generalised random walk model, for mobile user location management for personal communication services (PCS) networks. This model is used towards the study of the location management cost associated with the distance-based and movement-based policies as a function of the directional movement. Estrela et al. propose a comprehensive framework for the classification of IP existing mobility protocols and carries out performance comparisons via simulation studies using TIMIP/aMIP, a terminal's independent mobility architecture. Loukatos et al. describe an enhanced mobility mechanism in terms of a typical system and user oriented handover scenarios. A beyond 3G (B3G) platform is implemented for the study of mobility requirements and performance evaluation through experimentation. Pichon et al. develop a new execution environment framework involving a service provider and a mobility manager in order to react to mobility events. An adequate overlay network is constructed by adding a controlled proxy to match mobile network capacities and an application is described involving scalable video coding multimedia flows.

In *Part Two* 'Optimal Admission Control', Garcia-Roger et al. propose a new methodology and associated algorithms, based on the solution space concept, for the computation of the optimal configuration of the so-called Multiple Fractional Guard Channel (MFGC) admission control policy in multi-service mobile wireless networks. Pla et al. assess the impact of incorporating handover prediction information into the process of call admission control in mobile cellular networks and employ dynamic programming and reinforcement learning techniques to compute the optimal admission policy with performance gains.

In *Part Three* 'Performance Modelling Studies', De Vuyst et al. apply the generating functions approach and the spectral decomposition theorem to analyse the stop-and-wait Automatic Repeat Request (ARQ) protocol over a bursty and correlated transmission error-prone channel, modulated by a two-state Markov Chain. Wang et al. address the issue of differentiated QoS for the design and development of the next generation telecommunication networks and propose a new analytic model based on an efficient hybrid-scheduling scheme, which integrates Priority Queueing (PQ) and Weighted

Fair Queueing (WFQ) for QoS differentiation. De Turck and Wittevrongel apply the probability generating functions approach to obtain robust analytic expressions for the unfinished work and the packet delay in order to investigate the impact of the Go-Back-N ARQ protocol over a wireless channel with correlated error probabilities. Song et al. propose a new network topology and an open queueing network model (QNM) consisting of Erlang loss systems representing an integrated system of cellular and wireless LANs (WLANs) with handoff.

In *Part Four* 'Access Network Coverage', Xing et al. devise an optimized cell size, subject to motorway width and overlapping length between adjacent motor-pico cells. The frequency of evolved handoffs amongst remote antenna units (RAUs) is taken into account and the novel concept of time slot distributor (TSD) is introduced to avoid co-channel interferences and make the system cost-effective.

In *Part Five* 'Quality of Service (QoS)', Agharebparast and Leung propose a novel methodology, based on a min-plus system theory, to derive statistical delay bounds for efficient stochastic QoS evaluations and cross-layer design in wireless networks. De Vogeleer et al. address the users' demand for interactive multimedia services over the Internet with end-to-end QoS and employ the concept of overlay networks to design and develop the new route discovery protocol (RDP) as part of a framework towards efficient overlay QoS routing. Belzarena et al. focus on end-to-end delays in QoS enabled networks and propose a hierarchical per-node delay assignment method to distribute the delay requirement to 'homogeneous zones' that use different scheduling policies and to assign the distributed delay to each node within each zone. Moscholios and Logothetis review an extension of the Erlang Multirate Loss Model under a bandwidth reservation policy and derive approximate recursive formulae for the calculation of basic performance measures, subject to QoS guarantees at call level for each service class. Aspirot et al. focus on QoS estimation based on end-to-end active measurements of Internet performance and propose an extension of Ferraty et al work on functional nonparametric regression to include non-stationary Internet traffic. Finally, Bertrand and Texier develop, for a path computation element (PCE) based architecture, an efficient routing algorithm consisting of high flows with end-to-end QoS guarantees, based on the Ad-hoc Recursive Inter-domain Path Computation method (ARPC).

I would like to end this preface by expressing my thanks to the EU Networks of Excellence Euro-NGI and Euro-FGI for sponsoring in part the publication of this research book and to the members of HET-NETs

Advisory Boards and Programme Committees as well as the external referees worldwide for their invaluable and timely reviews. My thanks are also due to Professor Ramjee Prasad, Director of the Center for TeleInFrastruktur (CTIF), Aalborg University, Denmark, for his encouragement and valuable advice during the preparation of this book.

Demetres D. Kouvatsos

Participants in the Peer Review Process

Samuli Aalto
Ramon Agusti
Sohair Al-Hakeem
Eitan Altman
Jorge Andres
Vladimir Anisimov
Laura Aspirot
Salam Adli Assi
Tulin Atmaca
Zlatka Avramova
Frank Ball
Simonetta Balsamo
Ivano Bartoli
Alejandro Beccera
Monique Becker
Pablo Belzarena
Andre-Luc Beylot
Andreas Binzenhoefer
Jozsef Biro
Pavel Bocharov
Miklos Boda
Sem Borst
Nizar Bouabdallah
Richard Boucherie
Christos Bouras
Onno Boxma
Chris Blondia
Alexandre Brandwajn

George Bravos
Oliver Brun
Herwig Bruneel
Alberto Cabellos-Aparicio
Patrik Carlsson
Fernando Casadevall
Vicente Casares-Giner
Hind Castel
Llorenc Cerda
Eduardo Cerqueira
Mohamad Chaitou
Ram Chakka
Meng Chen
Stefan Chevul
Tom Coenen
Doru Constantinescu
Marco Conti
Laurie Cuthbert
Tadeusz Czachorski
Alexandre Delye de Clauzade
   de Mazieux
Luc Deneire
Koen De Turck
Danny De Vleeschauwer
Stijn De Vuyst
Felicita Di Giandomenico
Manuel Dinis
Tien Do

Jose Domenech-Benlloch
Rudra Dutta
Joerg Eberspaecher
Antonio Elizondo
Khaled Elsayed
Peder Emstad
David Erman
Melike Erol
Jose Oscar Fajardo Portillo
Fatima Ferreira
Markus Fiedler
Jean-Michel Fourneau
Rod Fretwell
Wilfried Gangsterer
Peixia Gao
Ana Garcia Armada
David Garcia-Roger
Georgios Gardikis
Vincent Gauthier
Alfonso Gazo
Xavier Gelabert Doran
Leonidas Georgiadis
Bart Gijsen
Jose Gil
Cajigas Gillermo
Jose Manuel Gimenez-Guzman
Stefano Giordano
Jose Gonzales
Ruben Gonzalez Benitez
Annie Gravey
Klaus Hackbarth
Slawomir Hanczewski
Guenter Haring
Peter Harrison
Hassan Hassan
Dan He
Gerard Hebuterne
Bjarne Helvik
Enrique Hernandez

Robert Hines
Helmut Hlavacs
Amine Houyou
Hanen Idoudi
Ilias Iliadis
Dragos Ilie
Paola Iovanna
Andrzej Jajszczyk
Lorand Jakab
Sztrik Janos
Robert Janowski
Terje Jensen
Laszlo Jereb
Mikael Johansson
Hector Julian-Bertomeu
Athanassios Kanatas
Tamas Karasz
Johan Karlsson
Stefan Koehler
Daniel Kofman
Vangellis Kollias
Huifang Kong
Kimon Kontovasilis
Rob Kooij
Goerge Kormentzas
Ivan Kotuliak
Harilaos Koumaras
Tasos Kourtis
Demetres Kouvatsos
Udo Krieger
Koenraad Laevens
Samer Lahoud
Jaakko Lahteenmaki
Juha Leppanen
Amaia Lesta
Hanoch Levy
Wei Li
Yue Li
Fotis Liotopoulos

Renato Lo Cigno
Michael Logothetis
Carlos Lopes
Johann Lopez
Andreas Maeder
Tom Maertens
Thomas Magedanz
Sireen Malik
Lefteris Mamatas
Michel Marot
Alberto Martin
Jim Martin
Simon Martin
Jose Martinez-Bauset
Ignacio Martinez Arrue
Martinecz Matyas
Lewis McKenzie
Madjid Merabti
Bernard Metzler
Geyong Min
Isi Mitrani
Nicholas Mitrou
Is-Haka Mkwawa
Hala Mokhtar
Sandor Molnar
Edmundo Monteiro
Ioannis Moscholios
Harry Mouchos
Luis Munoz
Maurizio Naldi
Victor Netes
Pal Nilsson
Simon Oechsner
Sema Oktug
Mohamed Ould-Khaoua
Antonio Pacheco
Michele Pagano
Zsolt Pandi
Panagiotis Papadimitriou

Stylianos Papanastasiou
Nihal Pekergin
Izaskun Pellejero
Roger Peplow
Paulo Pereira
Gonzalo Perera
Jordi Perez-Romero
Rubem Perreira
Guido Petit
Maciej Piechowiak
Michal Pioro
Jonathan Pitts
Vicent Pla
Nineta Polemi
Daniel Popa
Adrian Popescu
Dimitris Primpas
David Remondo-Bueno
David Rincon
Roberto Sabella
Francisco Salguero
Sebastia Sallent
Werner Sandmann
Ana Sanjuan
Lambros Sarakis
Wolfgang Schott
Raffaello Secchi
Maria Simon
Swati Sinha Deb
Charalabos Skianis
Amaro Sousa
Dirk Staehle
Maciej Stasiak
Panagiotis Stathopoulos
Bart Steyaert
Zhili Sun
Kannan Sundaramoorthy
Riikka Susitaival
Janos Sztrik

Yutaka Takahashi
Sotiris Tantos
Leandros Tassiulas
Luca Tavanti
Silvia Terrasa
Geraldine Texier
David Thornley
Florence Touvet
Phuoc Tran-Gia
Chia-Sheng Tsai
Thanasis Tsokanos
Krzysztof Tworus
Rui Valadas
Rob Van der Mei
Vassilios Vassilakis
Vasos Vassiliou
Sandrine Vaton
Tereza Vazao
Speros Velentzas

Dominique Verchere
Pablo Vidales
Nguyen Viet Hung
Manolo Villen-Altamirano
Bart Vinck
Jorma Virtamo
Kostas Vlahodimitropoulos
Joris Walraevens
Xin Gang Wang
Wemke Weij
Sabine Wittevrongel
Mehti Witwit
Michael Woodward
George Xilouris
Mohammad Yaghmaee
Bo Zhou
Stefan Zoels
Piotr Zwiezykowski

PART ONE
MOBILITY MANAGEMENT

PART ONE

1

IPv6 over DVB-T: Mobility Issues, Challenges, and Enhancements

Daniel Negru[1], Yassine Hadjadj-Aoul[2], Ahmed Mehaoua[3] and
Anastasios Kourtis[4]

[1]*LaBRI Laboratory, CNRS, University of Bordeaux, 351 Cours de la Libération,
Talence, France; e-mail: daniel.negru@labri.fr*
[2]*School of Computer Science & Informatics, University College Dublin, Belfield,
Dublin 4, Ireland; e-mail: hadjadj@ieee.org*
[3]*CRIP5 Laboratory, Faculty of Mathematics and Computer Science, University
Paris Descartes, 45 Rue des Saints Peres, 75006 Paris, France;
e-mail: ahmed.mehaoua@math-info.univ-paris5.fr*
[4]*Institute of Informatics and Telecommunications NCSR "DEMOKRITOS", Agia
Paraskevi Attikis, 15310 Athens, Greece; e-mail: kourtis@iit.demokritos.gr*

Abstract

Next generation networks will consist of the interoperation of digital video
broadcasting and Internet protocols. Internet protocols standards are IPv4
and IPv6 and concerning broadcasting, DVB is expected to be the leading
technology. As a matter of fact, integrating Internet over DVB is not a simple
task, due to the many differences of those two in the way of handling particular fields, such as flows' directionality, mobility, and quality of service.
There exist several kinds of DVB technologies, DVB-S for Satellite, DVB-
C for Cable, DVB-H for handheld, DVB-T for terrestrial, each one with its
own particular specifications. This paper focuses on the mobility aspects of
IPv6 over DVB-T mechanisms; it presents the mobility issues and challenges
arising from the integration of next generation Internet, IPv6, over the coming
terrestrial digital television standard, DVB-T. It presents mobility scenarios in a special DVB-T environment, along with a complete broadcasting

*D. D. Kouvatsos (ed.), Mobility Management and Quality-of-Service for Hetero-
geneous Networks, 3–31.*

architecture, into which all leading mobile technologies coexist, and provides solutions and enhancements for a better technology interoperation. This work has been partially accomplished within the context of the IST-funded European project ATHENA.

Keywords: Internet protocol, digital video broadcasting, mobility, multimedia services.

1.1 Introduction

Interoperation of digital video broadcasting standards and Internet protocols based services represents one main evolution in next generation of new services for digital television. Services range from normal web surfing to many other multimedia applications, like video streaming, distant learning, videoconference and TV-specific services.

Concerning broadcasting standards, Digital Video Broadcasting (DVB) [1] is expected to be prominent television broadcast standard for next decades, as well through a satellite-based technology (DVB-S), as in terrestrial television (DVB-T), or through cable (DVB-C) and handheld devices (DVB-H). The evolution in satellite digital broadcasting has lead to a wide variety of Internet and multimedia applications, even in the few first years of service launch. Even though being member of the same protocol family, terrestrial digital television standard (DVB-T) [2] sets quite different constraints on performance than satellite digital television (DVB-S).

Internet protocols standards are IPv4 and IPv6, respectively the former and next generation of prominent networks protocols. IPv6 [3], with its enormous number of addresses and other new features, is undoubtedly more adequate to multimedia applications. Therefore, integration of DVB and IPv6 seems to be an interesting objective to conceive for next generation services of digital television.

Additionally, the growth in wireless networks predicts an increasing role of wireless communication techniques in the future and consequently, mobility is taking an important role in next generation services. DVB-T provides a relatively high bandwidth data channel but it is only unidirectional. Mobile multimedia terminals require also a return channel through different techniques, such as WLAN or UMTS. Future evolution on portability and mobility has also effects on use and performance of Internet protocols. Mobile IP in IPv4, but especially in IPv6, offers solution to handle mobility at a network protocol layer.

In this paper, we present the mobility issues and challenges on the integration of the new version of the Internet, IPv6, and the terrestrial television broadcasting standard, DVB-T. A complete inter-working architecture is presented, for which any mobility scenario is depicted, from horizontal handovers to vertical ones, from one designed technology to another, featuring wireless IP based technologies (WLAN IEEE 802.11) and cellular ones (3GPP UMTS). Novel mobile solutions are developed and enhancements are suggested in order to optimize such mechanisms. Also, the European project ATHENA through which the validation and integration of the approaches are being achieved is outlined.

1.2 Context and Motivations

1.2.1 Importance of Mobility in DVB-T

Mobility support in DVB-T represents an important parameter for the deployment of this technology. In countries where a broadband wired backbone has already been set up, there is not really a will to promote DVB-T. On the contrary, if the support of mobility by DVB-T is to be achieved then this technology will have an important added value compared to xDSL. With the combination between mobile technologies such as WLAN and UMTS, people will be able to receive their broadcasted favorite programs or an Internet access wherever they are (inside a Broadcasting Area settled by a DVB-T transmitter) and no matter how they move. Above all, this feature is not addressable by the wired xDSL technologies.

1.2.2 Next Generation IP Networks: IPv6

IPv6 [3] is the new version of IP which is designed to be an evolutionary step from IPv4. There are numerous reasons why IPv6 could be appropriate to the next generation of networks. It solves the Internet scaling problem, provides a flexible transition mechanism for the current Internet, and was designed to meet the needs of new markets such as nomadic personal computing devices, networked entertainment, and device control. It provides these features in an evolutionary way, which reduces the risk of architectural problems.

The advantages of incorporating IPv6 support in an interactive DVB-T network are mainly focused on: (i) Increased security – mandatory in a broadcast network, where the downlink data, for all users, are available to everyone, (ii) Mobility support, which can enhance the DVB-T interactive services to mobile users, (iii) QoS classification, which is of increased importance as the downlink data are not provided via multiple narrowband channels, but all users are simultaneously served by a multi-megabit broadband trunk.

1.2.3 Internet over DVB-T

DVB [1] is designed to be a broadcast transmission system over Terrestrial, Satellite or Cable link. On one hand, this technology leads to a very high degree of efficiency regarding transmission robustness, simplicity and spectral efficiency in the broadcast scenario. On the other hand, it does not provide any radio resource management facilities: it has no power control functionality and cannot dynamically assign channels to individual receivers. Since this radio system is by design unidirectional, resource negotiation and any form of QoS control need to be performed via a second, uplink-capable radio system.

The DVB-T [2] system has first been designed to provide terrestrial broadcasting of MPEG-2 coded TV signals. But with the emergence of Internet multimedia applications, the integration of digital television standards and Internet protocols based services became an evolutionary step. Transporting IP packets over DVB can be done in several ways depending on applications, bandwidth and reliability needs.

Multi-Protocol Encapsulation (MPE) is the recommended way of transporting IP datagrams in DVB transport streams [4]. Through this method, encapsulation is done as DSM-CC sections like several MPEG and DVB tables. Transport scheme limits maximum section size to 4 KB (4096 Bytes). IP encapsulation in MPE packets is the most widespread method for transporting datagrams, it is well standardized and supported among manufacturers. Also, MPE is well-suited to IPv6, since some IP bindings to Multi-Protocol Encapsulation rely on ATSC (US) standards of transporting IP over MPEG and these limit Maximum Transfer Unit (MTU) to size 1008 B. This would typically cause IP fragmentation in IPv4, but would prevent use in IPv6, because IPv6 assumes that all links can transfer at least 1280 B datagrams.

Ultra-Lightweight Encapsulation [5] (ULE) is a standard for efficient encapsulation of IPv4/IPv6 datagrams into MPEG-2 Transport Streams. It has been proposed by the IETF IP over DVB Working Group for optimizing the encapsulation process. ULE is targeted to get 10% efficiency regarding to MPE and is supposed to replace it. The ULE concept is based on Protocol Data Units (PDUs); they represent packets coming from any kinds of networks (IP datagrams, Ethernet, LLC/SNAP, etc.). Those are sent to the encapsulator for the header and the CRC trailer. Then, they are inserted in

SubNetwork Data Units (SNDUs), which are mapped in MPEG-2 packets payload.

1.2.4 Mobility Management in Heterogeneous Networks

The mobility management in multi-networked environments has mainly been driven by the IPv6 protocol as the central unifying layer.

The European research project OverDRiVE [6] (Spectrum Efficient Uni- and Multicast Over Dynamic Radio Networks in Vehicular Environments) is a research and technology development project, which aims at UMTS enhancements and coordination of existing radio networks into a hybrid network to ensure spectrum efficient provision of mobile multimedia services. An IPv6 based architecture enables inter-working of cellular and broadcast networks in a common frequency range with Dynamic Spectrum Allocation (DSA). The project's objective is to enable and demonstrate the delivery of spectrum efficient multi- and unicast services to vehicles. Due to the heterogeneous world of access systems, the connection of a vehicular system to Internet will not be limited to a single access system. Instead, several different communication networks will be used, having different characteristics and providing different services. For example a vehicle might be connected to cellular telephone networks (GSM, UMTS) for voice communications, WLAN for local hot spot access, DVB-T for multicast distribution of multimedia streams. OverDRIVE addresses these issues and proposes solutions focussing essentially on IPv6 mobility. In [7], the OverDRIVE consortium presents the analysis and comparison of approaches based on standard MIPv6 [8], HMIPv6 [9], and Prefix Scope Binding Updates [10]. They strongly emphasize on network mobility but they do not take into consideration the diversity of access networks, especially DVB-T.

DVB-T, although primarily developed for fixed reception with roof-top directive antenna and portable reception, may also be used for mobile TV services. Investigations and trials [11] on this domain have already confirmed the feasibility of mobile DVB-T services provided a robust modulation scheme as well as a suitable receiving system is adopted. Studies have been elaborated for using DVB-T standard to deliver broadcast services to mobile receivers. They have shown that the 8k mode is only viable with slowly moving terminals. For the others, a 2k mode is required. Referenced studies indicate that a 2k mode using 16-QAM and a code rate of 1/2 or 2/3 provide sufficient error performance up to high speeds. It has to be noted that both studies were looking at mobile reception of MPEG-2 coded video streams (TV program-

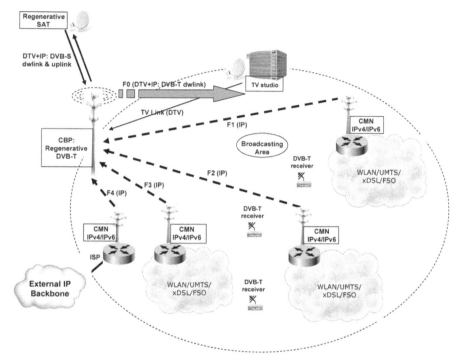

Figure 1.1 Overall IP/DVB-T network configuration.

ming). This type of transmission imposes very strict quality requirements: since transmission is real-time, errors, which are not recoverable by channel coding, cannot be corrected through retransmission.

In a context where the radio spectrum is extremely busy and where the availability of a UHF channel is so complicated and expensive, transmitting only few programs, even if receivable in mobile reception is definitely not acceptable for broadcasters. To overcome this problem, taking into account that for mobile reception, small LCDs are usually adopted (for example mounted in the cars head-rests), the encoding process can be done at low bit-rate (i.e. less than 500 kbps, in CIF format 352×288), and IP encapsulated over DVB Transport Stream, using the Multi-Protocol Encapsulation (MPE), and transmitted on DVB-T channel. Hence, there exists the possibility of using DVB-T for the delivery of IP-based multimedia services to mobile terminals. Such experiments have been held in [11] and resulted in a better service with IP over DVB-T mechanism.

1.3 Overall Configuration of the DVB-T Environment

The environment herein consists of an infrastructure which uses regenerative DVB-T streams for the interconnection of distribution nodes, enabling access to IP services and digital TV programs in wide areas such as big cities. This is certainly the most interesting case for DVB-T features. Such a configuration enables multi-service capability, as regenerative DVB-T creates a single access network physical infrastructure, shared by multiple services (i.e. TV programs, interactive multimedia services, Internet applications, etc.). In this approach, the DVB-T stream is used in a backbone topology and thus creates a flexible and powerful IP broadband infrastructure, thus permitting broadband access and interconnection of all local networks. Figure 1.1 shows an overall representation of such an environment. The Broadcasting Area is provided with regenerative DVB-T stream by the Central Broadcasting Point (CBP). Cell Main Nodes (CMNs) enable a number of simple users (geographically neighbouring the CMN) to access IP services hosted by the network. Each CMN constitutes the 'physical interface' to the common Ethernet backbone of users/citizens of a local network (i.e. IEEE 802.11b/g), customers of a mobile network operator making use of 3G and B3G technology (i.e. UMTS), individual users and service providers. In such configuration, both reverse and forward IP data traffic are encapsulated into the common DVB-T stream, thus improving the flexibility and performance of the network.

The IP data stemming from the CMNs, consisting of either requests/acknowledgements or of useful data, are forwarded to the CBP to be included in the common broadcast downlink. This traffic will be conveyed via unidirectional point-to-point wireless links, acting as return channel trunks. The technology adopted for the implementation of the return channel can be any point-to-point wireless data transmission technique, without any need for additional link-level procedures, like multiple access schemes or error resilience via retransmissions.

The important components of such an architecture are the IPv6 to MPEG-2 encapsulator, which is located at the CBP level (and de-encapsulator at CMN level) and the IP routing module. Further information on modules description of all the components can be found in [12, 13].

No component dealing with mobility is present at the CBP side, since it is just a broadcasting point. They are all located at CMNs level as described in the next section.

As the predominant mobile technologies nowadays are UMTS and WLAN, the architecture and the scenarios proposed are based on them.

1.4 Mobility in DVB-T Environment

In order to support all the mobility features of UMTS, WLAN and DVB-T technologies, we present all the possible mobility issues and scenarios, as well as solutions, in such an environment. Since the intention is to deploy a fully mobile DVB-T access network, all the eventual features are being outlined, from layers of mobility to specific handovers.

1.4.1 Mobility Problems Statement

In this study, we try to incorporate the widest range of scenarios on mobile terminals in a DVB-T environment. There exist different types of mobile terminals; we will distinguish them through their reception technology:

- Terminals with a DVB-T reception antenna, which can be divided in:
  - Terminals that can receive the MPEG-2 programs directly thanks to their DVB-T reception device, such as set-top boxes;
  - Terminals that can receive IP audio/video streams, such as personal computers equipped with a DVB-T board;
- Terminals with a UMTS interface, such as mobile phones;
- Terminals with a WLAN interface, such as laptops and PDAs.

Of course, a mobile terminal can as well be equipped with several reception devices, e.g. a UMTS and a WLAN (IEEE 802.11b/g) interface or other combinations. A complete mobile terminal could be a car equipped with a DVB-T antenna and a UMTS interface, in which several laptops or PDAs are linked through a WLAN network.

The mobility aspects underneath represent the interesting points in this work. Mechanisms are studied for seamless reception of IPv6 data, when transition from one UHF channel (DVB-T stream) to another is required, or when terminals move from one point to another, experiencing handovers. Mobility is treated at two levels: layer-2, or layer-3 mobility, at the link layer or at the network layer.

Mobility scenarios can be exposed through different ways. This can be made according to the layer they depend, either layer-2 or layer-3, whether a change at the link level occurs or at the network level. This interferes directly to the mechanisms to use and opens issues for proposals.

At network layer, the IP routing module of each CMN is in charge of all the mobility aspects. Therefore, this module has to deal not only with routing but also with mobility actions.

1.4.2 Mobility Scenarios

Mobile terminals of different technologies would be present in the Broadcasting Area (BA). Several mobility scenarios can happen. Horizontal handovers can occur, a mobile terminal switching from a BA to another or, inside a BA, from a WLAN to another. Also, there could be vertical handovers, when a mobile terminal switches from a DVB-T stream reception to a UMTS one or from its WLAN interface to its UMTS one. In each case, seamless mobility must be achieved.

1.4.2.1 Horizontal Handovers

Horizontal handovers happen in several cases and mobility scenarios arising can be presented through these three: UMTS mobility, WLAN mobility, and DVB-T mobility.

From one cellular area to another: UMTS mobility   A mobility aspect with a horizontal handover could be performed when a mobile user is switching from a cellular area to another, still inside the Broadcasting Area. Mobile users equipped with a UMTS interface may move across cellular areas, as shown in Figure 1.2. Two sub-scenarios can happen then: (i) a layer-2 handover between base-stations; (ii) a layer-3 handover between GGSN which will be performed in very rare cases.

The access technology handles all the layer-2 mobility and the IP network layer is unaware of changes in the point of attachment. Current 3G networks, like General Packet Radio Service (GPRS) and Universal Mobile Telecommunications System (UMTS) provide an IP-mobility solution that is access technology-specific. Such a model is applicable when the mobile device is within the scope of GPRS/UMTS networks.

Another important point resides in the fact that the release 00 of UMTS specifies IPv6-exclusively support, especially for IP Multimedia Service (IMS). It is solely designed to use IPv6 and no more IPv4. Also, based on release 99 specifications, which describes an IPv6 client support for native IP mobility and multicast, this release provides enhancements to Mobile IPv6 relative items to be considered for layer-3 handovers.

In conclusion, for UMTS mobility scenarios, solutions that already exist do not need any change in order to make them applicable in this environment.

From one WLAN to another: WLAN mobility   Another case of mobility is performed when a mobile user is switching from a WLAN to another, in-

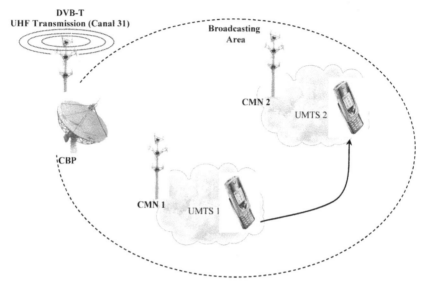

Figure 1.2 UMTS mobility scenario.

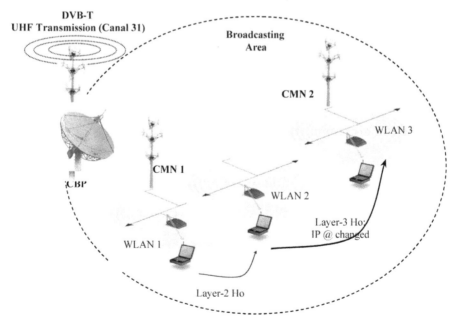

Figure 1.3 WLAN mobility scenarios.

side the Broadcasting Area. Mobile users, equipped with an IEEE 802.11b/g card, may move across WLANs, as shown in Figure 1.3. As for UMTS mobility, two sub-scenarios happen then: (i) a layer-2 handover between access points, covered by a CMN; (ii) a layer-3 handover, when switching from one CMN coverage area to another.

The case when the mobile device moves from an access point to another but keeps its IP address can be seen as WLAN mobility through switched networks. It essentially happens within a CMN coverage area. This case is undertaken by WLAN technology very efficiently and does not depend at all on the environment. On the contrary, when the mobile terminal switches from one CMN coverage area to another and performs a layer-3 handover, its IP address changes, it represents WLAN mobility through routed networks. In order to achieve seamless mobility in such scenarios, known approaches are not sufficient. Hence, a novel mechanism is proposed, based on Mobile IPv6. The IP routing module of the CMN the mobile terminal has just joined will permit packets destined to the terminal's former address to pass thanks to a binding procedure (as described in MIPv6 specifications [8]). Details on this proposed mechanism are explained in Section 1.4.3.3.

From one broadcasting area to another: DVB-T mobility A DVB-T receiver moving from one DVB-T area to another has to keep the service continuity. In a MFN (Multi Frequency Network), the DVB-T receiver has to tune a new frequency and, eventually, to get the new transport stream providing the same service when it performs a handover. This represents DVB-T mobility.

The issue of DVB-T mobility has to be taken into account, especially when the mobile user is moving from one Broadcasting Area to another (both utilizing the same infrastructure), or when switching from one UHF channel to another (within the same area) due to traffic loaded DVB-T streams. This scenario is depicted in Figure 1.4. In this respect, a traffic policy mechanism is required, which, in collaboration with the bandwidth management system, will route/direct the data traffic destined to a specific citizen, via the appropriate DVB-T stream (proper UHF channel within the same area or among different territories), in order to provide seamless access to the targeted services, besides enabling for any-time, any-where ubiquitous services distribution. This represents the proposed solution for mobility issues based on this scenario and it is fully explained in Section 1.4.3.1.

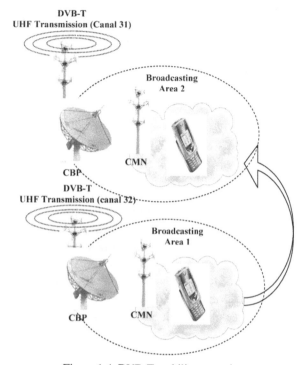

Figure 1.4 DVB-T mobility scenario.

1.4.2.2 Vertical Handovers

The cases when vertical handovers occur are much more complex and interesting. There are two principal scenarios of vertical handovers; in each case the mobile terminal has at least two possible receptive interfaces: (i) when the mobile terminal switches from UMTS to WLAN and back, and (ii) when it switches from DVB-T to UMTS.

Switching from UMTS to WLAN and vice versa   A mobile terminal equipped with two interfaces, a UMTS one and a WLAN one, can perform a vertical handover, switching from one technology to another. A roaming mobility aspect is then addressed.

As the mobile user moves across networks, the Mobile Terminal Device's (MTD) IP address may change. The mobile user, equipped with a WLAN network interface, initially moves within the Metropolitan CMN 1 accessing IP multimedia services, via the WLAN interface with address IP0, while the appropriate reply signals are forwarded to him via the regenerative DVB-T

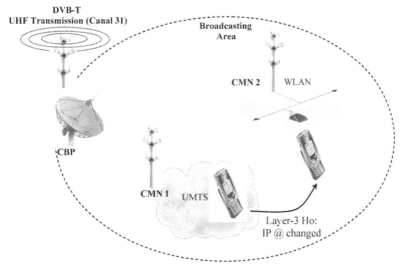

Figure 1.5 Vertical handover UMTS/WLAN mobility scenario.

stream. As this user moves out of the WLAN range, his MTD switches to the UMTS interface and assumes the – dynamically assigned by the UMTS (or GPRS) provider – address IP1. Next, he may move into the range of a second WLAN CMN (CMN2) or into an area served by another UMTS provider, assuming the address IP2. This scenario is depicted in Figure 1.5.

Knowledge of the IP address change is necessary in order to avoid inter-ruption of the DVB-T service. MIPv4 support is specified by UMTS release 99 (R99) to allow vertical handover towards UMTS. In this context, a vertical handover mechanism allows users to move from one external IP subnet (e.g. WLAN) to one UMTS network (considered as an IP subnet from the external PDN point of view) during an active session. Some strong driving factors for deploying IPv6 and MIPv6 on a wide-scale have been seen in the 3GPP community. Mobile IPv6 has been identified by 3GPP as a solution for pro-viding mobility control between wireless LAN and GPRS/UMTS networks to support 3G services including IP Multimedia Services (IMS). Supporting IPv6/MIPv6 in GPRS/UMTS networks has become an imminent and impor-tant issue to 3G community. So, Mobile IPv6 stands out as the solution for vertical handovers problems from WLAN to UMTS.

An IPv6 client support has been elaborated in UMTS Releases. The way of acquiring an IPv6 address for a mobile host visiting a UMTS network is done as follows. The most common one would be through a stateless auto-

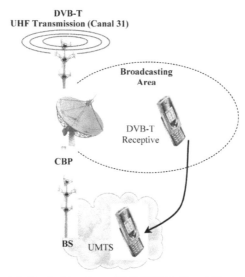

Figure 1.6 Vertical handover UMTS/DVB-T mobility scenario.

configuration method, where no external agent is used for address allocation. The Mobile Station (MS) sends a classical PDP context activation message and requests a dynamic IPv6 address allocation. In the response to the PDP activation, the MS creates a "link-local" address from the interface identifiers provided by the GGSN according to the RFC2373. After that the GGSN has sent a "Create PDP context response", it starts sending router advertisements periodically. These messages contain some parameters allowing the mobile to create a global or site-local unicast address. So, the MS uses an IPv6 prefix combined with the identifier, distributed by the GGSN, to create its IPv6 address. Afterwards, instead of the classical Mobile IPv6 operations, the specific MIPv6-based solution proposed in this paper (Section 1.4.3.3) will be undertaken, in order to achieve seamless mobility when a vertical handover from WLAN to UMTS is performed by the mobile station.

Switching from DVB-T to UMTS and back   This vertical handover scenario, when a mobile terminal switches from DVB-T to UMTS, will occur in a special case: when the mobile user moves out of a Broadcasting Area and access a no broadcasting coverage area, provided there is a cellular coverage, through a UMTS infrastructure.

In this particular case, the Mobile Station has two interfaces, a DVB-T reception antenna and a UMTS device; it is able to switch from one to the other. When accessing a BA, it naturally receives the streams through its DVB-T antenna. Meanwhile, if the streams are transmitted through a UMTS network, it may also receive them through its UMTS device (Figure 1.6). The switching from one to the other represents an important issue and needs the introduction of new concepts of services deployment, based on complementariness. Indeed, the same service available through UMTS and through DVB-T will not be deployed the same way according to the technology. The complementariness of the service and also of the coverage, which seems essential for the interoperation of DVB-T and UMTS, represents an important concept to deliver and is explained in-depth in Section 1.4.3.2.

1.4.3 Proposed Solutions

In this section, solutions are presented and explained thoroughly for mobility scenarios that do not have proper mechanisms working efficiently. First, concerning layer-2 handovers, approaches for DVB-T mobility and for UMTS/DVB-T vertical handover are exposed. Next, for all the scenarios in which the IP address needs change when the mobile terminal is moving, assuming a layer-3 handover, a Mobile-IPv6 based solution is brought out.

1.4.3.1 DVB-T Mobility

As discussed above in Section 1.4.2, the issue of DVB-T mobility has to be taken into account in two cases:

- When the mobile user is moving from one Broadcasting Area to another.
- When the mobile user is switching from one UHF channel to another within the same area.

The handover issue is depicted in Figure 1.4, where a mobile user, initially located in city 2, is accessing IP multimedia services via its UMTS device, while the appropriate reply signals are forwarded to him via the regenerative DVB-T stream. As this user moves towards a new BA, he leaves behind the first broadcasting area, and enters the broadcasting area of city 1.

For seamless reception and uninterrupted access to the provided services (i) the user must switch his DVB-T receiver device to the new UHF channel, and (ii) the core infrastructure must redirect the IP traffic (targeted to him) from regenerative DVB-T 2 to the regenerative DVB-T 1 platform (located in city 1). In this respect, a handover policy mechanism is required for enabling efficient redirection of the IP traffic and fast transition from one UHF channel to another.

Towards this, we propose the following mechanism: a Location Aided DVB-T Handover (LADH) policy mechanism, capable of providing DVB-T mobility and able to interconnect cities that make use of regenerative DVB-T platforms. This mechanism will monitor the geographical position of the mobile user, i.e. by making use of the location and direction data that will be available at the user's Mobile Terminal Device. The Location and Direction Information (LCDI) data will be provided via the UMTS uplink to the LADH module, where it will be processed and compared with Coverage Information (CI) and Road Topology (RT) data, related to the coverage area of each regenerative DVB-T platform stored in an appropriate GIS database. The LADH module will decide if traffic redirection is required or not. In case that traffic redirection is essential, i.e. mobile user is leaving city 2 and enters city 1, the LADH module will signal the Traffic Policy Mechanism (TPM) that the mobile user needs to switch to the new UHF channel. The TPM will accomplish this by providing the mobile user with the appropriate control data encapsulated in the DVB-T stream of the appropriate channel. At the same time, the TPM will reroute all IP traffic targeted to this user, from regenerative DVB-T 2 to regenerative DVB-T 1. As a result, the mobile user's receiver/demodulator will switch to the new UHF channel (upon reception of control data), and the regenerative DVB-T 1 will broadcast the IP services (targeted to this user) via its regenerative DVB-T stream.

This mechanism based on a LADH and with the use of a TPM represents the solution for achieving DVB-T mobility in the presented architecture, when a mobile terminal switches from one BA to another or from a UHF frequency to another within the BA and is depicted in Figure 1.7.

1.4.3.2 UMTS/DVB-T Vertical Handover Issue

When switching from UMTS technology to DVB-T and back, a novel concept has to be taken into account. Since those two technologies do not have the same characteristics at all and have not been created to be incorporated and to work together, we propose the concept of complementarities activities for delivered services. We consider a dedicated service is available via UMTS radio access and via DVB-T TV broadcasting. Complementariness is achieved through the DVB-T channel, which carries general information and through the UMTS delivery, which transports more detailed information. Complementariness is applicable at two levels: service and coverage.

Service Complementariness is when two or more links within a single or different access system are needed simultaneously to provide the service. The broadcast network carries general interest information, without going in

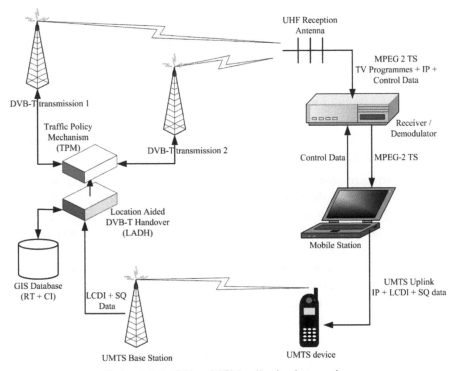

Figure 1.7 LADH and TPM policy implementation.

details (history of the town, significant photo shots, small video clips), real-time messages, Service Announcements and live content (documentary on some selected events). The information given by DVB-T is also specially targeted to crowded areas (cities) where it is convenient to send a basic set of information with the broadcast technology. The UMTS network provides detailed information (History of a certain palace, details on an event), the opportunity of active interaction (buying tickets for an event, booking hotels or restaurants) and for Service Subscription.

Coverage Complementariness is needed when the terminal is in an area not covered by DVB-T or UMTS. In areas without the DVB-T network, the UMTS network shall provide some (i.e. partial coverage complementary) of the information usually present in the broadcast network:

- General information (on request of the user).
- Optionally, a low bit-rate version of the video content (e.g. MPEG-4).

The transition, when the DVB-T network is not available, cannot be "transparent", because of the different cost of the service and the available bit-rate. The user shall be notified when there is a network switching. After the notification, the switching can be automatic or manual, and this feature has to be set during the initial configuration of the terminal. There are two options:

- After a switching notification, the terminal asks the user to explicitly allow the network switching.
- After a switching notification, the terminal automatically switches network and the service keeps running seamlessly.

3GPP is currently standardizing Multimedia Broadcast/Multicast Services (MBMS). The use of the multicast/broadcast extensions of the UMTS protocol will facilitate the transition, allowing the use of the same network (and upper) layer protocols. When the UMTS network is not available two alternatives exist: The user will not be able to access detailed information and interact. Links present in the broadcasted content will be disabled. Alternatively, other bi-directional networks will be used (e.g. the GSM/GPRS network for simple interactions and to request a small amount of detailed information). In this case, the terminal shall inform the user of the network swap.

When dealing with vertical handovers between DVB-T and UMTS, setting up service and coverage complementariness would permit to handle efficiently such an issue.

1.4.3.3 Efficient MIPv6 Based Solution for Layer-3 Handovers in DVB-T Environment

Layer-3 handovers, with IP address switching, occur in two special cases, as described above and depicted in Figure 1.3:

- When a mobile terminal connected via a IEEE 802.11b/g interface to an access point in a WLAN switches from one CMN to another; it then performs a horizontal handover between CMNs with WLAN technology;
- When a mobile terminal connected via a IEEE 802.11b/g interface to an access point in a WLAN switches to its UMTS interface in a cellular area (and vice versa) either inside the same CMN or between two CMNs; it then performs a vertical handover from WLAN to UMTS or vice versa;

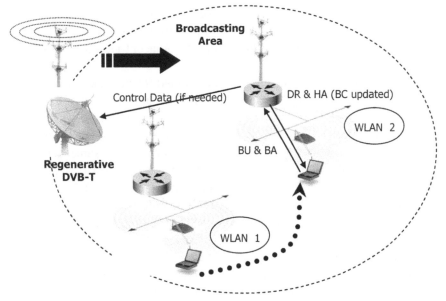

Figure 1.8 IP Mobility mechanism in DVB-T context.

In order to support the mobility aspects described above in the context of DVB-T networks, an IP layer solution has been taken into consideration.

Mechanism Description: The proposed solution is based on Mobile IPv6 approach. Mobile IP facilitates node movement by permitting a Mobile Node to communicate with other (stationary or mobile) nodes, after changing its point of attachment from one IP subnet to another, yet without changing the MN's address. A MN is always addressable by its HoA and receives a CoA each time it visits a different network. A HA in the Home Network of the MN is responsible for performing the appropriate re-routing process, according to its Binding Cache.

Herein, we have a Broadcasting architecture; therefore all the CMNs receive all the flows (TV+IP) transmitted by the Regenerative DVB-T CBP. After the de-multiplexing and de-encapsulation process of the IP over DVB-T streams, the routing module of each CMN decides either to redirect the packets to the terminals behind it or to destroy them according to their IP addresses.

The proposed mechanism for achieving seamless layer-3 handovers in a DVB-T environment consists of the following steps (see Figure 1.8):

1. First, the MN has a HoA and receives a CoA when accessing a new subnet, covered by a new CMN. It moves from WLAN 1 to WLAN 2 and acquires its new IPv6 address from the adjacent Designed Router of CMN2.
2. At that point, the MN will inform its Designed Router (the routing module of the adjacent CMN) that it has joined its subnet and that a binding must be established between its home address and its care-of-address.
3. A Binding Update message is sent to the Designed Router (DR) and the DR replies with a Binding Acknowledgement, the same it is done in Mobile IP. The adjacent DR is considered as a pseudo Home Agent.
4. The DR establishes a Binding Cache and since it receives all the IP traffic, it can retransmit the packets destined to this entry in the Binding Cache to the respective care-of-address, instead of dropping them.
5. Timers should be set and signalling messages between the DR and the mobile terminals that correspond to an entry in the Binding Cache need to be addressed. For example, every 2 seconds, a discovery message from the DR is sent to the care-of-addresses of the MNs present in the binding cache. If there is no response from the MNs, it means that the MN is not in the subnet anymore and consequently, the corresponding entry in the BC is deleted and the packets are not re-transmitted.
6. If mapping of PIDs and IP addresses is performed at the Regenerative DVB-T (CBP) side, control data is sent by the CMN in order to set dynamically a new appropriate IP address/PID mapping entry.

Multicast Support: Multicast is essential in such a multimedia-based environment since most of the relevant applications use this transmission technique. In order to be coherent to the former status of DVB environments, in which by definition broadcast is the leading technology, there is a need to deploy the corresponding delivery technique, by means of multicast.

In such environment, all the CMNs will receive all the multicast flows delivered in the network. No multicast routing process, as in native IP networks, needs to be present. Each CMN has to decide either to pass the multicast packets to the adjacent access network behind it or to block them, depending on the requests of the users. Thanks to IGMP or MLD messages, the users inform their corresponding CMN of their will to register to the desired multicast stream.

Concerning mobile nodes, a Mobile IP Remote Subscription process is thus established for multicast receivers, as shown in Figure 1.9. In order to receive packets sent to a given multicast group, a mobile receiver needs to

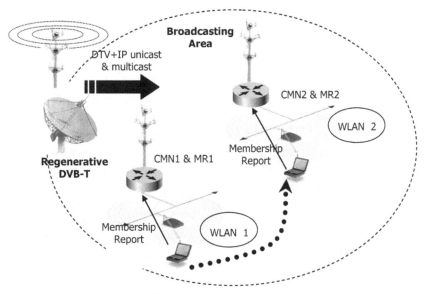

Figure 1.9 Mobile multicast support in DVB-T context.

first join that multicast group. With the remote subscription approach, the mobile receiver joins the multicast group via a local multicast router on the foreign network. To join a multicast group, an MN sends its membership report message to the local multicast router located on the visited network. The local multicast router intercepts this membership report message and joins the requested multicast group. Following this approach, the MN uses its CoA as the IP source address when sending its membership report message to the multicast router MR1. After handover to the foreign network 2, MN again sends a new membership report message to MR2 by using its new CoA. While the MR2 constructs a new multicast branch for the MN, the MR1 may stop re-transmitting the multicast flows to the access network behind it if it has no other receivers.

1.5 Large-Scale Evaluation and Demonstration

1.5.1 Athena IST Project

The EU-funded IST Project ATHENA (ATHENA – Digital Switchover: Developing Infrastructures for Broadband Access [12]), which started in January 2004, takes into consideration the mobility concepts defined in this paper, as well as the overall network architecture.

ATHENA proposes the use of the DVB-T in regenerative configurations and exploits the networking capabilities of the television stream for the creation of a powerful backbone that interconnects distribution nodes within a city. As these distribution nodes (local networks) make use of broadband access technologies (i.e. the "local loop", WLAN, LMDS, MMDS, Optical) they enable all citizens to have broadband access to the entire network and to be interconnected. Such a configuration enables for multi-service capability, as the regenerative DVB-T creates a single access network physical infrastructure, shared by multiple services (i.e. TV programs, interactive multimedia services, Internet applications, etc.). In such approach, the DVB stream is used in a backbone topology and thus creates a flexible and powerful IP broadband infrastructure.

Among its objectives, the ATHENA European project is conducting research activities in DVB-T system and mobility. One of the goals of this research project is to set the proposed architecture described in this paper and make feasible such scenarios of mobility, with the proper solutions and enhancements presented thereby. Therefore, the ATHENA project consists of a perfect support for integrating and developing at a large scale, mobility aspects, issues and proposed solutions.

1.5.2 The ATHENA Demonstrator

Within the framework of the IST project ATHENA, the deployment of an inter-working IP/DVB-T demonstrator has been setup at the premises of the Centre of Technological Research of Crete (CTRC), in Heraklion, Crete.

The networking configuration has been established in IPv4 and in IPv6, through the addressing and routing processes, as well as end-user autoconfiguration mechanisms. An IPv6-to-MPEG2 encapsulator (the AMBER equipment) has been implemented at the Regenerative DVB-T side and IP routing modules at the Cell Main Nodes. Figure 1.10 shows the IPv6 networking configuration of the ATHENA demonstrator with all the addressing and routing features.

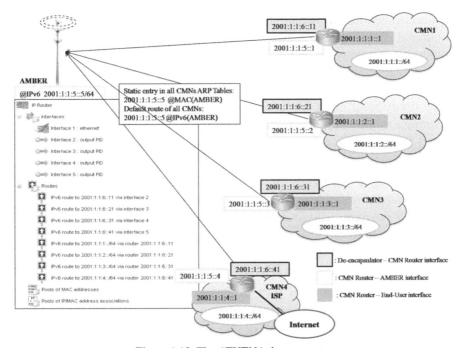

Figure 1.10 The ATHENA demonstrator.

The modules of the mobility solution for IP mobile users in a hybrid IP/DVB-T environment have been developed for IPv6 only. The use of IPv6 addresses and Router Advertisements for the end-users permits a successful integration of the modules inside the overall platform. The work being held so far includes the development of most of the points of the mechanism, all those that are located at CMN level. As well, the multicast support is efficient since no cooperation with the Regenerative DVB-T CBP is needed for this point. Only the MIPv6-based solution for layer-3 mobility has been implemented so far inside the demonstrator. The handling of DVB-T mobility and the provision of an UMTS access inside the area constitute a work in progress. Only laboratory simulations could have been performed for these cases.

This ATHENA demonstrator makes use of available software and hardware modules/tools for enabling passive custom user/citizens, apart from receiving digital TV programmes via a UHF channel, also and most predominant to access Internet, e-mail, VoD, AoD, IP-TV and IP-Radio services via the same UHF beam. In this context, and by making use of a regenerative configuration for the DVB-T stream, this UHF channel constitutes a broadband access neutral infrastructure that is commonly shared and exploited by all citizens (Broadcasters, telecom operators, active users, potential content providers, e-businessmen, etc.).

One Regenerative DVB-T Central Broadcasting Point has been developed inside the University of Heraklion, as well as four CMNs, two inside the University (CMN1 and CMN2) with WLAN access, one in the city center providing a minimal ISDN/PSTN access to citizen (CMN3), and the last one is used for providing an Internet access (CMN4). The possible services to access are:

- Three digital MPEG-2 TV programs (one of them is a satellite TV program retransmitted in real time);
- IP-TV programs (distributed by an active user/student) in IPv6;
- IP-Radio in IPv6;
- Common Internet (at the rate of 2Mbps);
- VoD and AoD services in IPv6.

All the IP users are attached to a CMN through a WLAN IEEE 802.11g hotspot and receive an IPv6 address automatically thanks to the Router Advertisement program in each CMN. This way, they have a direct access to IPv6 services. They may move from one CMN to another and experience layer-3 handovers. Further details on the demonstrator can be found in [14].

1.5.3 Measurements and Evaluations

The first measurements were performed in order to observe the delays induced by such a configuration, especially since these types of networks are based on a one-way communication. We first analyzed through simple messages (ping6) the difference in delays between a network composed of three classic IPv6 routers and one ADN (ATHENA Demonstrator Network) with four entities (the Regenerative DVB-T and the CMNs). For these measurements, the network was heavily loaded with significant multimedia streams and data requests. Figure 1.11 represents the results obtained.

We can see through the ping6 process that there is undoubtedly a delay induced by such networks. However, this delay is not significant; it is at most twice as long as the classical networks. Consequently, it would not affect the important multimedia services.

As stated earlier, the environment presented herein is mostly dedicated to multimedia applications. Therefore, we made the following measurements based on a multicast IP TV service. We chose two different videos in MPEG-

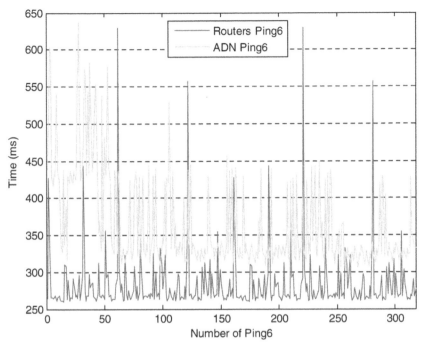

Figure 1.11 Delays induced by the IP/DVB-T environment.

4 format (at 700 and 1800 kbps, respectively) to be multicasted in IPv6 into the broadcasting area by two different active users' servers, located behind CMN4.

The testing scenario was established as follows. A mobile user was connected to CMN1 (through WLAN) and registered to receive the IP TV1 stream first. He then moved to the CMN2 coverage area and got connected to the local hotspot, experiencing a layer-3 handover between CMN1 and CMN2 (Ho1). Afterwards, he came back to his original coverage area, behind CMN1, experiencing another layer-3 handover (Ho2). We reiterated the experience with the IP TV2 stream, making the mobile user experience another layer-3 handover between CMN1 and CMN2 (Ho3).

We focused on delay and jitter, the two most important metrics for multimedia applications. Figure 1.12 represents the delay induced by the treatment process for each UDP media packet. The Y-axes represent the delays in seconds and the X-axes represent the time scale related to this communication session. The figure shows a graphical representation of the two video streams sent to the following IPv6 addresses: ff1e::e002:201 and ff1e::e002:202. The

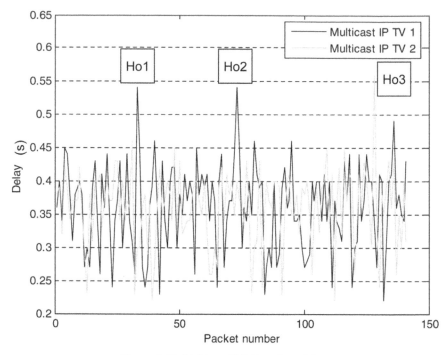

Figure 1.12 Multicast IP TV packets delays.

measurements obtained from our performance evaluation reveal relatively short delays and smoothness in the delays variation. This will particularly be in favor of a good perceived video quality at the client side. The three handovers (Ho1, Ho2 and Ho3) are represented on the delay graph. A longer delay is induced at those times due to the processing of our mechanism, which introduce the latency of the treatment of new exchange messages (BU, BA, and Membership Report) between the mobile user and the newly attached CMN. However, this increased delay is not significant and does not introduce complications towards the video quality at the receiver side.

Figure 1.13 is a graphical representation of the jitter in the same environment as described above. We observe that the delays variation (jitter) remains under an acceptable threshold, the maximum being 0.38 s, and the average around 0.07 s. Therefore, real-time services for which jitter is primordial, will not be affected. When handovers are processed, we also notice a more important jitter (Ho1, Ho2 and Ho3). However, the delay variation is not high enough to perturb the quality of the received streams. Our mechanism

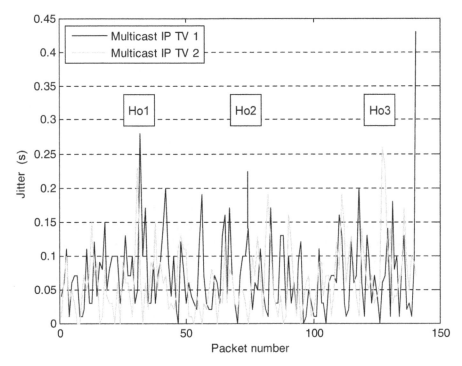

Figure 1.13 Multicast IP TV packets jitter.

performs well under these conditions. Further performance evaluations and analyses are available in [15].

1.6 Conclusions and Further Work

The interoperation of next generation IP networks technology, IPv6, and digital terrestrial television standard, DVB-T, is undoubtedly an important step to deploy for Next Generation Networks. Integrating IP packets into DVB-T streams can be realised through the Multi-Protocol Encapsulation (MPE) process or the Ultra-Lightweight Encapsulation (ULE) one.

In this paper, we propose an architecture that takes into consideration the use of regenerative DVB-T streams for the transport of both TV programs and IP flows at the same time and through the same frequency channel. It is based on a Central Broadcasting Point (CBP), which transmits and regenerates the streams and Cell Main Nodes (CMN), which group different wireless and fixed access networks to permit users to be interconnected and to receive services.

In order to make the above feasible and since wireless access networks are primordial nowadays, an important and most innovative issue is to handle mobility. The paper focuses on this part. It presents all the possible mobility scenarios in such a DVB-T environment, covering all the wireless access networks from WLAN to UMTS and depicting all the types of handovers, vertical and horizontal, from a technology to another, layer-2 or layer-3 based. We propose solutions and enhancements for achieving seamless mobility in each case. First, a Location Aided DVB-T Handover (LADH), along with a Traffic Policy Mechanism (TPM), is proposed for DVB-T mobility issues. Next, a concept of complementariness, for services and for coverage, is suggested for UMTS/DVB-T vertical handover scenarios. Finally, when layer-3 handovers occur, an efficient Mobile IPv6 based mechanism with multicast support is designed to undertake proper actions.

Inside the framework of the ATHENA project, we set-up and developed such an inter-working IP/DVB-T architecture through a powerful demonstrator. Furthermore, it has been conceived to support all IPv6 features and to evaluate in a real environment the efficiency of our solutions, besides enabling the transition from the analogue to digital TV and the provision of a powerful backbone to citizens. Future work will consist of exploiting this large-scale demonstrator for setting up enhanced mobility solutions, such as the UMTS/DVB-T possible handover, for improving the evaluations processes and for introducing the DVB-H complementary solution.

Acknowledgements

This work has been partially performed within the context of the European research project IST ATHENA (http://www.ist-athena.org). The authors would like to thank the participants for their contributions.

References

[1] ETSI: Digital Video Broadcasting (DVB); DVB specification for data broadcasting, European Standard EN 301 192.
[2] ETSI: Digital Video Broadcasting (DVB); Framing structure, channel coding and modulation for digital terrestrial television, European Standard EN 300 744.
[3] S. Deering and R. Hinden, Internet Protocol, Version 6 (IPv6) Specification, RFC 2460, December 1998.

[4] M. J. Montpetit, G. Fairhurst, H. D. Clausen, B. Collini-Nocker and H. Linder, A framework for transmission of IP datagrams over MPEG-2 Networks, RFC 4259, November 2005.

[5] G. Fairhurst and B. Collini-Nocker, Unidirectional Lightweight Encapsulation (ULE) for transmission of IP datagrams over an MPEG-2 Transport Stream (TS), RFC 4326, December 2005.

[6] R. Tönjes, K. Moessner, T. Lohmar and M. Wolf, OverDRiVE, Spectrum efficient multicast services to vehicles, in *Proceedings IST Mobile Telecommunications Summit 2002*, Thessaloniki, Greece, June 2002.

[7] M. Wolff, Evaluation of mobility management approaches for IPv6 based mobile car networks, KiVS 2003, Leipzig, 25–28 February, 2003.

[8] D. Johnson, C. Perkins and J. Arkko, Mobility support in IPv6, RFC 3775, June 2004.

[9] H. Soliman, C. Castelluccia, K. El Malki and L. Bellier, Hierarchical Mobile IPv6 Mobility Management (HMIPv6), RFC 4140, August 2005.

[10] V. Devarapalli, R. Wakikawa, A. Petrescu and P. Thubert, Network Mobility (NEMO) basic support protocol, RFC 3963, January 2005.

[11] A. Bertella, M. Rossini, P. Sunna and L. Vignaroli, Mobile DVB-T reception: Quality of streaming over IP of audiovisual services, IBC 2002 Conference Papers, September 2002.

[12] E. Pallis, C. Mantakas, G. Mastorakis, A. Kourtis and V. Zacharopoulos, Digitical switchover in UHF: The ATHENA concept for broadband access, *European Transactions on Telecommunications*, vol. 17, no. 2, March 2006.

[13] D. Negru, A. Mehaoua and E. Pallis, ATHENA: A large-scale testbed for the next generation of interopable networks and services, in *Proceedings 2nd International Conference on Testbeds and Research Infrastructure for the Development of Networks and Communities, TRIDENTCOM*, March 2006.

[14] ATHENA Deliverable D11.1, ATHENA demonstrator set-up, www.ist-athena.org.

[15] ATHENA Deliverable D12.1, ATHENA demonstrator evaluation, www.ist-athena.org.

2

Hidden Markov Modeling and Macroscopic Traffic Filtering Supporting Location Based Services

Theodore S. Stamoulakatos, Sofoklis Kyriazakos and
Efstathios D. Sykas

*Telecommunications Laboratory, Department of Electrical and Computer
Engineering, National Technical University of Athens, 9 Heroon Polytechniou
street, 15773, Athens, Greece; e-mail: {tstamoul, skyriazakos}@telecom.ntua.gr,
sykas@cs.ntua.gr*

Abstract

Location Based Services (LBS) is a new type of services for mobile phone
users based on mobile terminal (MT) location. A large number of service providers is developing LBS, however, each service has different requirements
on accuracy, response time, signaling overhead and number of subscribers
that can be localized at the same time. Therefore, the operators are trying to
make use of such position location technologies that can bring the best results,
also considering the cost. In this study it is presented a technique that combines pattern recognition techniques with cellular signaling measurements
and more precisely information extracted from Abis/Iub air interfaces in
GSM and UMTS networks respectively. The pattern recognition is performed
by Hidden Markov Model (HMM) which is trained with downlink prediction
data modeling the strength of the received signals for specific areas employing K-means (KM) as the clustering method. The accurate results from a
single probe vehicle show the potential of the method when applied to large
scale of MTs for vehicle load estimation in main city routes providing in that
way Traffic Information Service to mobile phone users. Another important

*D. D. Kouvatsos (ed.), Mobility Management and Quality-of-Service for Hetero-
geneous Networks, 33–53.*

issue is that this technique can be easily integrated in a cellular system and it also fulfils the requirements of a reliable localization technique.

Keywords: Location based services, traffic information, pattern recognition, hidden Markov model.

2.1 Introduction

One of the most powerful ways to personalize mobile services is based on location. Utilizing this information enables the user to experience value-added services and the cellular network provider to offer differentiation and incremental profitability by increasing its subscribers base [1]. As shown in the literature [2], LBS includes various types of services, increasing by that way the subscriber base available to operators, making LBS market growth to look promising.

This study has been focused on a particular class of LBS which is Traffic Information Service (TIS). A network architecture for supporting TIS applications has already been proposed [3]. Already various techniques for collection, reduction and reporting of travel time data [4] can support estimation of traffic density, traffic flow and speed. Among these techniques, one makes use of equipped vehicles as sensors or probe vehicles. In this study vehicles with MTs are converted into probe vehicles while trying to estimate traffic basic parameters. Traffic reporting is performed already by companies with either low accuracy levels [5] or not applicable to standard MTs, since they require the use of a GPS kit [6]. After the successful implementation of the proposed method in [7] while examining urban environments, the proposed method is expanded in order to cover also rural environments while trying to estimate vehicle volume estimation by utilizing already proposed traffic models. Key aspect in this study is that no hardware modifications in the cellular network are needed, making it possible for the cellular network provider to deploy LBS without any additional implementation cost.

This paper is organized in six section. In Section 2.2 the mathematical background of the proposed method is presented, which includes Hidden Markov Modeling, the examined Clustering techniques, as well as the definition of the RSSI vector. The examined road traffic model is analytically described in Section 2.3, while in Section 2.4 the vehicle volume estimation schema architecture is constructed step by step. In Section 2.5 the simulation of the proposed method for individual vehicle speed estimation and com-

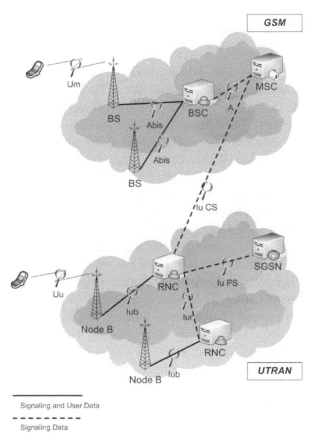

Figure 2.1 Air interfaces in GSM/UMTS networks.

ments on the results are presented. The paper closes with conclusions along with currently work in progress in Section 2.6.

2.2 Mathematical Background

Depending on the cellular network infrastructure, different air interfaces have been investigated like Abis interface of GSM network and Iub interface of UMTS networks (Figure 2.1). Based on RSSI measurements the "radio environment" seen by the mobile can be described. Since MT located inside cars are investigated, Rayleigh fading is also examined. Thus, the "environment" for a specific area can be imprinted and by this way a database of predefined "environments" can be created which will be later used for location and speed estimation of the MT.

In a GSM network, Abis interface is monitored. Abis interface is defined in the BSS (Base Subsystem Station) between the BTS (Base Transceiver Station) and the BSC (Base Station Controller). The data extracted from Abis interface include serving Cell ID, TA (Timing Advance) and NMR (Network Measurement Results). TA indicates the TOA between a base station (BS) and MT where NMR contains information about RSSI among others. Received Signal Strength Indicator (RSSI) is the measured wide-band received power within the relevant channel bandwidth. Measurement shall be performed on a GSM BCCH carrier, having as a reference point the antenna connector at the MT. Based on Received Signal Strength (RSS) measurements, the "radio environment" seen by the MT can be described.

Various techniques examining signaling load can be found in literature for efficiently calculating the position of MT in cellular networks. Due to ambiguity and signal degradation, the position estimation from the comparison process must be filtered. Thus, it is essential to use an effective filter for database comparison of measured "location sensitive" parameters. Two different techniques are suitable for such filtering; Hidden Markov Models filtering (HMM) [8] and Kalman filter (KF) [9]. Comparing HMM with KF, HMM parameters are fewer and easier to estimate, while HMM estimator seems to be less sensitive to errors during model parameters determination. Last but not least, HMM multi-hypothesis processing is much simplier. On the other hand in KF technique, wrong estimation of the model parameters, could lead the position estimation to extreme values. Thus HMM approach is more suitable for performing signal processing of a DB comparison process.

2.2.1 Hidden Markov Models

The fundamental assumption of an HMM is that the process to be modeled is governed by a finite number of states and that these states change once per time step in a random but statistically predictable way. To be more precise, the state at any given time depends only on the state at the previous time step. This is known as the Markovian assumption.

Hidden Markov Models (HMM) were used with success for speech recognition [8] as well as for other applications that are based on pattern comparison and detection. A discrete Markov process is characterized by a finite or non-numerable infinite number of states $S_1, S_2, S_3 \ldots, S_N$.

According to Rabiner and Juang [10], suppose T observation times exist. At each time $1 \leq t \leq T$, there is a discrete state variable q_t where $q_t \in \{S_1, S_2, S_3, \ldots, S_N\}$. The probability distribution function q_{t+1} depends only on the value of q_t since an HMM is examined. This is described as a state transition probability matrix A whose elements a_{ij} represent the probability that q_{t+1} equals j given that q_t equals i. Since it has been used for modeling the hidden Markov process, the states can not be observed. What can be observed is the data O_t that are generated according to a PDF which depends on the state at time t. PDF of O_t under state j is denoted as $b_j(O_t)$.

The formal constituents of the discrete HMM are the following:

1. N, the number of the states in the model.
2. M, the number of distinct observation symbols per state, i.e. the discrete alphabet size.
3. A, the state transition probability distribution matrix; defined as $A = \{a_{ij}\}$, where $a_{ij} = P(q_{t+1} = S_j \mid q_t = S_i)$.
4. B, the observation symbol probability distribution matrix; defined as $B = \{b_j(k)\}$, where $b_j(k) = P(v_k \text{ at } t \mid q_t = S_j)$.
5. π, the initial state distribution vector; defined as $\pi = \{\pi_i\}$, where $\pi_i = P(q_1 = S_i)$.
6. T, the length of the observation sequence.
7. $O = O_1, O_2, O_3, \ldots, O_T$, the observation sequence.

Given appropriate values of N, M, A, B and π the HMM can be used as a generator to give an observation sequence $O = O_1, O_2, O_3, \ldots, O_T$, where each observation O_t is one of the symbols and T is the number of observations in the sequence.

The symbols are denoted as $V = V_1, V_2, V_3, \ldots, V_M$ in the discrete case.

A complete description of the model could be given as $\lambda = \{A, B, \pi)$. For the vehicle speed estimation problem, given an observation sequence, it is possible to decide in which state the model is.

Having defined HMM parameters, the following problems can be solved according to Rabiner [8]:

1. Speed estimation where given an observation sequence $O = O_1, O_2, O_3, \ldots O_N$ and a model $\lambda = \{A, B, \pi)$, the probability $P(O \mid \lambda)$ that the sequence results from the model. This problem is referred to as the evaluation problem (problem formulation 1).
2. Training and optimization where given an observation sequence $O = O_1, O_2, O_3, \ldots, O_N$ and a model $\lambda = \{A, B, \pi)$, the optimal state sequence $O' = O'_1, O'_2, O'_3, \ldots, O'_N$ would result in training and op-

timization. This problem is referred as the decoding problem (problem formulation 2).

3. Deriving the models from prediction data where given an observation sequence $O = O_1, O_2, O_3, O_N$, maximize $P(O \mid \lambda)$ by adjusting the model parameters $\lambda = \{A, B, \pi\}$. This would result in derivation of the models from prediction data. This problem is referred to as the learning or estimation problem (problem formulation 3).

In order to find the single best state sequence $q = (q_1, q_2, q_3, \ldots, q_T)$ for the given observation sequence $O = O_1, O_2, O_3, \ldots, O_T$ Viterbi algorithm has been used. For a detailed description, see [11].

During the simulation of the proposed model, street segments that follow varying speed distribution have been modeled individually. A given pattern, i.e. observation sequence, belongs most likely to the model that yields the greatest value for $P(O \mid \lambda_i)$. The step of street modeling is one of the most basic steps in this technique.

2.2.2 Clustering Methods

Cluster analysis is a technique for grouping data and finding structures in data. In real applications there is very often no sharp boundary between clusters creating the need for developing various approaches to solve the problem. This study has been focused on the statistical based clustering techniques; KM and FCM clustering techniques.

2.2.2.1 K-Means (KM)

The KM minimization problem is a widely used method with a computational complexity of $O(n)$, where n is the number of data points to be clustered. A detailed description of KM method can be found in [12].

Although it can be proved that the procedure will always terminate, the KM method does not necessarily find the global optimal solution. It often terminates at a local optimum. The global optimum may be found using techniques such as deterministic annealing and genetic algorithms. KM is also significantly sensitive to the initial randomly selected cluster centers. KM can be run multiple times to lessen this affect. Techniques that improve the initial starting centers can be found in [13].

The advantage of KM method is its efficiency, having $O(tkn)$, where n is the number of objects, k is the number of clusters and t is the number of iterations with normally $k, t \leq n$. A weakness of KM method is that it is applicable only when mean is defined. Also it is needed to specify k, the number

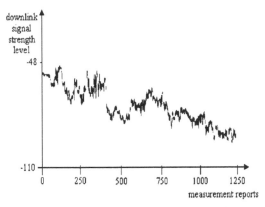

Figure 2.2 Progression of downlink level.

of clusters, in advance. Also KM is relatively sensitive to noise perturbation in the data as mentioned in [14]. These outliers can distort centroid positions and ruin the clustering). Last but not least, clusters are forced to have convex shapes.

2.2.2.2 Fuzzy C-Means (FCM)

KM partitions the data set of feature sequences into disjoint crisp, or hard, clusters. A feature sequence can be a member of one cluster only. Fuzzy clustering allows feature sequences to have membership of multiple clusters, each to varying degrees.

Consider a data set with two known clusters and a data point which is close to both clusters but also equidistant to them. Fuzzy clustering gracefully copes with such dilemmas, by assigning this data point equal but partial memberships to both clusters. A detailed definition of FCM can be found in [15]. FCM method performs better than KM at avoiding local minima but is not immune from the problem [15]. Hoppner et al. detail more specialized fuzzy clustering algorithms, some derived from the C-Means algorithm [16].

2.2.2.3 RSSI Vector

In GSM, MT registers RSS measurement reports every 480 ms. Within this interval the MT moves a distance that depends on the unknown speed. The radio level of the downlink burst that the MT measures is in the interval -110 and -48 dBm [17]. Thus, based on real radio maps of the examined area, they have been generated sequences of downlink signal strength level measurements of moving vehicles that travel having varying speed. In Fig-

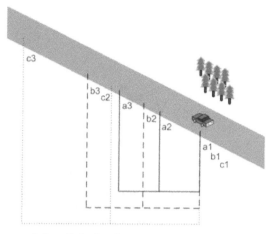

Figure 2.3 Sequences of downlink signal strength level measurements of MT moving at different speed.

ure 2.2 the progression of downlink signal strength level measurements of a MT travelling at 40 km/h is illustrated.

The number of RSSI values that a sequence of downlink signal strength level measurements contains during our modeling is 50 elements (total duration of 24 sec). In Figure 2.3, three (3) sequences of downlink signal strength level measurements of MT are illustrated. Sequence $a = \{a1, a2, a3\}$ corresponds to MT speed φ_1, sequence $b = \{b1, b2, b3\}$ corresponds to MT speed υ_2 while sequence $c = \{c1, c2, c3\}$ corresponds to MT speed υ_3, where $\upsilon_1 < \upsilon_2 < \upsilon_3$.

Suppose that there exists a sequence of downlink signal strength level measurements z of a MT moving at speed υ_z. During HMM modeling and after compared it with the sequences a, b, and c, suppose that the higher likelihood value is achieved after the comparison of the sequence z with the sequence b. Then, the MT is very likely to travel at speed $\upsilon_2 (\upsilon_z \approx \upsilon_2)$.

In the general case, the unique downlink signal strength level vector p^i, at position i, having dimensions $M \times 2$, is defined as follows:

$$p^i = \begin{bmatrix} \text{RSSI}_0^i & \text{cell}_0^k \\ \text{RSSI}_1^i & \text{cell}_1^k \\ \dots \\ \text{RSSI}_j^i & \text{cell}_j^k \end{bmatrix},$$

where RSSI^i_j is the received downlink signal power level at position i, cell^i_j is the jth BS. $o \leq j, k \leq 6$ for the serving BS and the six neighbouring BSs.

During the simulation of the proposed Vehicle Volume Estimation Schema, we process RSS measurements only from the serving BS. The downlink signal strength level vector p^i, at position i, is defined as follows:

$$p^i = [\text{RSSI}^i_0 \quad \text{cell}^i_0].$$

Sequence $S = \{31, 32, 34, 31, 35, 35, 36, 37, 39, 38, 33, 34, 33\}$ becomes S_1 in case the MT listens to only one BS, to BS #8, during the entire examination period:

$$S_1 = \{(31\ 8), (32\ 8), (34\ 8), (31\ 8), (35\ 8), (35\ 8), (36\ 8), (37\ 8), (39\ 8),$$

$$(38\ 8), (3\ 8), (33\ 8), (34\ 8), (33\ 8)\}.$$

2.3 Road Traffic Modeling

During the simulation of the proposed Vehicle Volume Estimation Schema, a traffic model has been employed. This traffic model has been based on an expanded macroscopic model capable to describe microscopic activity. This hybrid approach has as an objective the macroscopic simulation of a large network with the availability to focus at a microscopic level on some parts of the network, where information is available about MT movement (LA boundaries). First though, it is needed to prove that the existing sample is adequate to represent vehicles volume for the examined road.

2.3.1 Characterizing Existing Vehicles Sample

A vehicle can be characterized as probe if the MT that is located inside the vehicle is in a certain state. In GSM networks, MTs can be found in two states; idle or active whether in UTRAN networks, MTs support two modes; idle and connected. In idle state, the MT does not communicate with the network, but it listens to the information sent by the system on common radio operator channels. In active/connected state, the MT is in communication with the infrastructure of the network, and a great number of information is exchanged between the MT and the system. In order RSSI measurements from MTs located inside vehicles (probe vehicles) to result in a valid report, it is needed to be proved that the existed probe vehicles are adequate to support a traffic report estimation.

Table 2.1 MT distribution located inside vehicles.

	# MTs	%	# One phone in "active mode"	%
No phone	105	13.11	–	–
1 phone	542	67.67	373	68.82
2 phone	154	19.23	125	81.17
Total	801	100.00		

To estimate the percentage of MTs in probe vehicles, a survey has been conducted. The survey was carried out in one of the main routes in the city center of Athens, Alexandras street, on 20th, 21st and 22nd of October 2004.

The results illustrated in Table 1 show that there is at least one phone for 86.89% of the vehicles, and more precisely, there are 106 cellular phones for 100 vehicles. In Greece currently exist 3 cellular providers, having equally shared the market. 28.67% can be assumed that belongs to each cellular provider, a percentage much higher than what is needed for the theoretical 5% of probe vehicles needed to obtain a 95% accuracy of travel times estimates [18]. Since the survey was contacted on midweek days, the results were the outcome of a representative sample of the usual city traffic congestion.

2.3.2 Proposed Road Traffic Model

For the modeling of road traffic, the use of a dynamic traffic model is necessary [19]. Dynamic models are able to generate realistic time series of the simulation scenario, which is essential for a proper characterization of the transmission channel. Currently, two major approaches that differ in their level of resolution are available. They are termed "Macroscopic" and "Microscopic" traffic models [20].

Following the macroscopic formulation presented in [21] along with the given experimental data of the volume-density diagram, it is clear that volume is first increasing with density and then decreasing. The increasing part is the result of an increasing demand level, the decreasing part is the consequence of saturation (demand does not have anymore influence on volume). From the resulting diagrams (Figure 2.4) vehicle volume estimation for specific routes can be obtained resulting route vehicle volume estimation.

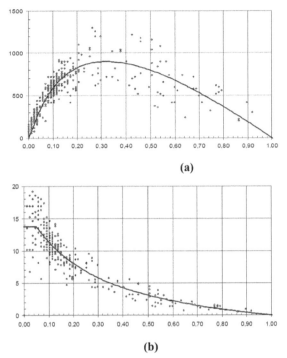

(a)

(b)

Figure 2.4 (a) VD (volume-density) diagram, (b) SD (speed-density) diagram (V: number of vehicles per hour, D: number of vehicles per length unit, S: m/sec).

2.4 Vehicle Volume Estimation Schema

In the previews, paragraphs were presented the basic components of our vehicle volume estimation model. The architecture for such estimation is illustrated in Figure 2.5. A GSM cellular network simulator has been developed, from where sequences of downlink signal strength measurements have been generated (see Figure 2.2) which are RSS measurements, as has been described earlier and are based on real radio coverage maps.

In order to process the sequences of downlink signal strength level measurements, first is needed to be constructed an infrastructure that would be used to identify in which speed class the examined sequence belongs to. This infrastructure would be consisted of all possible speed classes for the examined route. The construction of the infrastructure would consist of the following stages illustrated also in Figure 2.5:

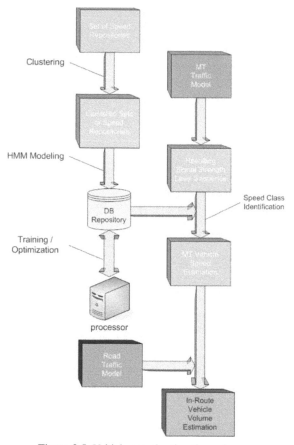

Figure 2.5 Vehicle speed estimation schema.

1. The construction of sets of speed repositories containing sequences of downlink signal strength level measurements based on real radio-map, while simulating MT located inside moving cars grouped by vehicle speed.

2. The clustering of the sequences in step 1 using KM method before moving to HMM modeling.

3. The computation of HMM models based on clustered sequences of step 2 that is carried out by Baum–Welch algorithm, creating in that way the DB repository with k HMM models $\lambda = \{A, B, \pi\}$ that correspond to k vehicle speed classes (problem formulation 3).

4. The training and optimization using Viterbi algorithm of the produced k vehicle speed classes in step 3 (problem formulation 2).
5. Given a sequence of signal strength level measurements, its log-likelihood is calculated against the k different HMM models. The speed class of the HMM model that gives the higher likelihood value when it is compared with the examined sequence, is the estimated speed of the MT (problem formulation 1).
6. Having an adequate sample size of vehicles, traveling at the examined route, use of the "Road Traffic Model" takes place in order to extract "Vehicle Volume Estimation" for the specific route, based on the vehicles speed of the sample.

Sets of sequences of signal strength level measurements, for every 5km/h from 20km/h to 110km/h for five (5) TA zones in the form smp_S_spx_TAy_z, have been generated, where:

- $S \in [001, \ldots, 100]$ corresponds to 100 sequences of signal strength level samples,
- $x \in [001, \ldots, 19]$ corresponds to speed values from 20 to 110 km/h,
- $y \in [0, 1, 2, 3, 4]$ corresponds to TA0, TA1, TA2, TA3 and TA4 zones,
- $z = \begin{cases} E \\ W \end{cases}$ corresponds to the two directions (East/West).

The generated 100 samples for each speed class. Thus a sample of the form smp_050_sp12_TA1_E corresponds to the speed class of 75 km/h traveling inside TA-1 with direction East. It should be noted that the number of zones could exceed the number of 5 (TA-0, TA-1, TA-2, TA-3, TA-4) while rural environment is examined contrary to urban environments, due to the frequent handovers that take place in the latter environments.

In step 2, pre-processing of the sequences of downlink signal strength level measurements by clustering takes place. Among the presented clustering methods KM and FCM, KM has been chosen for clustering due to faster convergence tested on the same data set and initial conditions (see Figure 2.6).

In step 3, HMM model computation of the sequences of downlink signal strength level measurements can be carried out by either Baum–Welch algorithm [22] or Segmental K-Means algorithm. Both algorithms performed almost identical, with the Baum–Welch to provide always slightly higher accuracy levels than the Segmental K-Means. The computation of the Segmental K-Means and Baum–Welch algorithms was done on the same initial values reached from KM clustering method. Table 2.1 presents likelihood values per sequence sample of 5 random cases.

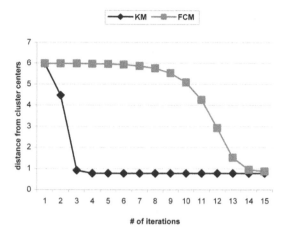

Figure 2.6 KM *versus* FCM.

Figure 2.7 TA-zones 0 and 1.

Then, in step 4 and after having derived for the various speed classes the k HMM models $\lambda = \{A, B, \pi\}$, they are trained. This is achieved by finding the optimum state sequence via the Viterbi algorithm, producing an updated set of k HMM models for the k speed classes.

In similar problems training has not been used at all [23], or has been performed in problems like position estimation [24]. Additionally it has been used in vehicle load estimation with Macroscopic [25] or Microscopic vehicle traffic modeling [26]. It should be noted that necessary filtering by the TA zones in GSM is also taking place as shown in Figure 2.7.

In step 5, the resulting (from the MT traffic model) sequence of downlink signal strength measurements is categorized into an HMM model class by calculating its likelihood value ($P(O \mid \lambda)$). Thus estimation of MT speed can be achieved.

Finally, in step 6, for the "Road Traffic Model" Mobility Model presented in Section 2.3.2 is followed.

As has been stated in Section 2.3.1, the existing vehicle sample is more than adequate to support vehicle volume estimation reports.

In the following example it is constructed the speed class sp04 that corresponds to 35 km/h based on 100 sequence samples as mentioned earlier. It took 249 iterations to converge. Then comparison between the log likelihood of three samples takes place (that could be a random sequence of signal strength level measurements); smp_022_sp04_TA0_W, smp_031_sp03_TA0_W and smp_099_sp05_TA0_W that correspond to 35, 30 and 40 km/h:

- loglik(smp_022_sp04_TA0_W) = −652.934431
- loglik(smp_031_sp03_TA0_W) = −664.810879
- loglik(smp_099_sp05_TA0_W) = −672.848484

It has been found (as expected) that the sample smp_022_sp04_TA0_W gives a higher log likelihood value than the other two samples.

2.5 Implementation

Based on the bibliography, many wireless traffic models that propose various approaches can be found. Since the accuracy of a simulation model is judged by its level of consistency compared to analytic techniques, the formulation proposed in [27] has been followed, which has also being used with success in modeling highway mobile networks, even though wireless traffic models that require less computational effort, do exist, based on the random walk model [28]. It has been assumed that vehicles are moving along the highway while their speed is a function of time and location [29]. Based on the original formulation, mobile communication networks can be simulated by investigating all their different aspects, like dynamic periodic location area update schemes [30].

Unlike other wireless traffic models that can be found in the literature, as has been stated earlier, the examined vehicle speed is a function of time and location. By following this assumption, simulation of both major and minor speed variations can be achieved (like accidents and traffic jams or vehicle acceleration and deceleration). Even though the traffic model it has

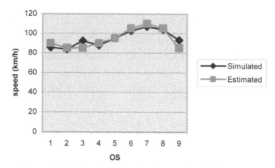

Figure 2.8 Simulated and estimated vehicle speed.

been employed simulates realistically the movement of MT (traffic lights, acceleration, deceleration), the simulated vehicle was moving in Alexandra's street for approximately 5.67 km (see Figure 2.7). Thus more complex mobility models can be found in [31]. The speed variation of the examined vehicle is illustrated in Figure 2.8.

The street segments have an approximate range from 558 to 714 m and have been modeled for both directions. In this way errors have been avoided, that could occur while measuring a vehicle that was traveling along the wrong direction.

In Table 2.2, the results are given by utilizing the measurement reports of a MT being in a vehicle traveling for a distance of 5.67 km and nine observation sequences. As mentioned, each observation sequence consists of 50 measurement reports, recorded every 480 ms. The specific route crosses five TA zones (TA-0, TA-1, TA-2, TA-3 and TA-4) and had duration of 214 seconds. The comparison of the test sequences with the vehicle speed classes in DB repository, are giving in column 3 ("Closest Repository Models") where for example sp08_TA0_W is the model of downlink signal trained with mean speed 55 km/h, driving in West direction of the street segment of Alexandra's street that belongs to TA-0. According to Table 2.2, the vehicle speed is approximated by the upper limit of the selected repository models. It has been achieved successful estimation in 77.78% of the cases using K-means as the clustering method (the bold entries indicate the occurrence of successful approximation by the highlighted vehicle speed class). Under the same conditions KM clustering gave accuracy level of 72.73%.

Moreover it is very helpful to find the places where the user is not moving. This can be estimated with high precision from the parts of the curve of the downlink signal power level where the level is almost constant (±3 dBm).

Table 2.2 Comparison of the estimated and real probe vehicle speed.

OS	Measurement Reports Models	Closest Repository	Time Duration	Distance Covered	Successful Estimation
1	85.52	sp14_Ta4_E **sp15_Ta4_E (90)**	24	570	Yes
2	83.76	sp13_TA3_E **sp14_TA3_E (85)**	24	558	Yes
3	92.42	sp13_TA2_E sp14_TA2_E	24	616	No
4	87.66	sp14_TA1_E **sp15_TA1_E (90)**	24	584	Yes
5	95.09	sp15_TA0_E **sp16_TA0_E (95)**	24	634	Yes
6	102.56	sp17_TA0_E **sp18_TA0_E (105)**	24	684	Yes
7	107.11	sp18_TA2_E **sp19_TA2_E (110)**	24	714	Yes
8	103.76	sp17_TA3_E **sp18_TA3_E (105)**	24	692	Yes
9	93.13	sp13_TA4_E sp14_TA4_E	24	621	No
		Total	216 sec	5.67 km	77.78%

The assumption of staying at the traffic lights in combination with a database, can limit the considered area of the user even more. Also an aid in speed estimation is the identification of the kind of road of the examined area. Containing the DB accurate models for each individual street segment enables us to estimate with higher accuracy the vehicle speed value.

In Table 2.3, ten different scenarios are illustrated, including the one that has already been described analytically. The rates of a successful estimation are all between 70 and 80%, reaching almost 82% in one case. More scenarios needed to be run in order to highlight under which conditions our approach results in good estimates and thus emphasize its strengths and weaknesses.

Table 2.3 Different simulation scenarios.

Scenario	# OS	Time Duration (sec)	Distance Covered (m)	Successful Estimation
1	11	264	3310.02	81.82
2	10	240	3102.04	80.00
3	14	336	2810.45	78.57
4	12	288	2947.32	75.00
5	12	288	2503.67	72.73
6	13	312	2893.87	76.92
7	11	264	3312.45	72.73
8	10	240	2998.61	70.00
9	9	216	2894.49	77.78
10	13	312	2932.15	69.23
			Mean	75.48%

2.6 Conclusions and Further Work

Adaptation of the proposed technique can be realized without modifications neither to the MT nor to cellular networks, making it an attractive solution to cellular network providers that offer LBS applications. Cluster analysis for this particular case is also a complex but challenging task. A clustering algorithm can always provide as output either a partition into k clusters or a hierarchical grouping, and does not answer the question whether there is actually structure in the data, and if there is, what are the clusters that best describe it. In fact, the "correct" number of clusters in a data-set often depends the external perspective about the data, and sometimes equally good answers can be obtained for the same data. A widely used solution is cluster validation technique as the assessment of clustering quality to determine the proper number of the clusters in the data set. A cluster validation index is the number which estimates the quality of clusters. Future work could be directed into this direction. A worth to mention clustering algorithm with high validation index is proposed in [32] where it is expected to lead to more credible clustering in the context of sequence clustering during HMM modeling.

It is also worth mentioning that speed estimation of a MT has not been carried out by examining sequences of downlink signal strength level measurements. The usual way to perform such a task was to calculate two of the possible locations of the MT traveling along a specific route as well as the distance between two of these locations. By this way though inaccuracies during location estimation had a big impact resulting in wrong MT speed estimation. Also during traffic modeling, MT speed does not remain constant. In contrary, in the traffic model that has been used, the examined vehicle speed is a function of time and location. This modeling could reflect real traffic conditions like traffic lights. Last but not least processing of the downlink signal data has been extensively studied. Results so far concerning speed estimation of a probe vehicle have been quite optimistic [33], but need to be applied in a larger scale.

Special attention should be given to the efficient filtering of possible measurement errors. First of all the length of street segments would influence the vehicle volume distribution and thus the result accuracy. Also attention is needed while constructing the repositories by measurements of the downlink signal strength values. The latter values correspond to different street and are directly affected by the speed distribution of the vehicles in the examined area.

Adaptation of the proposed technique can be realized without modifications neither to the MT nor to cellular networks. Thus the proposed technique appears to be an attractive solution to cellular network providers that offer LBS applications. At the moment construction of the UMTS simulator takes place. It is expected to offer more precise results that would lead to less effort during error filtering, thus more accurate traffic report estimations.

Acknowledgments

The authors wish to thank Professor Michali Theologou and Professor Vassili Loumo as well as the anonymous reviewers for their valuable comments that considerably improved the quality of the manuscript.

References

[1] C. Drane, M. Macnaugtan and C. Scott, Positioning GSM Telephones, *IEEE Commun. Mag*, vol. 36, no. 4, pp. 46–59, April 1998.

[2] T. Stamoulakatos and E. Sykas, A review on cellular location methods targeting location based services, in *Proceedings of the IASTED International Conference on Communication Systems and Networks (CSN 2003)*, Benalmadena, Spain, September 8–10, pp. 118–124, 2003.

[3] T. Stamoulakatos and E. Sykas, A network architecture to obtain traffic information for Location Based Service applications, in *Proceedings of ConTEL 2003, 7th International Conference on Telecommunications*, Zagreb, Croatia, June 11–13, pp. 197–204, 2003.

[4] S. Turner, W. Eisele, R. Benz and D. Holdener, *Travel Time Data Collection Handbook*, Report No. FHWA-PL-98-035, Texas Transportation Institute, March 1998.

[5] D. Lapidot, Blazing new trails in mobile traffic information, in *Proceedings of Mobile Venue '02*, Athens, Greece, May 30–31, pp. 136–140, 2002.

[6] SnapTrack, *Location Technologies for GSM, GPRS and WCDMA Networks*, SnapTrack White Paper, 2001.

[7] T. Stamoulakatos and E. Sykas, Hidden Markov modeling and macroscopic traffic filtering supporting location based services, *Wireless Commun. Mobile Comput*, vol. 7, no. 4, pp. 415–429, March 2006.

[8] L. Rabiner, A tutorial on hidden Markov models and selected applications in speech recognition, *Proc. IEEE*, vol. 77, no. 2, pp. 257–286, February 1989.

[9] R. Brown and P. Hwang, *Introduction to Random Signals and Applied Kalman Filtering*, 3rd edition, John Wiley & Sons, 1997.

[10] L. Rabiner and B. Juang, *Fundamentals of Speech Recognition*, Prentice Hall, Englewood Cliffs, NJ, 1993.

[11] A. J. Viterbi, Error bounds for convolutional codes and an asymptotically optimal decoding algorithm, *IEEE Trans. Informat. Theory*, vol. IT-13, pp. 260–269, April 1967.

[12] A. K. Jain, M. N. Murty and P. J. Flynn, Data clustering: A review, *ACM Comput. Surveys*, vol. 31, no. 3, pp. 264–323, 1999.

[13] P. S. Bradley and U. M. Fayyad, Refining initial points for K-Means clustering, in *Proceedings of 15th International Conference on Machine Learning*, Morgan Kaufmann, San Francisco, CA, pp. 91–99, 1998.

[14] G. Chen et al., Evaluation and comparison of clustering algorithms in analyzing ES cell gene expression data, *Statistica Sinica*, vol. 12, pp. 241–262, 2002.

[15] F. Hoppner, F. Klawonn, R. Kruse and T. Runkler, *Fuzzy Cluster Analysis, Methods for Classification, Data Analysis and Image Recognition*, John Wiley & Sons, 1999.

[16] X. Chang, W. Li and J. Farrell, A C-means clustering based fuzzy modeling method, in *Fuzzy Systems, 2000, The Ninth IEEE International Conference on Fuzzy Systems*, vol. 2, pp. 937–940, 2000.

[17] GSM 05.08, Digital cellular telecommunication system (Phase 2+); Radio subsystem link control.

[18] J. L. Ignace, Travel time estimates from cellular phone positioning, in *Proceedings of Mobile Venue '02*, Athens, Greece, May 30–31, 2002.

[19] W. R. McShane and R. P. Roess, *Traffic Engineering*, Prentice Hall, Englewood Cliffs, NJ, 1990.

[20] A. D. May, *Traffic Flow Fundamentals*, Prentice Hall, Englewood Cliffs, NJ, 1990.

[21] M. Papageorgiou and G. Schmidt, Freeway traffic modeling, (Technische Universität München, Germany) – in *Concise Encyclopedia of Traffic & Transportation Systems*, M. Papageorgiou (Ed.), Pergamon Press, pp. 162–167, 1991.

[22] A. P. Dempster, N. M. Laird and D. B. Rubin, Maximum likelihood from incomplete data via the EM algorithm, *J. Roy. Stat. Soc*, vol. 39, no. 1, pp. 1–38, 1977.

[23] T. Nypan and O. Hallingstad, A cellular positioning system based on database com-

parison – The hidden Markov model based estimator versus the Kalman filter, in *Proceedings 5th Nordic Signal Processing Symposium (NORSIG 2002)*, Norway, October, , 2002.

[24] S. Mangold and S. Kyriazakos, Applying pattern recognition techniques based on hidden Markov models for vehicular position location in cellular networks, in *Proceedings of 50th IEEE Vehicular Technology Confererence, VTC 1999*, pp. 780–784, Fall 1999.

[25] T. Stamoulakatos and E. Sykas, Signal pattern recognition, hidden Markov modeling and traffic flow modeling filters applied in existing signaling of cellular networks for vehicle volume estimation, in *Proceedings of ISICT '03, International Symposium on Information and Communication Technologies*, Trinity College, Dublin, Ireland, September 24–26, pp.397–403, 2003.

[26] T. S. Stamoulakatos, A. Yannopoulos, T. Varvarigou and E. D. Sykas, Hidden Markov filtering with microscopic traffic modeling for vehicle load estimation in cellular networks, in *Proceedings of IASTED International Conference on Communication Systems and Networks (CSN 2003)*, Marbella, Spain, September 1–3, pp. 349–355, 2004.

[27] Y. B. Lin, Modeling techniques for large-scale PCS networks, *IEEE Commun. Mag*, vol. 35, pp. 102–107, February 1997.

[28] Y. C. Tseng and W. N. Hung, An improved cell type classification for random walk modeling in cellular networks, *IEEE Commun. Lett*, vol. 5, no. 8, pp. 337–339, August 2001.

[29] B. Cvetkovski and L. Gavrilovska, A simulation of a mobile Highway traffic, in *IEEE VTC*, pp. 1429–1433, 1998.

[30] Y.-B. Lin, P.-J. Lee and I. Chlamtac, Dynamic periodic location area update in mobile networks, *IEEE Trans. Vehicle Technol*, vol. 51, no. 6, pp. 1494–1501, 2002.

[31] P. I. Bratanov and E. Bonek, Mobility model of vehicle-borne terminals in urban cellular systems, *IEEE Trans. Vehicular Technol*, vol. 52, no. 4, pp. 947–952, July 2003.

[32] D. D. Kouvatsos and I. M. Mkwawa, Multicast communication in grid computing networks with background traffic, *IEE Proc. Softw*, vol. 150, no. 4, August 2003 (IEE Proceedings online no. 20030810, doi:10.1049/ip-sen:20030810).

[33] T. S. Stamoulakatos, A. S. Markopoulos, M. E. Anagnostou and M. E. Theologou, Vehicle velocity estimation based on RSS measurements, *J. Wireless Pers. Commun*, vol. 40, no. 4, pp. 523–538, March 2007.

3

Mobile User Location Management under a Random-Directional Mobility Pattern for PCS Networks

Ignacio Martínez-Arrúe, Pablo García-Escalle and
Vicente Casares-Giner

*Departamento de Comunicaciones, GIRBA-ITACA, Universidad Politécnica de Valencia, Camino de Vera s/n, 46022 Valencia, Spain;
e-mail: imartinez201j@cv.gva.es, {pgarciae, vcasares}@upvnet.upv.es*

Abstract

In this paper, we propose a new random-directional model which generalizes the random walk model. Our proposal gathers mobility patterns with several degrees of randomness, so that both random walk and totally directional mobility patterns are modeled. This model is used as an input to study and compare the location management cost of the distance-based and the movement-based policies as a function of the directional movement.

Distance-based and movement-based location update (LU) strategies are usually selected when dynamic location management schemes are employed. The distance-based outperforms the movement-based mechanism, even though the former requires to compute the distance traveled by a mobile terminal (measured in cells). These strategies are often studied using the random walk mobility model, that yields a lower location management cost for the distance-based scheme. However, we point out that if a directional mobility pattern is considered the distance-based LU cost increases and it approaches the movement-based LU cost. In such cases, movement-based policies may be more suitable for its simplicity rather than distance-based schemes. Both one-step paging and selective paging strategies are considered.

D. D. Kouvatsos (ed.), Mobility Management and Quality-of-Service for Heterogeneous Networks, 55–78.

Keywords: Mobility models, location update, distance-based schemes, movement-based schemes, paging.

3.1 Introduction

In a personal communication services (PCS) system, location management is defined as the set of procedures that keep information about the location of a mobile terminal (MT) at any time. These procedures are called location update (LU) and call delivery (CD). The LU process consists of reporting the location of an MT to the system databases. This location information must be maintained up to date. Thus, the database entry of an MT is updated each time the MT sends an LU message or it accepts an incoming call – whenever the MT has a contact with the fixed network (FN). This updating operation is known as registration. During the CD procedure, the system searches for the called MT in order to deliver the incoming call. The CD may be further decomposed into interrogation (IG) and terminal paging (PG). Each time an incoming call arrives, the IG process starts. The system databases are queried in the FN to obtain the last known location of the MT. The query result is a set of cells where the MT is roaming. This set is called uncertainty region or registration area (RA). The IG process is followed by the PG procedure. By means of the PG procedure, an MT is searched by polling the mentioned set of cells where the MT is roaming.

It is well known there is a trade off between LU and PG procedures. The more frequently LU messages are delivered to the system, the LU cost becomes higher but the PG cost decreases because the MT position is known more accurately. On the other hand, the less frequently LU messages are triggered, the lower is the LU cost but the uncertainty of the MT position is higher and the PG cost increases.

The LU procedures may be classified into static or dynamic strategies. In static schemes, also called global, the whole coverage area is divided into several location areas (LAs). An LA is a fixed set of several neighboring cells. Whenever an MT crosses an LA border, it delivers an LU message. Nevertheless, this kind of strategies has several drawbacks. One of them occurs if an MT moves back and forth between two LAs. Then, it may trigger a large and unnecessary number of LU messages. This fact may be overcome through dynamic strategies, also called local, proposed in [1, 2]. These strategies are MT-dependent because the RA borders depend on the cell where the MT had its last contact with the FN.

In [1, 2], three LU schemes were proposed. LU messages are delivered according to the elapsed time, the performed number of movements or the traveled distance from the cell where the MT had contact with the FN for the last time. It is pointed out that the distance-based scheme outperforms both movement-based and time-based mechanisms. Furthermore, the movement-based strategy has a lower cost than the time-based scheme and it is very simple to be implemented. The distance-based mechanism is more complex as it requires to compute the distance traveled by the MT (measured in cells). Then, the MT stores all cell identifiers within its $RA - 3(D-1)^2 + 3(D-1) + 1$ cell identifiers for a distance threshold D, [3] – or it computes the traveled distance using specific algorithms [4]. Therefore, a movement-based scheme is more suitable for its simplicity rather than a distance-based scheme.

The PG process can be carried out according to two different policies. In the first strategy, hereinafter called one-step PG, all cells in the RA are polled simultaneously. In this case, the called MT is found with a minimum delay, but the PG cost is high. The second PG policy is called selective PG [3, 5]. The RA is divided into several PG areas (PAs) in this strategy. The PAs are polled sequentially so that PAs where the called MT is more likely to be located are polled first. In a shortest distance first (SDF) strategy [3, 5], the PAs are composed by rings of cells around the cell where the MT had contact with the FN for the last time (the center cell). Once the MT is located, the sequential polling process ends. This policy provides a lower cost than one-step PG, even though there is a delay on finding out the cell where the called MT is roaming.

Mobility models are employed to obtain the LU and PG costs in location management. The most extended mobility model is the random walk model. It is defined as a mobility model in which an MT moves from its current position to a new location by randomly choosing a direction and speed to travel [6]. Both of these parameters are chosen within the ranges $[0, 2\pi]$ and $[v_{min}, v_{max}]$ respectively.

In this paper, we propose a new random-directional mobility model that generalizes the random walk mobility model. In the proposed model, we compute transition probabilities as dependent on a directional movement parameter (denoted by α) in both cell transitions and ring transitions. All possible values of α gather both random mobility patterns and directional mobility patterns. Location management is studied through this new mobility model in both macrocellular and microcellular scenarios, just depending on the movement randomness to be considered in each scenario. Under the proposed mobility model, we study two dynamic schemes,

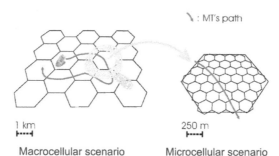

Figure 3.1 Mobility pattern in macrocellular and microcellular scenarios.

namely, the movement-based and the distance-based schemes, because the movement-based policy may be more suitable in some cases.

In Section 3.2, related work about mobility modeling and location management under directional mobility patterns is depicted. Section 3.3 describes the scenario and the proposed mobility model. In Section 3.4, we use this model to compute the LU and PG costs in both distance-based and movement-based schemes. Some numerical results related to these costs and their discussion are detailed in Section 3.5. Finally, Section 3.6 presents the conclusions of this paper.

3.2 Related Work

The random walk mobility model was proposed in [1, 2] under a one-dimensional (1D) approach. In this proposal, an independent and identically distributed movement and a Markovian model were developed. The 1D version was extended to a two-dimensional (2D) random walk model, whose properties were studied in [7, 8]. In these papers, the 2D random walk model is modeled by a 2D Markov chain, that is studied and simplified by using symmetry features in both mesh and hexagonal cell layouts.

However, the random walk mobility model was a common assumption for macrocellular scenarios. As the number of mobile users has increased over the last few years, cells have become smaller and macrocellular scenarios have been divided into microcellular scenarios. If we consider an overlaid microcellular scenario, the mobility rate is higher and MTs may move according to directional mobility patterns as it is shown in Figure 3.1.

Thus, there are some related works about location management under directional mobility patterns. In [9], the authors studied distance-based and movement-based LU strategies depending on the directional characteristics of the user mobility pattern. They supposed each mobile user has a privileged direction in its movement. Thus it is defined a parameter, a, that indicates the degree in which movement is directional. Assuming a hexagonal cell layout, transition probabilities between two cells are defined as $ma/(5+a)$ for moving towards a neighboring cell in the privileged direction, and $m/(5+a)$ for moving to a neighboring cell in any other direction, being m the probability that the MT crosses a cell boundary. Also, the authors feature mobility through the call-to-mobility ratio (CMR). The CMR is defined as $\theta = \lambda_c/\lambda_m$, where λ_c is the rate of call arrivals for the MT and λ_m is the rate of cell boundary crossings performed by that MT.

In [10], a distance-based scheme is studied considering a transitional directivity index. In this case, the directivity index does not influence movements between two cells but movements between two rings of cells. Hence, transition probabilities between rings are defined so that they depend on both the current ring of the MT and the directivity index.

Based on the previous ideas, we propose a model that generalizes the 2D random walk mobility model through a directional movement parameter. This parameter appears in expressions related to transition probabilities between two neighboring cells and transition probabilities between two neighboring rings. Thus, both mobility among cells and mobility among rings are gathered in our model. This fact allows us to study distance-based and movement-based schemes under the proposed model. Also we consider the MT mobility is dependent on the CMR, which is a common parameter to study location management strategies.

3.3 Proposed Mobility Model

3.3.1 Cell Dwell Time Distribution

A hexagonal cell layout has been chosen. Results obtained for the hexagonal and the mesh cell configurations are very similar though [5]. It is assumed that all cells have the same size. An MT is roaming within a cell during a random time which is featured through a generalized gamma distribution [11], with probability density function (pdf) denoted by $f_c(t)$ and mean cell dwell time value equal to $1/\lambda_m$ (hence, λ_m is the mobility rate of the MT),

$$f_c(t; a, b, c) = \frac{c}{b^{ac}\Gamma(a)} t^{ac-1} e^{-(t/b)^c}; \qquad t, a, b, c > 0 \qquad (3.1)$$

where $\Gamma(a)$ is the gamma function, defined as $\Gamma(a) = \int_0^\infty x^{a-1}e^{-x}dx$ for any real and positive number a. Values a, b and c are set up according to Zonoozi and Dassanayake [11], where it is shown through simulations that the values which best fit the cell dwell time distribution are $a = 2.31$, $b = 1.22R$ and $c = 1.72$. The parameter R is the equivalent cell radius assuming an average speed of 50 km/h and zero drift.

The Laplace transform (LT) of the generalized gamma distribution is denoted as $f_c^*(s)$. Its mean value is given by [12],

$$-f_c^{*'}(0) = b\frac{\Gamma(a + \frac{1}{c})}{\Gamma(a)} = \frac{1}{\lambda_m} \tag{3.2}$$

The parameters a and c are constant. There is a relation among b, R and λ_m. By adjusting the parameter λ_m to a given value, b and R can be obtained.

Analogously, let $f_{cr}(t)$ be the pdf of the residual cell sojourn time [13], and $f_{cr}^*(s)$ its respective LT. Both $f_c(t)$ and $f_{cr}(t)$ are related through their LTs as

$$f_{cr}^*(s) = \lambda_m\frac{1 - f_c^*(s)}{s} \tag{3.3}$$

The cell dwell time is an activity that is observed during a time period. If the model is observed during the time interval between two consecutive incoming calls to the MT, and we assume that interval is exponentially distributed with mean value $1/\lambda_c$ (λ_c is the call arrival rate), $f_c^*(\lambda_c)$ is the probability that the MT leaves its current cell before a new incoming call is received. Also, $f_{cr}^*(\lambda_c)$ is the analogous probability if a residual pdf is used.

3.3.2 Scenario Description and Transition Probabilities

In a 2D random walk model with a hexagonal layout, when the MT leaves its current cell, it moves to one of its six neighboring cells with probability 1/6. In the proposed model, a single parameter denoted by α provides directivity to the MT motion. To that end, cells are labeled as it is shown in Figure 3.2. Cells are divided into two sets to obtain the transition probabilities in the system. Cells labeled $(x, 0)$, with $x = 1, 2, \ldots$, are located in the corners of their ring and they limit with three neighboring cells in their contiguous outer ring. Cells labeled (x, i), with $x = 2, 3, \ldots$, $1 \leq i \leq \lfloor x/2 \rfloor$, are placed in the edges of their ring and they are bordered by two neighboring cells in their outer ring. The ring 0 consists of one cell. Notice that any other ring q, for $q > 0$, is a set composed by $6q$ cells, from which 6 are labeled $(x, 0)$ and $6(q - 1)$ cells are labeled (x, i), with $1 \leq i \leq \lfloor x/2 \rfloor$.

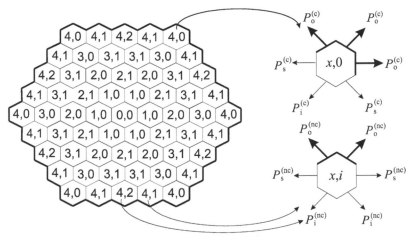

Figure 3.2 Transition probabilities between two cells.

The transition probabilities from one cell to another are defined according to the motion of a given MT implying a movement to the outer ring (P_o), a movement to the same ring (P_s) or a movement to the inner ring (P_i).

Then, for cells of type $(x, 0)$, with $x = 1, 2, \ldots$, that are located in the corners (c) of each ring, we define

$$P_o^{(c)} = \frac{\alpha}{3(1 + \alpha)}; \quad \text{three edges per cell}$$

$$P_s^{(c)} = \frac{1}{3(1 + \alpha)}; \quad \text{two edges per cell} \qquad (3.4)$$

$$P_i^{(c)} = \frac{1}{3(1 + \alpha)}; \quad \text{one edge per cell}$$

and for cells of type (x, i), with $x = 1, 2, \ldots$ and $i = 1, 2, \ldots, \lfloor x/2 \rfloor$, which are not placed in the corners (nc) of their ring, we define

$$P_o^{(nc)} = \frac{\alpha}{2(2 + \alpha)}; \quad \text{two edges per cell}$$

$$P_s^{(nc)} = \frac{1}{2(2 + \alpha)}; \quad \text{two edges per cell} \qquad (3.5)$$

$$P_i^{(nc)} = \frac{1}{2(2 + \alpha)}; \quad \text{two edges per cell}$$

Table 3.1 Transition probabilities between two cells for several values of α.

α	Cells $(x, 0)$			Cells (x, i)		
	$P_0^{(c)}$	$P_s^{(c)}$	$P_i^{(c)}$	$P_0^{(nc)}$	$P_s^{(nc)}$	$P_i^{(nc)}$
0	0	$\frac{1}{3}$	$\frac{1}{3}$	0	$\frac{1}{4}$	$\frac{1}{4}$
$\frac{1}{2}$	$\frac{1}{9}$	$\frac{2}{9}$	$\frac{2}{9}$	$\frac{1}{10}$	$\frac{1}{5}$	$\frac{1}{5}$
1	$\frac{1}{6}$	$\frac{1}{6}$	$\frac{1}{6}$	$\frac{1}{6}$	$\frac{1}{6}$	$\frac{1}{6}$
2	$\frac{2}{9}$	$\frac{1}{9}$	$\frac{1}{9}$	$\frac{1}{4}$	$\frac{1}{8}$	$\frac{1}{8}$
100	$\frac{100}{303}$	$\frac{1}{303}$	$\frac{1}{303}$	$\frac{100}{204}$	$\frac{1}{204}$	$\frac{1}{204}$

Notice that the above probabilities fulfill the following conditions respectively (see Figure 3.2),

$$3P_0^{(c)} + 2P_s^{(c)} + P_i^{(c)} = 1$$
$$2P_0^{(nc)} + 2P_s^{(nc)} + 2P_i^{(nc)} = 1$$

The parameter α is defined as a directional movement parameter that may take values within the range $[0, \infty[$. If α is equal to 1, the previous transition probabilities are equal to the probabilities in a random walk mobility model. As α increases from 1 to infinity, the movements towards the outer ring from the current one become more probable ($P_0^{(c)}$ and $P_0^{(nc)}$ grows). In the limit case of α tending to infinity, each cell boundary crossing is, in fact, a movement towards the outer ring. For decreasing values of α from 1 to 0, the probability of remaining within the same ring ($P_s^{(c)}$ and $P_s^{(nc)}$) or moving towards a cell of the inner ring from the current one ($P_i^{(c)}$ and $P_i^{(nc)}$) increases. If α is 0, the MT is always roaming within the same ring or it moves towards an inner ring from the current one after a cell boundary crossing. In Table 3.1, the transition probabilities are shown for some values of α.

From the previous transition probabilities, a two-dimensional (2D) Markov chain arises with states denoted by (x, i) (see Figure 3.3). In this chain, the horizontal dimension (label x) represents the number of ring where the MT is roaming, and the vertical dimension (label i) indicates the kind of cell where the MT is located. The transition probabilities of the 2D Markov chain are shown in Figure 3.3, and the general expressions of these probabilities are detailed in Table 3.2.

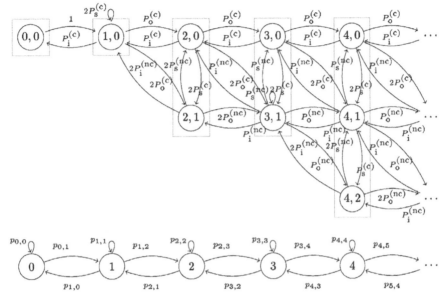

Figure 3.3 Two-dimensional and one-dimensional Markov chains in the proposed model.

The previous 2D Markov chain can be converted into a one-dimensional (1D) Markov chain, that simplifies our random-directional model without any significant degradation (see Figure 3.3). For this conversion, cells are grouped by rings, so that each state of the 1D Markov chain represents the number of ring where the MT is roaming. In the 1D Markov chain, we denote by $p_{i,j}$ the transition probability from a cell within ring i to a cell within ring j after a single cell boundary crossing. These probabilities are obtained by averaging the transition probabilities of the 2D Markov chain as

$$\left.\begin{array}{l} p_{0,0} = 0 \\ p_{0,1} = 1 \end{array}\right\}, \qquad q = 0 \qquad (3.6)$$

$$\left.\begin{array}{l} p_{q,q+1} = 6\frac{1}{6q}3P_o^{(c)} + 6(q-1)\frac{1}{6q}2P_o^{(nc)} = \frac{1}{q}\left(\frac{\alpha}{1+\alpha}\right) + \frac{q-1}{q}\left(\frac{\alpha}{2+\alpha}\right) \\ p_{q,q} = 6\frac{1}{6q}2P_s^{(c)} + 6(q-1)\frac{1}{6q}2P_s^{(nc)} = \frac{1}{q}\left(\frac{2}{3(1+\alpha)}\right) + \frac{q-1}{q}\left(\frac{1}{2+\alpha}\right) \\ p_{q,q-1} = 6\frac{1}{6q}P_i^{(c)} + 6(q-1)\frac{1}{6q}2P_i^{(nc)} = \frac{1}{q}\left(\frac{1}{3(1+\alpha)}\right) + \frac{q-1}{q}\left(\frac{1}{2+\alpha}\right) \end{array}\right\}, \ q > 0$$

$$(3.7)$$

As α becomes larger, the probability of moving towards an outer ring after a cell boundary crossing ($p_{q,q+1}$) increases and the probabilities of

Table 3.2 Transition probabilities between two states of the 2D Markov chain.

Transition probabilities for cells with label $(x, 0)$, $\alpha \geq 0$	
$P[(0, 0), (1, 0)] = 1$	
$P[(1, 0), (1, 0)] = 2P_{\text{s}}^{(\text{c})}$	
$P[(x, 0), (x - 1, 0)] = P_{\text{i}}^{(\text{c})}$	$\forall x \geq 1$
$P[(x, 0), (x, 1)] = 2P_{\text{s}}^{(\text{c})}$	$\forall x \geq 2$
$P[(x, 0), (x + 1, 0)] = P_{0}^{(\text{c})}$	$\forall x \geq 1$
$P[(x, 0), (x + 1, 1)] = 2P_{0}^{(\text{c})}$	$\forall x \geq 1$

Transition probabilities for cells with label (x, i), $\alpha \geq 0$	
$P[(x, \frac{x}{2}), (x - 1, \frac{x}{2} - 1)] = 2P_{\text{i}}^{(\text{nc})}$	x even, $\forall x \geq 2$
$P[(x, \frac{x}{2}), (x, \frac{x}{2} - 1)] = 2P_{\text{s}}^{(\text{nc})}$	x even, $\forall x \geq 2$
$P[(x, \frac{x}{2}), (x + 1, \frac{x}{2})] = 2P_{0}^{(\text{nc})}$	x even, $\forall x \geq 2$
$P[(x, \frac{x-1}{2}), (x - 1, \frac{x-1}{2})] = P_{\text{i}}^{(\text{nc})}$	x odd, $\forall x \geq 3$
$P[(x, \frac{x-1}{2}), (x - 1, \frac{x-1}{2} - 1)] = P_{\text{i}}^{(\text{nc})}$	x odd, $\forall x \geq 3$
$P[(x, \frac{x-1}{2}), (x, \frac{x-1}{2})] = P_{\text{s}}^{(\text{nc})}$	x odd, $\forall x \geq 3$
$P[(x, \frac{x-1}{2}), (x, \frac{x-1}{2} - 1)] = P_{\text{s}}^{(\text{nc})}$	x odd, $\forall x \geq 3$
$P[(x, \frac{x-1}{2}), (x + 1, \frac{x-1}{2})] = P_{0}^{(\text{nc})}$	x odd, $\forall x \geq 3$
$P[(x, \frac{x-1}{2}), (x + 1, \frac{x-1}{2} + 1)] = P_{0}^{(\text{nc})}$	x odd, $\forall x \geq 3$
$P[(x, i), (x - 1, i)] = P_{\text{i}}^{(\text{nc})}$	$\forall x \geq 4, 1 \leq i < \lfloor \frac{x}{2} \rfloor$
$P[(x, i), (x - 1, i - 1)] = P_{\text{i}}^{(\text{nc})}$	$\forall x \geq 4, 1 \leq i < \lfloor \frac{x}{2} \rfloor$
$P[(x, i), (x, i - 1)] = P_{\text{s}}^{(\text{nc})}$	$\forall x \geq 4, 1 \leq i < \lfloor \frac{x}{2} \rfloor$
$P[(x, i), (x, i + 1)] = P_{\text{s}}^{(\text{nc})}$	$\forall x \geq 4, 1 \leq i < \lfloor \frac{x}{2} \rfloor$
$P[(x, i), (x + 1, i)] = P_{0}^{(\text{nc})}$	$\forall x \geq 4, 1 \leq i < \lfloor \frac{x}{2} \rfloor$
$P[(x, i), (x + 1, i + 1)] = P_{0}^{(\text{nc})}$	$\forall x \geq 4, 1 \leq i < \lfloor \frac{x}{2} \rfloor$

moving inside the same ring ($p_{q,q}$) or towards an inner ring ($p_{q,q-1}$) decrease. Therefore, a movement in a certain direction is modeled for large values of α (Figure 3.4).

The effect of the movement directivity can be verified through computing the probabilities of an MT moving forward (P_{fw}) and backward (P_{bk}) from an

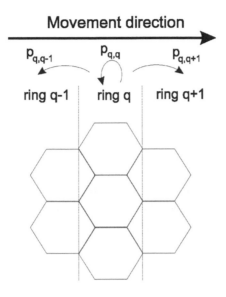

Figure 3.4 Directional movement in transitions between two rings.

specific ring q when a change of ring occurs. They are defined as

$$P_{\text{fw}}(q) = \frac{p_{q,q+1}}{p_{q,q+1} + p_{q,q-1}} = 3\alpha \frac{1 + q(1 + \alpha)}{\alpha - 1 + 3q(1 + \alpha)^2}; \quad q > 0 \quad (3.8)$$

$$P_{\text{bk}}(q) = \frac{p_{q,q-1}}{p_{q,q+1} + p_{q,q-1}} = \frac{-(1 + 2\alpha) + 3q(1 + \alpha)}{\alpha - 1 + 3q(1 + \alpha)^2}; \quad q > 0 \quad (3.9)$$

Obviously, $P_{\text{fw}}(0) = 1$ and $P_{\text{bk}}(0) = 0$. Also, notice that the transition probabilities of moving forward and backward from an specific ring q are complementary,

$$P_{\text{fw}}(q) + P_{\text{bk}}(q) = 1$$

3.4 Location Management Costs

3.4.1 LU Cost in a Distance-Based Scheme

In a distance-based scheme, the states of a 1D Markov chain represent the ring where the MT is roaming (see Figure 3.5). The 1D Markov chain is in state q, with $q \geq 0$, when the MT has currently traveled a distance q from the center cell of its RA [3]. The state labeled as *Abs* is an absorbing state. At

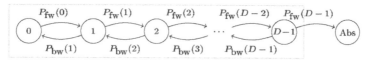

Figure 3.5 One-dimensional Markov chain in a distance-based LU scheme.

time $t = 0^+$, an incoming call arrives and an exponential observation interval begins. Then, the MT is roaming inside the center cell of its RA (ring 0). As the MT could have reached the center cell before the activity observation started, a residual sojourn time distribution [13] is considered in this case. The MT may move among different rings within its RA. Whenever the state *Abs* is entered, the distance threshold (D) is reached, an LU message is triggered, and the MT returns to state 0 instantaneously.

Let $f_q^*(s)$ denote the LT of the sojourn time distribution of an MT roaming in the ring q. Let $f_{qr}^*(s)$ denote its respective residual sojourn time distribution. If $q = 0$, $f_0^*(s)$ is given by $f_c^*(s)$ and $f_{0r}^*(s)$ is obtained as $f_{cr}^*(s)$. If $q \geq 1$, then

$$f_q^*(s) = f_c^*(s) \left[1 - p_{q,q}(1 - f_c^*(s)) \sum_{i=0}^{\infty} (p_{q,q} f_c^*(s))^i \right]$$

$$= \frac{f_c^*(s)(1 - p_{q,q})}{1 - p_{q,q} f_c^*(s)}; \quad q \geq 1 \tag{3.10}$$

In order to compute the LU cost, we define the sojourn time distribution in specific boxes of states as done in [14]. We denote by $r_{l,r}(i, j, t)$ the sojourn time distribution in a box that gathers all states between i and j ($i < j$), given that the MT enters the box through state i and it leaves the box through state j. Let $rr_{l,r}(i, j, t)$ be the analogous residual sojourn time distribution. Besides, we define $r_{l,l}(i, j, t)$ as the sojourn time distribution in a box composed by states between i and j ($i < j$), assuming that the MT enters the box through state i and it leaves the box through state i. The LTs of the previous distributions are denoted by $r_{l,r}^*(i, j, s)$, $rr_{l,r}^*(i, j, s)$ and $r_{l,l}^*(i, j, s)$ respectively.

Recalling the observation is exponentially distributed (see Section 3.3.1), with mean value $1/\lambda_c$, the LTs of the sojourn time distributions in the before mentioned boxes of states evaluated at $s = \lambda_c$ provide the probability of leaving the considered box of states before a new incoming call arrives. These probabilities can be evaluated recursively as done in [14],

$$
\begin{aligned}
rr_{l,r}^*(i, j, \lambda_c) &= P_{\text{fw}}(i) f_{ir}^*(\lambda_c)\big(r_{l,r}^*(i+1, j, \lambda_c) \\
&\quad + r_{l,l}^*(i+1, j, \lambda_c) r_{l,r}^*(i, j, \lambda_c)\big) \\
r_{l,r}^*(i, j, \lambda_c) &= \frac{P_{\text{fw}}(i) f_i^*(\lambda_c) r_{l,r}^*(i+1, j, \lambda_c)}{1 - P_{\text{fw}}(i) f_i^*(\lambda_c) r_{l,l}^*(i+1, j, \lambda_c)} \qquad (3.11) \\
r_{l,l}^*(i, j, \lambda_c) &= \frac{P_{\text{bk}}(i) f_i^*(\lambda_c)}{1 - P_{\text{fw}}(i) f_i^*(\lambda_c) r_{l,l}^*(i+1, j, \lambda_c)}
\end{aligned}
$$

where the following initial conditions must be used:

$$
\begin{aligned}
r_{l,r}^*(j, j, \lambda_c) &= P_{\text{fw}}(j) f_j^*(\lambda_c) \\
r_{l,l}^*(j, j, \lambda_c) &= P_{\text{bk}}(j) f_j^*(\lambda_c)
\end{aligned}
$$

From Figure 3.5, it can be noted that before the first absorption occurs in the system, the sojourn time is obtained through the residual distribution, $rr_{l,r}(0, D-1, t)$. Once entered the absorbing state for the first time, the non-residual sojourn time distributions are used. Thus, the probability that an MT performs no LU operation, between two consecutive incoming calls directed to that MT, is given by

$$
P(\text{no LU}) = 1 - rr_{l,r}^*(0, D-1, \lambda_c) \qquad (3.12)
$$

If we denote by m the number of LU messages triggered between two consecutive incoming calls, the probability of delivering m LU messages is

$$
P(m \text{ LUs}) = rr_{l,r}^*(0, D-1, \lambda_c)[r_{l,r}^*(0, D-1, \lambda_c)]^{m-1}[1 - r_{l,r}^*(0, D-1, \lambda_c)]
$$
$$(3.13)$$

From this expression, the average number of LU messages sent by an MT is given by

$$
\overline{m_{\text{LU}}} = \sum_{m=1}^{\infty} m P(m \text{ LUs}) = \frac{rr_{l,r}^*(0, D-1, \lambda_c)}{1 - r_{l,r}^*(0, D-1, \lambda_c)} \qquad (3.14)
$$

And finally, assuming the cost of a single LU operation is P_{U}, the total LU cost per call arrival, denoted by C_{LU}, is obtained as

$$
C_{\text{LU}} = \overline{m_{\text{LU}}} P_{\text{U}} \qquad (3.15)
$$

3.4.2 LU Cost in a Movement-Based Scheme

In a movement-based scheme, each MT has a movement counter. It is increased with one unit every time the MT crosses a cell boundary. Thus, the LU cost does not depend on wether the movement is directional or not. The LU cost is only dependent on the cell dwell time and the CMR. For computing the LU cost, it should be obtained the probability that an MT moves across k cells before the observation ends. According to Akyildiz et al. [5], this probability is defined as

$$a(k) = \begin{cases} 1 - f_{cr}^*(\lambda_c); & k = 0 \\ f_{cr}^*(\lambda_c)[f_c^*(\lambda_c)]^{k-1}[1 - f_c^*(\lambda_c)]; & k > 0 \end{cases} \qquad (3.16)$$

Let $\overline{m_{LU}}$ be the average number of LUs between two consecutive call arrivals directed to an MT. The probability that an MT delivers m LU messages depends on its number of performed movements, $a(k)$, so that the average number of LUs is obtained as

$$\begin{aligned} \overline{m_{LU}} &= \sum_{m=1}^{\infty} m P(m \text{ LUs}) = \sum_{m=1}^{\infty} m \left[\sum_{i=mD}^{(m+1)D-1} a(i) \right] \\ &= \frac{1 - f_c^*(\lambda_c)}{\theta} \frac{[f_c^*(\lambda_c)]^{D-1}}{1 - [f_c^*(\lambda_c)]^D} \end{aligned} \qquad (3.17)$$

where D is the movement threshold and θ is the CMR. In [5], no closed expression was given for $\overline{m_{LU}}$. The last equality of equation (3.17) was derived in [15]. Later on, it was derived independently in [16].

Assuming the cost of sending a single LU message is P_U, the total LU cost per call arrival in a movement-based scheme, denoted by C_{LU}, is given by

$$C_{LU} = \overline{m_{LU}} P_U \qquad (3.18)$$

Notice that the LU cost in this scheme does not depend on α, so that the movement-based scheme is not influenced by the motion directivity.

3.4.3 Paging Cost in Distance-Based and Movement-Based Schemes

In order to compute the PG cost, we note that the MT is always located in its RA (a set of rings between 0 and $D - 1$). Either a one-step PG or a selective PG strategy may be employed. In a one-step PG policy, all cells within the RA compose the PA and they are polled simultaneously. We denote by N_q the number of cells in each ring q. Assuming the cost of polling a cell is given by P_V, the PG cost in a one-step PG strategy is computed as

$$C_{PG} = P_V \sum_{q=0}^{D-1} N_q \qquad (3.19)$$

where $N_q = 1$ for $q = 0$ and $N_q = 6q$ for $q \geq 0$.

The one-step PG policy is not α-dependent. However, the directional movement influences the PG cost if a selective PG scheme is used, as it usually depends on the probabilities of an MT being roaming in a specific ring when a new incoming call arrives.

As a selective PG strategy, it has been chosen an SDF PG policy [3, 5]. The RA of the MT is divided into $l = min(\eta, D)$ PAs. Each PA is denoted by A_j, with $0 \leq j < l$, and it contains one ring at least. Let s_j and e_j be the numbers of the first ring and the last ring, respectively, that belong to the PA A_j. These values are defined as

$$s_j = \begin{cases} 0; & j = 0 \\ \lfloor \frac{Dj}{l} \rfloor; & j \neq 0 \end{cases} \qquad e_j = \left\lfloor \frac{D(j+1)}{l} \right\rfloor - 1 \qquad (3.20)$$

The MT is roaming in a ring q with probability $P(q)$ when an incoming call arrives. This probability is evaluated in Appendix A. Then, the probability that an MT is roaming in A_j when a new call arrival occurs is

$$P(A_j) = \sum_{q=s_j}^{e_j} P(q) \qquad (3.21)$$

Let N_{A_j} be the number of cells within the PA A_j. Given that an MT is roaming in A_j, the number of polled cells until the MT is located in that PA, $N(A_j)$, is obtained as

$$N(A_j) = \sum_{i=0}^{j} N_{A_i} = \sum_{i=0}^{j} \sum_{q=s_i}^{e_i} N_q \qquad (3.22)$$

From the previous equations, the total PG cost in an SDF PG policy, being P_V the cost of polling a single cell, is

$$C_{PG} = P_V \sum_{k=0}^{l-1} P(A_k) N(A_k) \qquad (3.23)$$

3.5 Numerical Results and Discussion

Once the LU and the PG strategies have been evaluated in the system, the total cost per call arrival in the distance-based and movement-based schemes is obtained as

$$C_T = C_{LU} + C_{PG} \tag{3.24}$$

In Figure 3.6, the total cost per call arrival is shown as a function dependent on the LU threshold for each strategy. This function is computed for different values of the directional movement parameter (α). As it is done in [5, 12, 17], $P_U = 10$ and $P_V = 1$ have been chosen. A high mobility environment has been considered to compare the distance-based and movement-based schemes. This choice has been done because if an MT remains static a few LUs would be performed, thus it would be difficult to compare both strategies. Furthermore, in a microcellular scenario, cells are smaller and the cell dwell time is lower than it is in macrocellular systems. This fact yields a large number of cell boundary crossings that provides a low CMR. Hence, the above mentioned costs have been reported for a CMR $\theta = 0.1$. Besides, both one-step PG (Figure 3.6a) and SDF PG (Figure 3.6b) strategies have been considered. In the SDF PG, a PG delay $\eta = 2$ has been assumed.

The total costs shown in Figure 3.6 are convex functions so that optimum values can be found for the threshold and the total cost of each scheme. We define the optimum cost per call arrival, C_T^*, as the minimum cost per call arrival that can be achieved by adjusting the threshold D. The optimum threshold, D^*, is the threshold that yields the optimum cost [5]. In both distance-based and movement-based strategies, as the threshold increases from D^*, the PG cost dominates the total cost, which results an increasing function of D. For values of D lower than the optimum threshold, the LU cost dominates the total cost [12]. The total cost also varies with the selected PG strategy. In a selective PG policy, the PG cost is lower than in one-step PG. This fact leads to achieve a lower value for C_T^* when SDF PG is used. This also means that the PG cost dominates for higher values of D. Hence, a larger value of D^* is obtained, as it can be observed in Figure 3.6 and Table 3.3.

Coming back to the LU cost, in the distance-based mechanism, the LU cost increases as α grows because an MT moving randomly crosses its RA border less times than an MT moving according to a directional mobility pattern. On the other hand, in the movement-based scheme, the LU cost is not α-dependent (see Section 3.4.2).

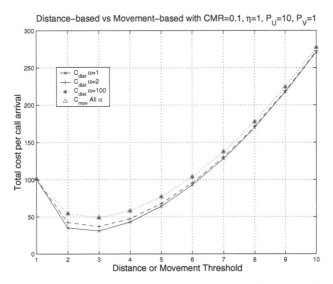

(a) Total cost per call arrival with one-step PG

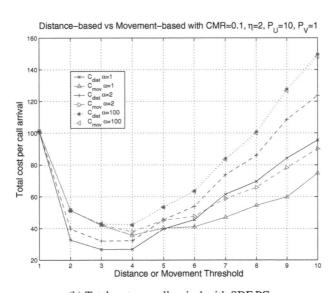

(b) Total cost per call arrival with SDF PG

Figure 3.6 Total costs per call arrival with CMR $\theta = 0.1$, $P_U = 10$ and $P_V = 1$.

Table 3.3 Optimum costs and thresholds in distance-based and movement-based schemes with CMR $\theta = 0.1$.

	PG delay $\eta = 1$			PG delay $\eta = 2$		
	$\alpha = 1$	$\alpha = 2$	$\alpha = 100$	$\alpha = 1$	$\alpha = 2$	$\alpha = 100$
Optimum distance-based cost C_T^*	31.0	36.8	48.7	26.7	32.1	42.2
Optimum distance threshold D^*	3	3	3	3	3	4
Optimum movement-based cost C_T^*		49.1		35.7	38.0	42.4
Optimum movement threshold D^*		3		4	4	4
% of cost outperforming in distance	58.4%	33.4%	0.8%	33.7%	18.4%	0.5%

In the results shown above, three values of the parameter α have been considered. The value $\alpha = 1$ represents the random walk mobility model, in which the distance-based outperforms the movement-based mechanism as it was expected (see [1, 2]). For a value $\alpha = 2$, a directional movement is being modeled so that the MT moves towards a cell of the outer ring with a greater probability than the random walk model ($\alpha = 1$). Under this mobility pattern, the distance-based cost is still lower than the movement-based cost, even though it approaches the movement-based cost as α grows. Thus the outperforming percentage of the distance-based scheme vanishes when α increases (see Table 3.3). In the limit case of α tending to infinity, the total cost of both strategies would be equal because an MT would always move towards a cell in the outer ring after a cell boundary crossing. Then, the LU messages would be triggered in the same locations in both strategies. This extreme case has been approached through $\alpha = 100$, that models a mobility pattern almost totally directional. For $\alpha = 100$, it is shown that the cost of both strategies is similar and the percentage of outperforming in the distance-based scheme is almost negligible. Values of α lower than 1 have not been considered because such values represent nomadicity in the MT mobility pattern. The MT tends to move within its RA so that a few LU messages are triggered in a distance-based strategy, while the LU cost is not α-dependent in a movement-based scheme. The evolution of the distance-based and movement-based costs as α increases is also shown in Figure 3.7, where it is represented the total cost per call a arrival as a function of the movement directivity (α) for both one-step PG (Figure 3.7a) and SDF PG strategies (Figure 3.7b).

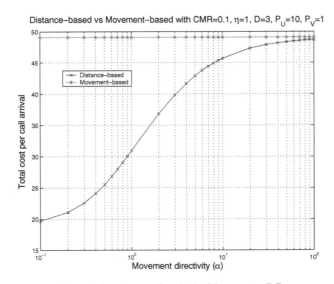

(a) Total cost per call arrival with one-step PG

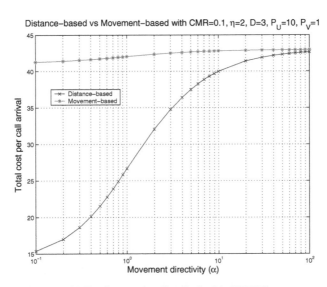

(b) Total cost per call arrival with SDF PG

Figure 3.7 Evolution of the total costs per call arrival depending on the movement directivity with CMR $\theta = 0.1$, and $D = 3$.

When one-step PG is used, the same location management cost is obtained in the movement-based scheme for all values of α. This fact occurs because the movement-based LU cost is not α-dependent (see Section 3.4.2) and neither it is the one-step PG (see Section 3.4.3). However, the distance-based LU cost is α-dependent. The distance-based optimum cost increases from 31.0 up to 48.7 between the cases $\alpha = 1$ and $\alpha = 100$.

When an SDF PG strategy is used, lower values for the optimum costs are obtained. The movement-based strategy performs worse than the distance-based scheme. Nevertheless, it can be observed that the movement-based optimum cost varies from 35.7 to 42.4 when α varies from 1 to 100, whereas the distance-based optimum cost increases from 26.7 to 42.2 for the same values of α. Besides, local changes can be noticed in the gradients of Figure 3.5. They are due to the way in which the PAs grow as the distance and movement thresholds increase. Hence, if the threshold D is a multiple of the PG delay η, all PAs have the same number of rings (D/η). If the threshold increases from D to $D + 1$, the new external ring of the RA (ring D) joins in the external PA according to Akyildiz and coworkers [3, 5]. Therefore, the outer PA increases its size in $6D$ cells, while the other PAs remain with the same size. Averaging the number of cells with the probability of an MT being roaming in each PA, it results a step in the cost curve.

When an MT moves directionally we should consider wether it is better to use a distance-based or a movement-based policy. The former achieves a lower optimum cost, but it requires to compute the distance traveled by the MT. On the other hand, the latter policy performs worse from the optimum cost point of view, although it needs no memory nor processing to be implemented, so that it becomes a simpler mechanism. Hence, a movement-based scheme can be used in these systems due to its similar cost and its simplicity.

3.6 Conclusions

In this paper, we have proposed a new random-directional mobility model for both macrocellular and microcellular scenarios. Movement can be modeled under several degrees of randomness depending on the value of a parameter α. It has been shown that the proposed random-directional model is consistent. For $\alpha = 1$, the random walk model is obtained. For increasing values of α, the movement becomes more and more directional and an MT tends to move towards a cell of the outer ring with a higher probability. When α tends to infinity, the probability of moving forward tends to 1. Also the transition probabilities maintain the dependence on the number of ring (q) and the movement directivity (α).

Our model has been used to obtain the optimum costs of the distance-based and movement-based LU strategies taking into account the MT mobility patterns. For $\alpha = 1$, the distance-based scheme outperforms the movement-based policy, as is well known [2]. However, for increasing values of α, the movement becomes directional and the percentage in which the distance-based outperforms the movement-based policy decreases. In the limit case of α tending to infinity, the cost of both strategies is the same.

The distance-based scheme outperforms the movement-based strategy. Nevertheless, the distance-based policy needs some storage capacity or processing to compute the distance traveled by the MT [4]. On the other hand, the movement-based mechanism only needs a movement counter. Therefore, in scenarios with high movement directivity, the movement-based policy may be more suitable. An example may be found in microcellular scenarios, where cells are smaller so that the number of cell boundary crossings is higher, and the MT does not change its movement direction so often.

Appendix

A Terminal Paging Probabilities

A.1 Distance-Based Schemes

An MT is initially roaming in ring 0. We look for the probability that the MT is roaming in ring q when a new incoming call arrives. There are two excluding cases to be taken into account (see Figure 3.5):

- First case: No LU message was triggered by the MT since the last call arrival. No absorption has occured ($m = 0$).
- Second case: The MT delivered an LU message at least since the last call arrival. Hence, one absorption happened at least ($m \geq 1$).

After visiting the ring q for the first time, the MT could leave and come back to ring q for a certain number of times (that we denote by k). Note that whenever an absorption occurs, the value of k is reset to 0.

Let $P(q, m, k)$ be the probability that an MT is located in ring q when a new incoming call arrives, given that m LUs have been triggered by the MT since the last call arrival and the MT has left and come back to ring q for k times since the last absorption. Then, for $m = 0$,

$$P(q, m = 0, k) = rr_{l,r}^*(0, q - 1, \lambda_c)[1 - f_q^*(\lambda_c)][f_q^*(\lambda_c)]^k$$

$$\times [P_{\text{fw}}(q)r_{l,l}^*(q + 1, D - 1, \lambda_c) + P_{\text{bk}}(q)r_{r,r}^*(0, q - 1, \lambda_c)]^k, \quad (3.25)$$

$$k = 0, 1, 2, \ldots$$

On the other hand, for $m \geq 1$,

$$P(q, m, k) = rr_{l,r}^*(0, D - 1, \lambda_c)[r_{l,r}^*(0, D - 1, \lambda_c)]^{m-1}$$

$$\times r_{l,r}^*(0, q - 1, \lambda_c)[1 - f_q^*(\lambda_c)][f_q^*(\lambda_c)]^k \quad (3.26)$$

$$\times [P_{\text{fw}}(q)r_{l,l}^*(q + 1, D - 1, \lambda_c) + P_{\text{bk}}(q)r_{r,r}^*(0, q - 1, \lambda_c))]^k$$

$$k = 0, 1, 2, \ldots, \quad m = 1, 2, 3, \ldots$$

Gathering all possible values of k, which constitute excluding events, the following probabilities are obtained:

$$P(q, m = 0) = \sum_{k=0}^{\infty} P(q, m = 0, k) \quad (3.27)$$

$$= \frac{rr_{l,r}^*(0, q - 1, \lambda_c)[1 - f_q^*(\lambda_c)]}{1 - f_q^*(\lambda_c)[P_{\text{fw}}(q)r_{l,l}^*(q + 1, D - 1, \lambda_c) + P_{\text{bk}}(q)r_{r,r}^*(0, q - 1, \lambda_c)]}$$

$$P(q, m) = \sum_{k=0}^{\infty} P(q, m, k) \quad (3.28)$$

$$= \frac{rr_{l,r}^*(0, D - 1, \lambda_c)[r_{l,r}^*(0, D - 1, \lambda_c)]^{m-1}r_{l,r}^*(0, q - 1, \lambda_c)[1 - f_q^*(\lambda_c)]}{1 - f_q^*(\lambda_c)[P_{\text{fw}}(q)r_{l,l}^*(q + 1, D - 1, \lambda_c) + P_{\text{bk}}(q)r_{r,r}^*(0, q - 1, \lambda_c)]}$$

$$m = 1, 2, 3, \ldots$$

Adding the previous probabilities for all possible values of m, the probability that an MT is roaming in the ring q, when a call arrival occurs, is given by

$$P(q) = P(q, m = 0) + \sum_{m=1}^{\infty} P(q, m) \quad (3.29)$$

$$= \frac{[1 - f_q^*(\lambda_c)][rr_{l,r}^*(0, q - 1, \lambda_c) + \frac{rr_{l,r}^*(0, D-1, \lambda_c)r_{l,r}^*(0, q-1, \lambda_c)}{1 - r_{l,r}^*(0, D-1, \lambda_c)}]}{1 - f_q^*(\lambda_c)[P_{\text{fw}}(q)r_{l,l}^*(q + 1, D - 1, \lambda_c) + P_{\text{bk}}(q)r_{r,r}^*(0, q - 1, \lambda_c)]}$$

A.2 Movement-Based Schemes

In a movement-based scheme, we consider a matrix that contains all possible transition probabilities between two rings after a single cell boundary crossing as it is done in [5]. It is also assumed that an absorption occurs whenever the movement threshold, D, is reached:

$$
\mathbf{P}(D) = \begin{pmatrix}
p_{0,0} & p_{0,1} & 0 & 0 & \cdots & 0 & 0 & 0 \\
p_{1,0} & p_{1,1} & p_{1,2} & 0 & \cdots & 0 & 0 & 0 \\
0 & p_{2,1} & p_{2,2} & p_{2,3} & \cdots & 0 & 0 & 0 \\
\vdots & \vdots & \vdots & \vdots & \ddots & \vdots & \vdots & \vdots \\
0 & 0 & 0 & 0 & \cdots & p_{D-1,D-2} & p_{D-1,D-1} & p_{D-1,Abs} \\
0 & 0 & 0 & 0 & \cdots & 0 & 0 & 1
\end{pmatrix}
\tag{3.30}
$$

The $p_{i,j}(D)$ element in the matrix represents the probability that an MT moves from a cell in ring i to a cell in ring j after one step – a single cell boundary crossing (see Section 3.3.2). For more than one step, being k the number of steps, we define

$$
\mathbf{P}^{(k)}(D) = \begin{cases}
\mathbf{P}(D); & k = 1 \\
\mathbf{P}(D) \times \mathbf{P}^{(k-1)}(D); & k > 1
\end{cases}
\tag{3.31}
$$

being $p_{i,j}^{(k)}(D)$ the probability of moving from a cell within ring i to a cell in ring j after k cell boundary crossings.

Thus, the probability that an MT is located in the ring q when a call arrival occurs is given by

$$
P(q) = \sum_{k=0}^{\infty} a(k) p_{0,q}^{(k \bmod D)}(D)
\tag{3.32}
$$

Acknowledgements

This work has been supported by the *Spanish Ministry of Science and Technology* under projects TEC2004-06437-C05-01 and TSI2007-66869-C02-02.

References

[1] A. Bar-Noy, I. Kessler and M. Sidi, Mobile users: To update or not to update?, in *Proceedings of INFOCOM'94*, IEEE, pp. 570–576, June 1994.

[2] A. Bar-Noy, I. Kessler and M. Sidi, Mobile users: To update or not to update?, *Wireless Communications Journal*, vol. 1, pp. 175–185, July 1995.

[3] J. S. M. Ho and I. F. Akyildiz, Mobile user location update and paging under delay constraints, *Wireless Networks*, vol. 1, pp. 413–425, December 1995.

[4] R. Vidal, J. Paradells and J. Casademont, Labelling mechanism to support distance-based dynamic location updating in cellular networks, *IEE Electronics Letters*, vol. 39, no. 20, pp. 1471–1472, October 2003.

[5] I. F. Akyildiz, J. S. M. Ho and Y.-B. Lin, Movement-based location update and selective paging for PCS networks, *IEEE/ACM Transactions on Networking*, vol. 4, pp. 629–638, August 1996.

[6] T. Camp, J. Boleng and V. Davies, A survey of mobility models for ad hoc network research, *Wireless Communications and Mobile Computing*, vol. 2, pp. 483–502, August 2002.

[7] K.-H. Chiang and N. Shenoy, A 2D random-walk mobility model for location-management studies in wireless networks, *IEEE Transactions on Vehicular Technology*, vol. 53, pp. 413–424, March 2004.

[8] I. F. Akyildiz, Y.-B. Lin, W.-R. Lai and R.-J. Chen, A new random walk model for PCS networks, *IEEE Journal on Selected Areas in Communications*, vol. 18, pp. 1254–1260, July 2000.

[9] A. Lombardo, S. Palazzo and G. Schembra, A comparison of adaptive location tracking schemes in personal communications networks, *International Journal of Wireless Information Networks*, vol. 7, no. 2, pp. 79–89, April 2000.

[10] T. Tung and A. Jamalipour, Adaptive location management strategy to the distance-based location update technique for cellular networks, in *Wireless Communications and Networking Conference*, IEEE, vol. 1, pp. 172–176, March 2004.

[11] M. M. Zonoozi and P. Dassanayake, User mobility modeling and characterization of mobility patterns, *IEEE Journal on Selected Areas in Communications*, vol. 15, pp. 1239–1252, September 1997.

[12] P. Garcia-Escalle, V. Casares-Giner and J. Mataix-Oltra, Reducing location update and paging costs in a PCS network, *IEEE Transactions on Wireless Communications*, vol. 1, pp. 200–209, January 2002.

[13] L. Kleinrock, *Queueing Systems*, vol. 1, John Wiley & Sons, New York, 1975.

[14] V. Casares-Giner, Variable bit rate using hysteresis thresholds, *Telecommunication Systems*, vol. 17, pp. 31–62, May 2001.

[15] V. Casares-Giner and J. Mataix-Oltra, On movement-based mobility tracking strategy-an enhanced version, *IEEE Communications Letters*, vol. 2, pp. 45–47, February 1998.

[16] J. Li, H. Kameda and K. Li, Optimal dynamic mobility management for PCS networks, *IEEE Trans. Networking*, vol. 8, no. 3, pp. 319–327, June 2000.

[17] Y. Fang, Movement-based mobility management and trade off analysis for wireless mobile networks, *IEEE Transactions on Computers*, vol. 52, pp. 791–803, June 2003.

4

Performance Evaluation of the TIMIP/sMIP Terminal Independent Mobile Architecture

Pedro Estrela[1], Teresa Vazão[1] and Mário Serafim Nunes[2]

[1] IST-TagusPark TU Lisbon/INESC-ID, Av. Prof. Cavaco Silva, 2744-016 Porto Salvo, Portugal; e-mail: pedro.estrela@inesc.pt, teresa.vazao@tagus.ist.utl.pt,
[2] IST TU Lisbon/INESC-ID/INOV, 2744-016 Porto Salvo, Portugal;
e-mail: mario.nunes@inesc.pt

Abstract

All IP mobility protocols currently proposed by the IETF assume that the mobile nodes always have a mobility-aware IP stack, which is still a scenario that can seldom be found nowadays.

The deployment of a terminal independent mobility solution, which efficiently supports handover and resource optimisation, can increase mobile services offer, as it may be easily deployed by changing the network infrastructure, while using the existing terminals.

This paper proposes a classification framework based on efficiency for existing mobility protocols; describes and classifies a terminal's independent mobility architecture – TIMIP/sMIP – and evaluates the performance of this and alternative solutions via simulation studies and trial experiments, in a variety of movement and traffic scenarios.

Keywords: Micro-mobility, transparency, TIMIP, surrogate MIP, IP mobility, NS2.

D. D. Kouvatsos (ed.), Mobility Management and Quality-of-Service for Heterogeneous Networks, 79–109.

4.1 Introduction

The number of people that use wireless Local Area Networks (wLANs) is increasing at a very fast rate, so it is foreseen that wLANs will have a major impact in the Internet [1]. Today, Mobile IP (MIP) [2] is the standard mobility solution for heterogeneous networks, representing a good approach for the mobility in wide area networks, where handover performance is not a major issue. Although developed within the scope of IPv4, it is accepted now as the macro mobility solution for IPv6 networks in its correspondent version (MIPv6) [3].

As far as micro-mobility is concerned, quick and smooth routing changes are required in order to achieve seamless handover, resulting on the work of Cellular IP (CIP) [4], HAWAII [5] and Hierarchical MIP (hMIP) [6]. In spite of their differences, all of them lack support for legacy terminals. Today, most of the nodes that would benefit from mobility, like laptops and PDAs, still use legacy IPv4 stacks. This might represent a constraint that prevents the deployment of mobility in the short term, as an entire migration to MIPv4, or even IPv6 coupled with MIPv6 is not envisaged for the near future [7, 8].

A terminal's independent mobility architecture was already proposed in [9, 10], and implemented in a testbed prototype. This architecture comprises a micro-mobility solution, named Terminal Independent Mobile IP (TIMIP), coupled with a MIP compatible adaptation for macro-mobility scenarios, named Surrogate MIP (sMIP). The unique transparency support will be of major importance to future heterogeneous 4G networks, being TIMIP already appointed as part of the 4G mobile's architecture [11].

This paper presents a performance evaluation study of several mobility solutions. This evaluation study is based on an original framework used to classify these proposals, according to their handover and resource utilisation efficiency. As far as TIMIP and sMIP are concerned, their architectures are also presented and classified, as well as simulation studies, trial experiments and results.

The remaining part of this paper is organised as follows: Section 4.2 presents the framework and classifies the protocols according to it. Section 4.3 presents the TIMIP/sMIP architecture, its operations, and its classification relative to the framework. In Section 4.4 we present the simulation studies and in Section 4.5 the trial experiments. The paper ends with some conclusions and future work in Section 4.6.

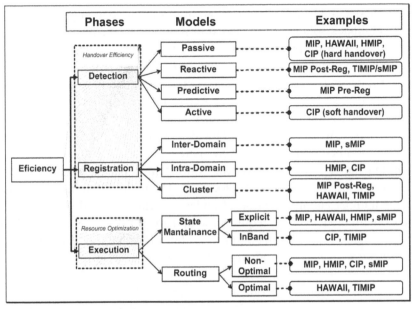

Figure 4.1 Classification framework.

4.2 Classification Framework

4.2.1 Framework Models Characterisation

Generally, efficiency is an essential and fairly well known metric used to compare and evaluate the mobility proposals suitability for each given scenario [7, 8]. For this, the efficiency level, as a whole, can be considered as the aggregation of simpler efficiency components, which can be combined into a single concept that enables the protocol evaluation for each specific utilisation scenario. In the proposed framework, the different phases of the handover procedure are analysed and several mobility solutions are classified accordingly (see Figure 4.1).

The detection phase considers the operations needed for the IP layer to become aware that the network has outdated routing information about the MN's location. The traditional IP layer is not aware of the MN's movements, as it is independent of the lower layers. However, on most radio technologies, the L2 is fully aware of these movements, and thus it is able to help L3 on their detection. The level of integration between L2 and L3 results in different detection phase models.

If a Passive model is used, no integration is required, being this either the sole or a fallback solution of the existing mobility protocols. In the Reactive model, there is a minimal cooperation between the two layers, as the L2 operations are simply exposed, without further interactions. In a Predictive model, L2 becomes responsible to predict handovers and inform L3 before they actually take place. Finally, the Active model requires a total cooperation and integration of L2, as L3 fully takes control on the L2 handover operations. Regarding this phase, L3 handover latency is expected to decrease as cooperation between the two layers increases.

During the registration phase the routing information is updated at all necessary nodes with the new MN location. The main differences are related with the number and the location of nodes associated with this process. The Inter-Domain model always requires the updating of nodes outside the MN's current domain, generally. In contrast, the Intra-Domain model only requires the notification of nodes belonging to the current MN domain, but which are located outside the shortest path between the old and new MN's locations. Finally, in the Cluster model, only the nodes belonging to the shortest possible path, between the old and new APs, must be notified. In this phase, handover latency is expected to decrease as long as the critical network nodes to be informed are kept closer to the MN.

The execution phase occurs between handovers, when the routing information is kept updated and data traffic reaches the MN in a stable way. The critical issues to consider are those that lead to an optimisation of the network resources during stationary periods, which occurs when the MN stays in the same AP. The efficiency of the mobility protocols is related to both the state maintenance overhead and the actual routing paths used to carry data traffic.

The Explicit State Maintenance model implies the existence of signalling messages, used to refresh MN information at the necessary nodes. Its dual is the Implicit State Maintenance model, which uses IP traffic to refresh MN information, at the relevant nodes. Concerning data paths, the Non-Optimal Route model uses paths that may be longer than needed. The Optimal Route model sends data through the shortest path, for the generality of situations. Regarding this phase, efficiency increases when fewer resources are needed both to maintain the MN active and to forward its data packets.

4.2.2 Classification of Existing Proposals

This section uses the previous defined models to classify existing protocols in the proposed framework, being this classification also summarised in Figure 4.1. The MIP protocol is currently the standard for macro-mobility, while CIP, HAWAII and hMIP are aimed to support micro-mobility. All of them have been previously described [13] and compared [15]. Additionally, several extensions to these proposals have also been made, and are also included in this classification.

The MIP protocol is the classical solution, suitable for large movements typical of nomadic computing. In its basic version, it defines two mobile agents – Home Agent (HA) and Foreign Agent (FA), being the packets always forwarded by the HA via a tunnel to the MN's current FA. Every time the MN moves, the terminal detects this event using MIP beacons, and starts a registration process, which always involves the HA, in order to re-establish the tunnel to the new location. As such, MIP features Passive Detection and Inter-Domain Registration models, with long latencies being expected for each handover. Regarding Execution phase, standard MIP has no optimisations, using Explicit State Maintenance and Non-Optimal Route models (also known as triangulation).

The extension proposed in [14] – low latency handovers – adds the notion of triggers to the MIP architecture, where standard MIP detection is aided by the link layer: for each handover event, an L2 trigger is sent to the MN and/or to the involved agents, either before or after the transition. In this framework, the former is an example of Predictive Detection, while the later represents a Reactive Detection. Although their advantages, there are still difficulties with the practical implementations of these L2 triggers, related to providing L3 information (e.g. IP addresses), which may not be available in L2 nodes.

Additionally, the post-reg trigger also presents an interesting optimisation to the Registration phase, by proposing a fast registration that removes the HA from the critical path. This extension creates a temporary tunnel between the involved FAs for MN data delivery, while the regular MIP registration runs in parallel. In this framework, this is a case of the Cluster model, and much lower latencies are expected in registration than those achieved by standard MIP.

Another MIP extension – hMIP – adds micro-mobility capabilities to MIP, by extending the single HA-FA tunnel to a hierarchy of FAs (usually with two single levels, for robustness purposes. When the MN moves, it only notifies the new FA and the Gateway Foreign Agent (GFA) of the current domain, which simply redirects the local tunnel, thus removing the HA from the critical path. Thus, hMIP has support for Intra-Domain Registration, resulting in smaller registration latencies. The remaining phases are classified as in standard MIP, as hMIP only optimises registration.

The CIP and HAWAII protocols are based on principles different from MIP's, being specially designed to support micro-mobility with higher efficiency. Both of them use an architecture fairly different from MIP's, being limited to domains structured in a tree topology.

On its basic form, the CIP protocol support both Passive Detection and Intra-Domain Registration models, as registration updates are propagated up the tree in direction of the GW. With the necessary L2 cooperation, CIP also supports a soft handover that provides an Active form of detection, where the network layer is able to control the link layer in order to start the handover while receiving packets through the old AP. When the update message reaches the crossover node, packets are sent to both APs at the same time (bicasting), resulting in marginal packet drops due to the L3 handover. However it should be noted that this mechanism can incur in significant increase of the L2 handovers latency, as the terminal may need to switch several times between the AP's frequencies.

Another optimisation of the CIP protocol is the support for Implicit State Maintenance model, as the regular IP data traffic is used to refresh routing entries for active MN, while explicitly generated signalling information is used only for idle terminals. However, CIP features Non-Optimal Route model, as all packets generated inside the network must always reach the domain's Gateway (GW), instead of being forwarded using the shortest path in the network tree.

By supporting MIP-compatible terminals, HAWAII inherits from it Passive detection and Explicit signalling models. However, signalling transparently follows the shortest path inside the network, between the two involved APs. Thus, a Cluster Registration model is used, which may increase the handover performance. In a similar way, for Execution phase this protocol also features Optimal Route model, as the data packets are routed inside the sub-tree through the shortest paths, without need to being routed through the domain's GW.

4.3 TIMIP/sMIP Mobility Solution

The mobile architecture evaluated in this paper is based on a dual mobility protocol, which uses TIMIP to support micro-mobility and an extended version of MIP (sMIP) to support macro-mobility.

The support of IP mobility requires changes and additions to the IP protocol, for both the network equipment and mobile nodes. This way, the existing standard IP terminals [16] cannot benefit of the new technology,

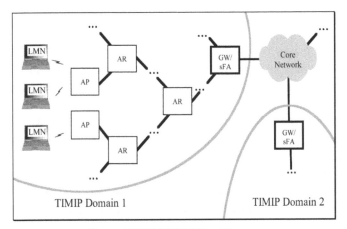

Figure 4.2 TIMIP/sMIP architecture.

unless their protocol stacks are changed with the addition of mobility support. However, this long migration cycle of all terminals is not currently envisaged for the near future.

To address this problem, TIMIP/sMIP provides a terminal independent mobility solution that explicitly supports Legacy Mobile Nodes (LMNs), with high efficiency. For this, the network is the sole responsible for the mobility actions that are typically executed by the terminals while roaming, implementing a "surrogate behaviour", originally defined and proposed in [17].

For Detection phase, the identification of legacy MN's movement is done at network-side only. Concerning the Registration phase, the network generates signalling information that triggers a routing update process. There are no requirements for the Execution phase concerning this support.

4.3.1 Architecture and Operations

4.3.1.1 Global architecture

The global TIMIP/sMIP network architecture proposed is outlined in Figure 4.2.

In this architecture mobile portions of the Internet are split into multiple domains. Inside each TIMIP Domain, the network elements are organised on a tree structure topology, being the IP routers that execute the TIMIP mobile protocol and cooperate among themselves to provide intra-domain mobility support to the terminals. At the top of the tree, a special network element called Gateway (GW) is used to centralise the management functions, while presenting to the Internet the domain as a classical IP network. On the other hand, the tree leaves contain the Access Points (APs), being connected via the Access Routers (ARs). The remaining components of a TIMIP domain are the Legacy Mobile Nodes (LMN), which can transparently roam between the APs. These terminals only have to be able to connect to the network via a suitable wireless interface, as it is the network that automatically reconfigures itself in order to provide the necessary IP connectivity.

Regarding the macro-mobility component, it is supported by the sMIP protocol, which is an adaptation of standard MIP with specific extensions to support both terminal independence and an optimised detection of the LMNs movements linked to the TIMIP operations. For this, the standard MIP agents are extended with extensions to detect movements and generate standard MIP signalling automatically for the terminals, on its behalf, resulting thus in surrogate agents sHA and sFA. When used in TIMIP domains, the sHA/sFA agent is to be unique and co-located with the GW.

4.3.1.2 TIMIP Operations

In TIMIP, the Detection phase is executed continuously by the APs connected to the network to track the movements of the LMNs between APs. To perform this task, the protocol uses a single primitive that signals the attachment of the LMN to the AP. This primitive uses all available L2 information for accurate localisation of terminal. If the required L2 information is not available, or is not enough, then a Passive Detection model is used, via a Generic Detection Algorithm [10]. As such, on the most efficient case, when the LMN attaches to an AP and performs an L2 explicit association, the L3 will be informed of the LMN's arrival via this primitive.

During the Registration phase, the LMN location is dynamically updated by the network itself, keeping that consistent with the LMN movements inside the domain. For this, the AP that detected the LMN in the previous phase is responsible for starting the registration operation, in order to inform the other interested nodes of the network about the LMN movement.

When the LMN arrives at a TIMIP domain, a Power-up operation is issued in the network, as depicted in Figure 4.3a. After detection, the AP reconfigures its own routing table (step 1) and issues a TIMIP update message that travels node-by-node up the tree to the GW, informing the necessary routers about the next hop to the terminal. Each node that receives this update message reconfigures its own routing table, and sends the message up the tree (steps 1 to 3). Reliability mechanisms were embedded into TIMIP sig-

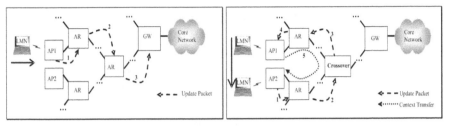

Figure 4.3 TIMIP registration: (a) power-up, (b) handover reconfiguration.

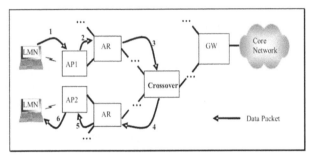

Figure 4.4 TIMIP execution: Intra-domain routing.

nalling messages, by using acknowledge packets and clock synchronisation procedures.

The outlined method for creating the initial routing path is also used to reconfigure it when the terminal roams inside the domain (Figure 4.3b). In these subsequent handover operations, TIMIP's signalling is again generated by the new AP, but is directed to the old AP via the local sub-tree that connects them. As no information about the LMN's old AP is available at the network-side, the network has to infer about its location using the previous outdated routing paths. Signalling is then sent using these paths (steps 3 and 4), in a process that also updates them with the new LMN's next-hop information. Outside this local subtree, the other TIMIP nodes remain with the same routing paths, resulting in faster reconfigurations. After this reconfiguration, the LMN's context present at the old AP is transferred to the new, using the network's routing paths (step 5).

The Execution phase is divided in two parts, downlink and uplink. When a network node receives a data packet, its destination is checked against the node's routing table. If a match is found, the packet is forwarded to the next hop defined in that entry (downlink routing); in the opposite case, the packet is forwarded to the upstream node, towards the GW (uplink routing). From these forwarding rules results that the packets destined to a terminal located in the same TIMIP domain, only in the worst case reach the domain's GW, as illustrated in Figure 4.4.

This phase also optimises the routing tables refresh, by using data packets as a proof of living for active terminals, avoiding additional overhead to perform state maintenance. On the other hand, the inactive terminals routing paths are refreshed by forcing the terminal to reply to standard signalling messages, subject to a backoff procedure, as described in [10].

4.3.2 Characterisation in the Proposed Framework

TIMIP primarily uses the Reactive Detection model, as the network tracks LMN movements by the use of a single L2 primitive that signals the attachment of LMN to the AP. Thus, there is a minimal cooperation between the two layers, in which the L2 operations are simply exposed without further interactions. If this reactive interaction is not possible, the TIMIP protocol can also use a Passive Detection, via the Generic Detection Algorithm already referred.

After detection, the LMN location is dynamically updated inside the domain, using a Cluster Registration model, where the update messages always follows the shortest path inside the network, between the two APs involved.

In the Execution phase, TIMIP uses the Implicit State Maintenance model, as regular data traffic is used to refresh routing entries for actives LMNs. TIMIP also use the Optimal Route model, as packets are always forwarded using the shortest path in the network tree. For intra-domain traffic, packets go up only until they reach the crossover node (uplink routing), being then directly forwarded down to the destination's AP (downlink routing).

4.4 Simulation Studies

The efficiency of existing micro-mobility solutions were evaluated by simulation and compared to TIMIP, by measuring the latency of the handover and the optimisation of network resources for a variety of traffic types, locations and movements. As TIMIP's macro-mobility solution is MIP-compliant, its performance is similar to the one presented by MIP and so it will not be addressed in this section.

The simulations were carried on Network Simulator (NS), version 2.26, enhanced with CIMS v1.0 mobility additions that provide support to hMIP (1-level), CIP and HAWAII, which were upgraded to the latest version of the

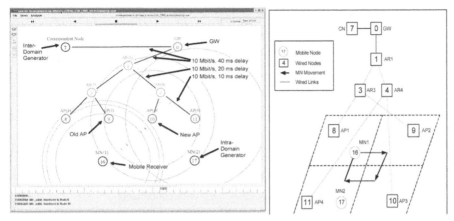

Figure 4.5 Simulation scenario: (a) logical deployment, (b) physical deployment.

simulator. Additionally, NS was modified both to support the TIMIP protocol, and to simulate 802.11 infra-structured behaviour with multiple channels. This forces L2 hard handover operations, where a station can only receive packets via its current associated AP, being this behaviour much closer to real 802.11 networks [18] than the soft-handovers paradigm that were considered in previous simulations studies, as in [15].

From the set of CIMS options, a sample subset was selected: CIP hard-handover option, due to L2 hard-handovers being the ones used in real 802.11 networks and HAWAII Multiple-Stream-Forwarding (MSF), due to the interesting packet buffering possibility.

Several sets of simulations were performed, as described in the following sections, using the network described in Figure 4.5. This network contains a single domain with 4 APs, that independently manage each one a 1 Mbit/s 802.11 cell, physically organised as depicted in Figure 4.5b. The APs are interconnected to the single GW by a series of internal nodes, organised on a tree structure; all internal connections are point-to-point wired links of 10 Mbit/s and increasing delays, for each hierarchical level closer to the GW (respectively 10, 20 and 40 ms).

The network features 2 MNs connected to it, which benefit from the mobility service. The first MN will roam inside the domain, performing circular movements to connect to each AP, in sequence, while receiving data. In each test, such traffic can either be originated inside the network, by the second MN located in AP4, or outside the network, by the correspondent node connected to the GW.

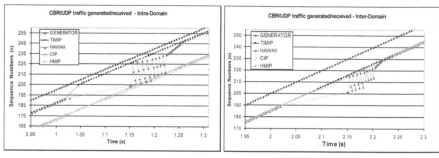

Figure 4.6 Handover latency: CBR intra and inter-domain traffic.

Table 4.1 Handover latency for different protocols.

Handover latency [ms]	Intra-Domain	Inter-Domain
TIMIP	27.3	67.3
HAWAII	196.8	184.5
CIP	69.7	75.4
hMIP	151.4	155.2

4.4.1 Simulation Experiment A – UDP Discrete Handover

The first experiment evaluated both handover latency and resource optimisation. For this, a Constant Bit Rate (CBR) traffic source, generating 200 pkt/s of 30 bytes, is allocated either outside the domain (Node 7), or inside it (Mobile Node 17), respectively for intra-domain and inter-domain traffic types. The traffic receiver is MN 16, which roams between AP 2 (Node 9) and AP 3 (Node 10).

The respective results are shown on Figure 4.6, which represents the packet sequence number at generation and reception for each protocol, versus time.

4.4.1.1 Evaluation of the handover efficiency

The handover latency represents the time gap between the first packet received by the new AP (AP 3) and the last packet received by the old AP (AP 2). This gap may be estimated by the number of missing packets during a roaming situation. Table 4.1 contains the handover latency, obtained from the data series used to generate Figure 4.6.

Considering intra-domain traffic situations, all protocols feature packet drops at the old AP, as the terminal must switch to the new AP channel; in HAWAII there are also out-of-order packets, immediately after the handover, for a short period of time. While the terminal is stationary at an AP, TIMIP and HAWAII protocols have similar results for routing delay, both lower than those of hMIP and CIP.

A more detailed analysis, shows that TIMIP takes 27.3 ms to perform handover, as the last packet received by AP2 (Node 9) is received at time 1.0177 s (packet 191) and the reception of the first one by AP3 (Node 10) occurs at time 1.0449 s (packet 204). Similarly, HAWAII, CIP, hMIP, have handover times of 196.8, 69.7 and 151.4 ms, respectively (summarised in Table 4.1). TIMIP has the lowest latency, as it only has to inform Node 4 to re-establish routing (crossover node); CIP must also inform Node 1, as all packets must pass through the GW, they will only be diverted on the first node common to both old and new paths (Node 1); hMIP must inform the gateway, in order to reconfigure the local tunnel; finally, although HAWAII must inform the same nodes as TIMIP, its routing updating starts at the old AP. Due to this, out-of-order packets are received, which for a real-time CBR flow counts as packets lost.

This experiment shows that, concerning the efficiency of Registration phase defined previously in Section 2, both TIMIP and HAWAII limit their updates to the minimum set of nodes that must be informed, and thus, can have the most efficient registration model – Cluster. However, by introducing methods that cause out-of-order packets, HAWAII can cancel this benefit, as showed above. Regarding the hMIP protocol, and also CIP (in a lesser degree), there are nodes inside the domain, further away from the MN's location, that need also to be informed about its new location – the domain's GW – resulting in higher handover times, caused by an Intra-Domain Registration model.

Concerning inter-domain traffic situations, packets are also dropped by the old AP while the terminal switches to the new AP channel; in HAWAII there are also out-of-order packets, immediately after the handover, during a short period of time. While the terminal is stable at an AP, all protocols have similar results for routing delay.

A more detailed analysis also shows that TIMIP takes 67.3 ms to perform handover, as the last packet received from AP2 (Node 9) is received at time 2.0178 s (packet 189) and the reception of the first one from AP3 (Node 10) occurs at time 2.0851 s (packet 202). Similarly, HAWAII, CIP and hMIP, have handover times of 184.5, 75.4 and 155.2 ms, respectively. TIMIP has

Table 4.2 Average packet delay per location.

Delay [ms]	Intra-Domain		Inter-Domain	
	AP2 (Node 9)	AP3 (Node 10)	AP2 (Node 9)	AP3 (Node 10)
TIMIP	62.4	22.3	72.5	72.5
HAWAII	62.4	22.3	72.5	72.5
CIP	142.7	142.7	72.5	72.5
hMIP	142.6	142.6	72.5	72.5

the lowest latency, as it only has to inform Node 4 and Node 1 to re-establish routing (crossover node); in the same group, CIP has a similar handover time, as the first node common to both paths is also Node 1. Like the previous case, hMIP must always inform the GW, in order to reconfigure the tunnel, thus resulting in an higher time; also as before, HAWAII must inform the same nodes as TIMIP, but as routing updating starts at the old AP, out-of-order packets are received, resulting in the highest handover time results.

This experiment shows that, concerning the efficiency of Registration phase, again both TIMIP and HAWAII limit their updates to the minimum set of nodes that must be informed, and thus, have the Cluster registration model, although HAWAII can cancel its benefit by causing packets to arrive out-of-order. As CIP and hMIP force all packets to pass through the domain's gateway, the registration classification defined previously for these protocols remain unchanged with inter-domain traffic.

4.4.1.2 Evaluation of the Resource Utilisation

Experiment A can also be used to evaluate the resource utilisation of the different protocols, for both traffic types, by checking the end-to-end latency that the packets suffer when passing through the network. This metric, of key importance to real-time services deployment, can be checked during stable conditions that happen before and after the handovers, by measuring the time interval between generation and reception of data packets. Table 4.2 contains these delay values, obtained from the data series used to compose Figure 4.6.

In the intra-domain case, during stable conditions, the gap between sending and receiving packets is smaller in TIMIP and HAWAII than in the other solutions. This gap represents the end-to-end delay. In TIMIP the average delay is 62.4 ms on AP2 (Node 9), being this value reduced to 22.3 ms on

AP3 (Node 10). Regarding the other protocols, HAWAII, CIP, hMIP have, respectively, 62.4, 142.7 and 142.6 ms delay on AP2, and 22.3, 142.7, 142.6 ms, respectively, on AP3. From these values, it is quite clear that both CIP and hMIP impose much higher delays for intra-domain type of traffic.

These results show that, for intra-domain traffic, both TIMIP and HAWAII can route packets efficiently inside the network, while both CIP and hMIP force all packets to pass through the domain's gateway. Thus, considering the efficiency of routing at the Execution phase, both TIMIP and HAWAII feature Optimal Intra-Domain Route model, by using the shortest paths inside the network, which optimises both the resource utilisation and the end-to-end delay. On the other hand, by forcing all traffic to pass at the domain's gateway, both hMIP and CIP have a non-optimal routing execution, resulting in longer delays and higher link usage inside the network.

In the inter-domain situation, during stable conditions, all end-to-end delays incurred by the protocols have similar values, in all MN's locations. For TIMIP, the average routing delay is 72.5 ms, for both AP2 (Node 9) and AP3 (Node 10). This value is also the same for the other protocols – HAWAII, CIP and hMIP, in both locations.

These results show that, for inter-domain traffic, all protocols route packets in the same way inside the network, without penalties associated with the requirement for traffic to pass through the domain's gateway (case of CIP and hMIP). This causes all protocols to have optimal routing for inter-domain traffic.

4.4.2 Simulation Experiment B – Multiple UDP Discrete Handovers

The second experiment evaluated resource utilisation, using a different metric – the sum of all packets forwarded inside the network, in all network nodes, per time interval. Several sequential handovers are simulated, where the MN starts at the first AP (AP1 – Node 8), and moves up to the last (AP4 – Node 11). A CBR traffic source generating 100 pkt/s of 30 bytes, is allocated either outside the domain (Node 7), or inside it (Mobile Node 17), in order to study the effect of intra and inter domain traffic in all locations in the network. This CBR source starts its transmission at time 1 second.

The average number of packets forwarded in the wired network is measured by the number of packets forwarded at each hop, being this summed for all network nodes, for each time interval of 50 ms. The respective results are shown on Figure 4.7, with the handover time instants that were taken directly from the simulation dumps present in Table 4.3.

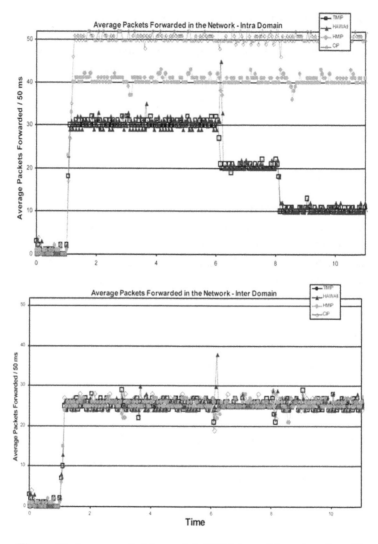

Figure 4.7 Average packet forwarded: CBR intra and inter-domain traffic.

4.4.2.1 Evaluation of the Resource Utilisation

Concerning intra-domain roaming situations, all protocols exhibit fairly constant resource utilisation of the network in the periods that the terminal is stable at each AP. The figure also shows that both TIMIP and HAWAII protocols have similar results for their network resource utilisation, which varies according to the present MN's location. However, this resource utilisation

Table 4.3 Instant of handovers at each AP.

Handover time [s]	AP2 (Node 9)	AP3 (Node 10)	AP4 (Node 11)
TIMIP	3.56	6.03	8.09
HAWAII	3.57	6.03	8.10
CIP	3.56	6.02	8.09
hMIP	3.52	6.09	8.45

is always lower than those of hMIP and CIP, as these have fairly constant results for all positions of the MN inside the network. Finally, it is possible to confirm the approximate instants of the handovers (present in Table 4.3), by observing the sharply changes incurred at these time instants in the graphs.

A more detailed analysis shows also that TIMIP and HAWAII forward an average of 30 packets per time interval, relative to 6 hops, when the MN is at the leftmost part of the domain (AP1 – Node 8); concerning hMIP, this protocol forwards 40 packets per time interval, relative to the 8 hops that the packets must transverse to reach the GW. CIP has the largest resource utilisation, by forwarding 50 packets per time interval, as its routing additionally forces packets to completely exit the domain, in order to being treated as coming from outside the domain. This results that packets must reach Node 7, summing thus a total of 10 hops per packet.

At time 3 s, the MN performs a handover to the next AP (AP2 – Node 9), but the resource utilisation of the protocols remains equal for all protocols. This happens because, in this topology, the new MN location is reachable by the same number of hops.

The situation changes at time 6 s, where a handover occurs to AP3 (Node 10). Here, both TIMIP and HAWAII reduce their network utilisation to 20 packets per time interval, as packets only need 4 hops to reach the new location of the MN. At this instant of time, there is a spike in HAWAII, which is a result of the out-of-order packets, due to the routing reconfigurations performed. As before, both hMIP and CIP maintain their high resource utilisation, as packets are forced to pass the gateway, maintaining the 7 hops minimum that was previously described.

The next change happens around time 8 s, where the last handover changes the MN point of attachment to the rightmost part of the domain – AP4 (Node 11). In this situation, both TIMIP and HAWAII reduce their utilisation to only 10 packets per time interval, relative to the two single hops needed to reach the MN1, through the AP4 (Node 11); again, CIP and hMIP must still force packets to reach the gateway, maintaining the minimum of 8 hops.

The results show, considering the efficiency of routing at the Execution phase, that both TIMIP and HAWAII feature optimal routing, by using the shortest paths inside the network, which optimises both the resource optimisation and the end-to-end delay. On the other hand, by forcing all traffic to pass at the domain's gateway (or even beyond), both hMIP and CIP have a non-optimal routing execution, resulting in longer delays and higher link usage inside the network.

Concerning inter-domain roaming situations, all protocols exhibit stationary resource utilisation of the network in the periods that the terminal is stable at each AP, being this utilisation always the same for all MN locations inside the domain. Again, it can be verified in the protocol behaviour the handover instants (presented in Table 4.3), by observing the sharp changes incurred in the number of packets forwarded per time interval.

A more detailed analysis shows also that all protocols forward an average of 25 packets per time interval, relative to 5 hops needed to reach all existing APs, including the source node (Node 7). These results show that, for inter-domain traffic, all protocols route packets in the same way inside the network.

4.4.3 Simulation Experiment C – TCP Discrete Handover

This experiment evaluates the effect of handover latency on reliable TCP flows. For this, a File Transfer Protocol (FTP) traffic flow, using TCP Tahoe agents, is established, using the same conditions of the CBR previous experiment (intra/inter domain). Again, the MN 16 will be the receiver of a TCP flow consisting of 1000 bytes packets, while roaming between AP2 (Node 9) and AP3 (Node 10). However, in contrast to the previous experiment, the agents will try to use as much bandwidth as possible, as determined by TCP-fairness. To ensure that all losses and retransmissions are due to the mobility protocols routing reconfigurations only, a 0.4 Mbit/s bottleneck is applied to the sender, which shapes the TCP flow to a sufficient low rate, avoiding random losses at the wireless links (that as their raw capacity is only 1 Mbit/s).

The respective results are shown in Figure 4.8, which represents the TCP and ACK sequence numbers at the mobile receiver for each protocol, versus time.

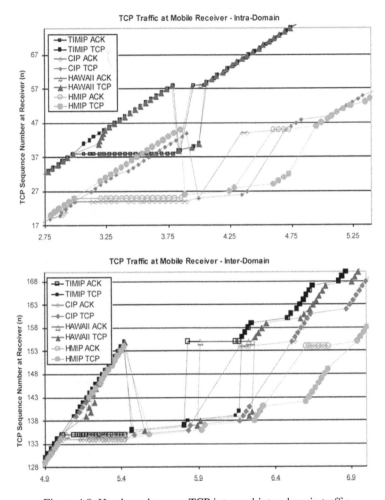

Figure 4.8 Handover latency: TCP intra and inter-domain traffic.

The handover latency represents the time gap between the last packet received by the old AP (AP2) and the first packet received by the new AP (AP3) after any necessary retransmissions of lost packets. This time gap measures the impact of handover in TCP flows, being influenced by the amount of lost packets and the Round Trip Time (RTT) that the TCP packets experience. Table 4.4 contains the handover latency, obtained from the data series used to generate Figure 4.8 using the method outlined above. The table also contains

Table 4.4 Handover latency and lost packets.

	Intra-Domain		Inter-Domain	
	Handover latency [s]	Lost packets	Handover latency [s]	Lost packets
TIMIP	1.08	3	1.18	4
HAWAII	1.07	3	1.26	4
CIP	1.71	3	1.50	5
hMIP	1.96	5	1.93	9

the number of lost TCP packets, which result of the routing reconfigurations, as studied in the previous simulations.

Considering intra-domain traffic situations, all protocols maintain the bottleneck link fully saturated before the handover, shaping the data flow at 0.4 Mbit/s. This is visible by the constant slope of both sequence number lines, immediately before the handover. Although the traffic flow was initiated at the same time for all protocols, (time 1 s), TIMIP and HAWAII feature an advance on their sequence numbers. This is a result of their lower RTT for intra-domain scenarios, which enables the TCP sender to saturate the bottleneck link earlier. During this time, each TCP packet received immediately originates an ACK packet, which is generated with the correspondent sequence number.

At time 3 s, all protocols perform a handover, which produces a series of misrouted / lost packets, verified by the missing TCP sequence numbers at the receiver. After the lost packets, a large series of packets is received, related to both the sender's TCP window and the existence of queued packets at the bottleneck link. The receiver buffers these packets for subsequent sorting, but generates ACKs requesting the first missed packet. This phenomena result in the horizontal ACK line in all protocols, one per received packet. When the sender receives three duplicate ACKs, it lowers its data rate and retransmits the requested packets; it should be noted that these packets only reach the receiver after passing through the saturated bottleneck's queue. After a complete iteration of this feedback cycle, the missed packets reach the receiver, which can then acknowledge multiple sequence numbers and sort the previously out-of-order received packets, ending the handover. As such time is fairly low for TIMIP and HAWAII, the sender can then regain the maximum rate immediately. Such is not the case of the other protocols, which require more RTT to regain full rate, further increasing the mentioned sequence number difference comparing to TIMIP or HAWAII. This problem is even more important in the hMIP case, where some packets are unnecessarily retransmitted (e.g., packet 32 at time 4.75 s).

A more detailed analysis shows that TIMIP takes 1.08 s to perform handover, as the last packet received by AP2 (Node 9) is received at time 2.98 s (packet 38) and the reception of the first one by AP3 (Node 10), after retransmissions, occurs at time 4.06 s (packet 59). Similarly, HAWAII, CIP and hMIP have handover times of 1.07, 1.70 and 1.96 s, respectively. Considering the lost packets only, all protocols, except hMIP, have similar packets lost. This happens because the TCP packet rate shaped by the bottleneck is too low to result in the drop of many in-transit packets, resulting only on the drop of further packets in hMIP, because of the additional time needed to update the GW.

Combining both metrics, it can be verified that the handover time is most influenced by the RTT value, as TIMIP and HAWAII help TCP to quickly recover from their low packet losses. On the other hand, as both CIP and hMIP impose similar RTT for intra-domain scenarios, it's the number of lost packets that determines the impact of the handover in TCP flows. In this case, CIP has a lower value, as the crossover node is located in Node 1, compared with the GW in hMIP.

The number of TCP lost packets in experiment shows that both TIMIP and HAWAII limit their updates to the minimum set of nodes between the APs, and thus, can have the most efficient registration model – Cluster. Comparing to the previous UDP experience, in this case HAWAII is not penalised for introducing out-of-order packets. On the other hand, as the full TCP rate is quickly regained, saturating the bottleneck, this experience confirms that both TIMIP and HAWAII feature optimal routing, using the shortest paths inside the tree, optimising both the resource utilisation and the end-to-end delay. Similarly, hMIP and CIP feature the Intra-Domain Registration model, as these protocols inform nodes further away from the MN's location, and feature the non-optimal routing execution, as all intra-domain traffic is forced to pass through the GW always.

Considering inter-domain traffic situations, the graph shows that all protocols impose the same RTT on the packets, as the bottleneck is saturated in the same way for all protocols, confirming the correspondent UDP experiences.

At time 5 s, all protocols perform a handover, which results in misrouted/lost packets, with out-of-order packets being reported for the HAWAII protocol additionally. After the lost packets, similar operations to the previous case occur, namely the buffering of packets, the request for dropped packets and their subsequent retransmission. A difference comparing to the intra-domain case is the larger feedback cycle, due to the higher RTT for inter-domain traffic scenarios. As the RTT is equal for all protocols, the differentiating factor in the handover impact will only be the number of lost packets. Thus, TIMIP and HAWAII have the shortest handover delay and convergence time, being followed by CIP and hMIP.

A more detailed analysis shows that TIMIP takes 1.18 s to perform handover, as the reception of the last packet by AP2 (Node 9) takes place at time 5.00 s (packet 135) and the reception of the first one by AP3 (Node 10), after retransmissions, occurs at time 6.18 s (packet 156). Similarly, HAWAII, CIP and hMIP have handover times of 1.26, 1.50 and 1.93 s, respectively.

Considering the lost packets only, all protocols, except hMIP, have similar packets lost, again because the TCP packet rate shaped by the bottleneck is too low to result in the drop of additional in-transit packets. As hMIP must always reconfigure the local tunnel located in the GW, it results in the highest handover time. TIMIP and HAWAII have the lowest convergence times, by featuring fewer drops than CIP.

This experiment shows that, concerning the efficiency of the Registration phase, again both TIMIP and HAWAII limit their updates to the minimum set of nodes that must be informed, and thus, have the Cluster registration model, without penalisation for out-of-order packets in the HAWAII case. For the same reasons as in the previous case, the CIP and hMIP are classified with the Intra-Domain Registration model, as these protocols inform nodes further away from the MN's location. As stated in UDP inter-domain traffic experiment all protocols route packets in the same way inside the network, without penalties associated to the requirement for traffic to pass through the domain's gateway (case of CIP and hMIP). This causes all protocols to have optimal routing for inter-domain traffic.

4.4.4 Simulation Experiment Set D – Continuous UDP Handovers

While the previous experiments evaluated the handover's effect of discrete handovers, this section presents simulation studies that consider the effect of continuous MN movement at varying speeds, resulting on increasing handover rates. For this, the achieved long-term throughput and loss rate, per handover rate, for a variety of traffic types and sources will be studied. In all simulation sets, the MN will be moving continuously in the circular motion illustrated in Figure 4.5b, being connected to each AP sequentially. This way, the MN performs a mixture of "short" and "long" handovers inside the logical tree. This procedure does not specifically benefit the protocols that most adequately support localised handovers. For this reason, it is expected that better results will be achieved by TIMIP and HWAII protocols, if more localised movements are performed by the MNs.

The first set considers the effect of continuous handovers on UDP packets, for both intra and inter domain traffic. For this, a CBR traffic source, generating 200 pkt/s of 100 bytes, is allocated either outside the domain (Node 7), or inside it (Mobile Node 17), for intra-domain and inter-domain traffic types, respectively. All cases were simulated with a sufficiently large simulation time, and the handover rate ranges from 0 to 14 handovers per minute (e.g., one handover each 4.2 s).

The respective results are shown in Figures 4.9, 4.10 and 4.11, representing the long-term throughput, packet loss rate and packet losses per handover rate, for both intra and inter domain traffic sources.

This experience shows that when the MN remains stationary (no handovers), in a intra-domain situation, there are no packet losses, and the full throughput is achieved. When the MN starts its movement, the protocols impose packet losses that increase linearly according to the MN's movement. Such losses are the sum of each individual handover studied in the corresponding discrete simulations; while one graph (Figure 4.11) presents the total packet losses per protocol, the other relates these values to the total number of received packets (Figure 4.10). These losses will then force a linear degradation on the total throughput received at the mobile node, as the CBR sender rate is constant. At maximum speed, there is a degradation of 0.84% using TIMIP. Similarly, CIP, HAWAII, and hMIP have throughput degradations of 0.88, 1.71 and 3.01%, respectively.

Considering the graph as a whole, the TIMIP protocol shows the smallest degradation for CBR traffic, being closely followed by CIP. Next, HAWAII shows an intermediate degradation ratio, because of its out-of-order phenomena; hMIP shows the worst performance, as all handovers must reach the GW. These results fully correspond to the previous discrete packet loss studies, and improve them by providing long term information, using a mixture of several types of handovers. This way, by periodically incorporating short handovers (namely, on the AP1 to AP2 and AP3 to AP4 movements), HAWAII improves its previous UDP result to an intermediate value between TIMIP/CIP and hMIP.

As studied before in the discrete case, the inter domain results are similar to the intra domain case. In these continuous long-term measurements, the ex-

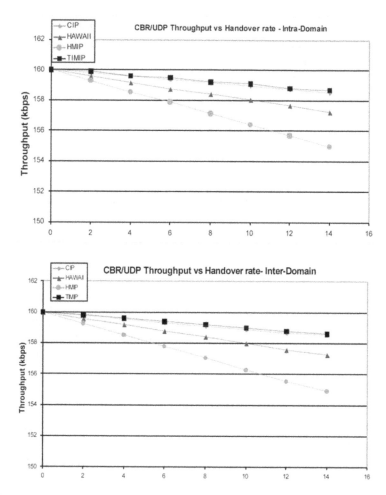

Figure 4.9 CBR/UDP throughput per handover rate: intra and inter-domain traffic.

isting differences are greatly shaped by the successful packets received; thus, almost similar results were experienced, which result on the same conclusions as in the previous case.

4.4.5 Simulation Experiment Set E – Continuous TCP Handovers

This simulation set investigates the effect of continuous handovers in TCP traffic, subject to the same scenario of the previous case. Again, a bottleneck link with the value of 0.3 Mbit/s is used to shape and limit the traffic at the

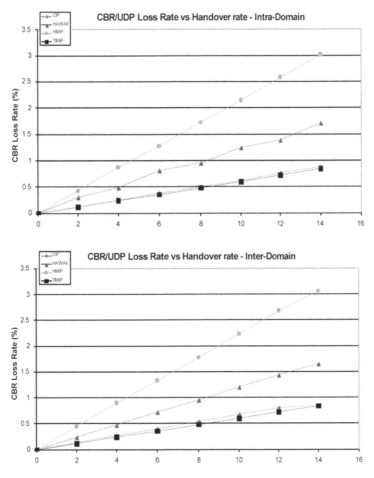

Figure 4.10 CBR/UDP throughput per handover rate: intra and inter-domain traffic.

sender, using 1000 byte packets. Comparing to the previous case, the TCP feedback mechanism will cause the sender to throttle back its transmission rate, based on the loss and RTT information received dynamically. Thus, for measuring the mobility impact of TCP flows, the long term throughput experienced by the receiver will be measured for each simulation. On the other hand, as TCP provides a reliable transfer, the necessary packets will be retransmitted and reordered by TCP. Taking this into account, the overhead necessary to ensure reliable communication will be measured, being defined as the ratio between transmitted and received unique ordered packets.

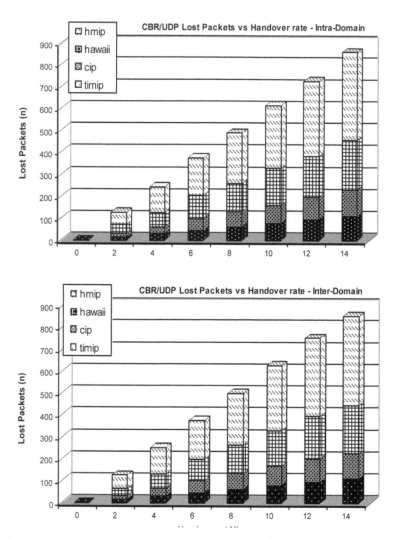

Figure 4.11 CBR/UDP throughput per handover rate: intra and inter-domain traffic.

The respective results are shown in Figures 4.12 and 4.13, representing the long-term bandwidth and overhead rate per handover rate, for both the intra and inter domain traffic sources.

When the MN is stopped (intra-domain situation), there are no packet losses, being the received throughput equal to the bottleneck value (0.3 Mbit/s). When the MN starts its movement, packet loss happened and

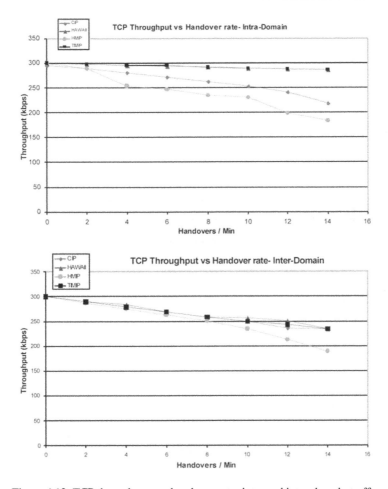

Figure 4.12 TCP throughput per handover rate: intra and inter-domain traffic.

the protocol's operations force the re-ordering phenomena previously studied. However, this is interpreted by the sender as a sign of congestion, which further degrades the data rate. This explains why the achieved long-term throughput has much lower values than in the previous UDP experience (Figure 4.12a vs Figure 4.9a). Such is the case because, after each handover, the sender stops emitting for a large period of time, and only raises it rate using packet bursts of increasing size.

For this reason, the subsequent handovers may occur at the same time as a packet burst, leading to further packet losses, or being unnoticed between

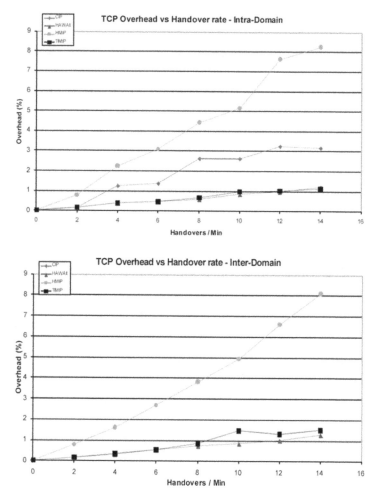

Figure 4.13 TCP overhead per handover rate: intra and inter-domain traffic.

TCP unshaped bursts. However, in practice this phenomenon exists for all protocols, and its effect is shaped by the long simulation time and multiple handovers in sequence. Thus, there is a fairly constant (although non-linear) throughput decrease while the handover rate increases. At maximum speed, there is a degradation of 4.72% in the TIMIP case. Similarly, HAWAII, CIP and hMIP have throughput degradations of 4.4, 27 and 39%, respectively (values derived from the data series of Figure 4.12a).

Considering the throughput graph as a whole, the TIMIP and HAWAII protocols show the least degradation as a result of their low packet losses and low RTT. Then, CIP has an intermediate degradation, because of its higher RTT which results in a slower feedback cycle; finally, hMIP shows the worst performance, because besides the high RTT, a large number of packets is also lost in all its handovers, as all updates must reach the GW, further reducing the sender's rate. These results fully correspond to the previous discrete packet loss studies, improving them with a long-term effect that considers a mixture of multiple handover types and situations.

The second graph gives a different view on the same experiment, by plotting the overhead necessary to ensure reliable TCP transfer. Their values are determined by to major factors: the amount of lost packets, which must be retransmitted, and the RTT time, which being too high, may lead to packets being unnecessarily retransmitted. In both cases, the number of retransmitted packets at the sender determines the TCP rate throttle back necessary to ensure TCP-fairness. At maximum speed, there is an overhead of 1.1% in the TIMIP case. Similarly, HAWAII, CIP and hMIP have overheads of 1.2, 3.2 and 8.3%, respectively. In particular, this high value for hMIP results from a very large number of retransmitted packets, both required and unnecessary.

When the MN is stopped (inter-domain case), there are no packet losses, being the received throughput equal to the bottleneck value (0.3 Mbit/s). When the MN starts its movement, all protocols, except hMIP, have similar results for throughput degradation and overhead increase, being these effects augmented while the handover rate increases (subject to the previous non-linear comments for intra-domain traffic). At maximum speed, there's a degradation of 21.9% in the TIMIP case. Similarly, HAWAII, CIP and hMIP have throughput degradations of 22.1%, 21.9% and 36.8%, respectively (values derived from the data series of Figure 4.12b).

Considering the throughput graph as a whole, the TIMIP, CIP and HAWAII protocols show the least degradation considering TCP traffic, as the crossover is in the same locations for each handover, and all protocols impose the same RTT. The clear difference is the hMIP protocol, which imposes a farther crossover node in all handovers (the GW). These results fully correspond to the previous discrete packet loss studies, improving them with a long-term effect of multiple handover types and situations.

The second graph shows the TCP overhead for reliable transfer for inter-domain traffic. At maximum speed, there is an overhead of about 1.4% for all the protocols, except hMIP that has 8.1% overhead, being again result of high number of packet retransmissions.

4.5 Conclusions

This paper presented the evaluation of a terminal independent mobility architecture composed of two protocols previously proposed – TIMIP/sMIP. This solution supports legacy terminals with high efficiency, for both aspects of handover latency and network resources utilisation.

To define the mobility solution requirements, a comprehensive and original classification framework featuring different models for the Detection, Registration and Execution phases was also proposed, being used to classify existing well-known mobility protocols (HAWAII, CIP and hMIP). Then, the architecture and operations of our solution were described, along their classification on the framework. In particular, TIMIP/sMIP uses the best models that are suitable for terminal independence, namely the Reactive Detection phase, the Cluster Registration model, and In-band State Maintenance coupled with Optimal Route, both for the Execution phase.

Simulations that compared TIMIP with the other micro-mobility protocols were performed, focusing on scenarios that, albeit their importance, were not addressed in previous research work. Thus, the simulations considering hard handovers in infra-structured 802.11 networks with multiple disjoint channels. During the tests, the handover latency, packet loss and resource utilisation for both intra-domain or inter-domain traffic, were evaluated, considering both CBR/UDP and TCP traffic under discrete and continuous handovers situations.

The results achieved have shown that TIMIP and HAWAII have the best performance in both topics of handover delay and resource optimisation, being this effect most impact in TCP intra-domain traffic situations of high speed mobile nodes. However, TIMIP has slightly better latency performance for UDP traffic, due to the existence of out-of-order packets in HAWAII during roaming situations. CIP and hMIP present the worst results, as all traffic must be forwarded through the domain's gateway, leading to high RTTs, and by having the crossover nodes in higher locations of the tree.

Future work comprises further optimisation of Optimal non-tree routes, Transparency improvements supporting Network Independency and incremental upgrading, and support of future IPv6 networks.

References

[1] K. Tachikawa, A perspective on the evolution of mobile communications, *IEEE Communications Magazine*, vol. 41, pp. 66–73, October 2003.

[2] C. Perkins (Ed.), *IP Mobility Support for IPv4*, RFC-3344, IETF, August 2002.

[3] D. Johnson and C. Perkins, *Mobility Support in IPv6*, RFC-3775, June 2004.

[4] A. Campbell et al., Design, implementation and evaluation of cellular IP, *IEEE Personal Communications*, vol. 7, no. 4, pp. 42–49, August 2000.

[5] R. Ramjee, K. Varadhan, L. Salgarelli, S. Thuel, S-Y. Wang and T. La Porta, HAWAII: A domain-based approach for supporting mobility in wide-area wireless networks, *IEEE/ACM Transactions on Networking*, vol. 10, no. 3, pp. 396–410, June 2002.

[6] E. Gustafsson et al., Mobile IP regional registration, draft-ietf-mobileip-reg-tunnel-09, June 2004.

[7] N. Banerjee, W. Wei and S. Das, Mobility support in wireless internet, *IEEE Wireless Communications*, vol. 10, no. 5, pp. 54–61, October 2003.

[8] J. Kristoff, Mobile IP, http://condor.depaul.edu/ jkristof/mobileip.html, DePaul University, Chicago, October 1999.

[9] A. Grilo, P. Estrela and M. Nunes, Terminal independent mobility for IP (TIMIP), *IEEE Communications*, vol. 39, no. 12, pp. 34–41, December 2001.

[10] P. Estrela, A. Grilo, T. Vazão and M. Nunes, Terminal Independent Mobile IP (TIMIP), draft-estrela-timip-01.txt, January 2003.

[11] D. Saha, A. Mukherjee, I. S. Misra and M. Chakraborty, Mobility support in IP: A survey of related protocols, *IEEE Network*, vol. 18, no. 6, pp. 34–40, November/December 2004.

[12] P. Eardley et al., A framework for the evaluation of IP mobility protocols, in *Proceedings of the 11th IEEE PIMRC*, London, September, pp. 451–457, 2000.

[13] P. Reinbold and O. Bonaventure, IP micro-mobility protocols, *IEEE Communications Surveys*, vol. 5, no. 1, 2003.

[14] K. El-Malki and H. Soliman, Low latency handoffs in mobile IPv4, draft-ietf-mobileip-lowlatency-handoffs-v4-09.txt, June 2004.

[15] A. Campbell et al., Comparison of IP microMobility protocols, *IEEE Wireless Communications*, vol. 9, no. 1, pp. 72–82, February 2002.

[16] R. Braden (Ed.), *Requirements for Internet Hosts – Communication Layers*, RFC 1122, October 1989.

[17] E. Gustafsson (Ed.), Requirements on mobile IP from a cellular perspective, Internet draft, draft-ietf-mobileip-cellular-requirements-02, June 1999.

[18] A. Mishra et al., An empirical analysis of the IEEE 802.11 MAC layer handoff process, *ACM SIGCOMM Computer Communication Review*, vol. 33, no. 2, pp. 93–102, April 2003.

5

Mobility in B3G Systems: Requirements and Performance Evaluation Through Experimentation

D. Loukatos, D. Kouis, N. Mitrou and M. E. Theologou

Electrical and Computer Engineering Department, Computer Networks Laboratory, National Technical University of Athens, 9 Heroon Polytechneiou Street, Zographou 15773 Athens, Greece e-mail: {dlouka; kouis}@telecom.ntua.gr

Abstract

The convergence of heterogeneous wireless networks, such as cellular mobile systems, wireless local area networks and other wireless broadband systems, leading to the so-called beyond 3G networks, introduces an increasingly strong trend, to margin monolithic approaches in favour of more flexible network architectures. Due to the diverse capabilities, in terms of bandwidth and delay characteristics, of each of the networks evolved in the formation of a beyond 3G system, the mobility functions (i.e. inter and intra-system handovers) present excessive complexity. It is apparent that new enhanced mobility mechanisms should be invented. In this direction, the typical system and user oriented handover scenarios, faced during the operation of such systems are described. The mobility mechanisms should be able to indicate the reasons (why) and the time instance (when) that a handover occurs, as well as its direction (where). Based on these scenarios and other QoS constraints, the generic requirements for mobility functionality are given. In order to strengthen more our arguments about the mobility complexity, we implemented an experimental B3G platform, studying this way various mobility scenarios. The analysis of the metrics tools used during the experiments

D. D. Kouvatsos (ed.), Mobility Management and Quality-of-Service for Hetero-geneous Networks, 111–127.

and the results taken, are included in this paper, presenting great interest and valuable conclusions.

Keywords: Performance evaluation, heterogeneous networks, IP metrics, DVB-T, WLAN, GPRS.

5.1 Introduction

The forthcoming beyond 3G systems or the so-called "fourth generation" (4G) systems combine diverse radio access technologies, working jointly, providing enhanced multimedia IP services. These types of environments exhibit enhanced performance in terms of network coverage and QoS guarantees [1]. It is probable that the forthcoming 4G systems will not be built as a monolithic structure, as the 3G systems, but as flexible interworking environments enabling new emerging technologies to interoperate, while offering new advanced services to users. Terms, such as software-defined radio (SDR, see [2]) and multi-interfaced terminal devices (PDA, laptops, etc.), are enabling means towards the realization of this concept.

The Network Provider (NP), either it possesses multiple licenses for operating different radio networks or just cooperates with other NPs, owning alternative wireless networks, is interest in the efficient coverage of the service area [3]. Efficient coverage means offering as high as possible Quality of Service (QoS) levels, at adequate capacity volumes, in a cost-effective manner. Therefore, a NP can choose, at a certain service area region and time zone in the day, instead of rejecting users or degrading their QoS levels, to direct them to an alternate radio technology, which may belong to an affiliated NP.

Regardless of the business model followed (i.e. common or different administrative domains), the mobility functionality plays a critical role for achieving seamless service delivery, during the user's transition to another wireless network. Towards this direction many solutions have been proposed for improving the performance, both for intra-system and inter-system handovers. More specifically, the intra-system handovers refer to the situation where a user's terminal chooses to access services through a different Radio Access Port (RAP) of the same wireless network. This type of mobility is usually faced from the networks themselves, based on their internal mobility functions (e.g. GSM Base Station Subsystem or IAPP Inter-Access Point Protocol [4] for IEEE WLANs). Also, micro-mobility protocols for improving the handover performance have been proposed, such as the Cellular IP [5] or HAWAII [6]. On the other side, the inter-system handovers refers to the case that user's terminal is attached to a completely new type of wireless network. This type of mobility (also known as macro-mobility) presents greater complexity than inter-system handovers, due to the fact that more parts are evolved (i.e. source and target network). The most common solution, for supporting macro-mobility functionality is the Mobile IPv4 protocol [7]. Also, nowadays the introduction of the IPv6 protocol, promises better support for macro-mobility operations [8, 9].

Nevertheless, none of the above presented solutions seems to be inadequate for providing uninterrupted connectivity and at the same time preserve the minimum QoS levels, during user's movement, through different wireless networks. The critical issues, faced during the mobility operation (concentrating on inter-system handovers), inside such environments are the reason and the time instance that the transition occurs (why and when), along with the selection of the most suitable target network (where) and the necessary adjustments from the radio up to application level at the terminal side.

Based on the above discussion, the rest of the paper is organized as follows: Section 5.2 presents the most typical mobility scenarios faced during B3G systems operation. Furthermore, the same section describes the main requirements for supporting enhanced inter-system handovers. Section 5.3 provides an overview of a prototype experimental platform, comprising three different wireless network, namely GSM/GPRS [10], IEEE 802.11b WLAN [11] and DVB-T [12]. Also, gives an overview of a Network Access Co-ordination Protocol (NACP), employed in order to improve mobility operations. Metrics tools used during the performance evaluation of the various types of inter-system handovers are presented and analyzed in Section 5.3.1. Moreover, indicative results, concerning delay aspects and packet losses phenomena, studying this way various mobility scenarios, are included in Section 5.4. Finally, Section 5.5 gives some concluding remarks and future steps within the context of B3G mobility requirements.

5.2 Typical Mobility Scenarios and Requirements in B3G Systems

5.2.1 Typical Mobility Scenarios

This section presents the typical network models for interworking different types of Radio Access Networks (RANs). Figure 5.1 depicts the most common architectures for coupling heterogeneous wireless segments.

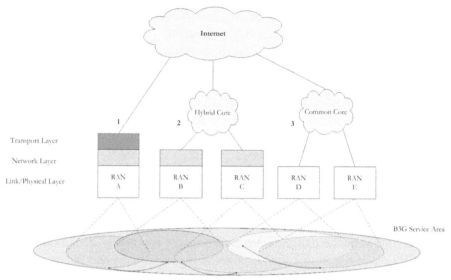

Figure 5.1 Network models for B3G integrated systems.

More specifically, models 1 and 2 refer to the "loose-coupling" archi-tectural style, where the interconnection between the different networks is achieved at the higher levels of the IP protocol (i.e. transport and session). The most common solution is the usage of MIP protocol. The "hybrid core" variation of the louse-coupling architecture (model 2), presents better per-formance during inter-RAN handovers, due to the immediate connection between the wireless segments, while at the same time avoids duplicating common network operations (e.g. billing, authentication, etc.). The third model refers to the "tight-coupling" architecture, where the common core operates as a single network, providing all network functionalities in a more integrated manner [13]. The first model is more suitable for networks co-operating through different administrative domains (DADs). The other two types of architectures, call for new technological advances and new business models [14] to be defined, as they require networks, operating under a com-mon administrative domain (CAD). It is noted that model 2 and 3 present better performance than model 1, especially during inter-system handovers.

Moreover, Figure 5.1 indicates three major types of inter-RAN (inter-system) handovers, depending on the adopted B3G network model (control plane mobility scenarios). In greater detail, the following cases exist:

- *Inter-RAN handover between networks belonging to DADs*: This type of handover is expected to present the worst performance, due to the intermediate role of the Internet.
- *Inter-RAN handover between networks belonging to a CAD, created following the "hybrid core" network model*: The common transport layer results in better performance, compared to the previous case. Still the inner operational layers of the coupled networks preserve different protocols for routing data and intra-system handovers.
- *Inter-RAN handover between networks belonging to a CAD, created following the "common core" network model*: The unified operation of the different wireless networks down to the link data level, results in optimal performance during this type of handovers. The different technology RAPs are faced as access points of the same network.

From a different point of view (still within control plane level), mobility scenarios can be categorized further, depending on the performance capabilities of the involved networks (source and target network). More specifically, the following cases exist:

- *Inter-RAN handover from "high" to "low" performance network*: This type of handover indicates the case where a user moves from a high performance network, in terms of bandwidth and delay characteristics, to a network with less capacity and increasing delay values.
- *Inter-RAN handover from "low" to "high" performance network*: This type of handover refers to the opposite case than the previously described.

Finally, from the user perspective, typical mobility scenarios (network or terminal initiated), in the context of B3G systems, include the following cases:

- *Service and cost oriented scenarios*: It is possible strict service requirements or cost related issues to trigger a handover to a different wireless network (e.g. user demands for a new service, like video streaming, result in his transfer to a high performance network).
- *Traffic congestion scenarios*: In the attempt of a B3G management system to resolve hot-spot situations, appearing at a part of the system, inter-RAN handovers may be triggered. This way a more balanced traffic distribution, among the co-operating wireless segments, is achieved.
- *Poor or loss of radio coverage scenarios*: This type of handover refers to case, where a user moves out from the coverage area of the serving

network. Services termination or QoS degradation (in the case where the user remains in an area with poor radio conditions) can be avoided through inter-system handover to an alternative network, operating in the same area.

The upcoming section, based on the previously described scenarios, presents an overview of the necessary requirements, in order to build reliable and well-behaved mobility support functionality for B3G systems.

5.2.2 Mobility Mechanisms Requirements

The complexity appearing during user mobility, inside B3G systems, demand for the satisfaction of a series of specific requirements, from the mobility support mechanisms. In more details:

- Mobility functions should exhibit robustness in case of network connection failure. The B3G system should be able to track the exact location of the terminal, even if it has been switched to a different radio technology.
- Singalling overhead should be kept to minimum.
- The adopted mobility framework must co-operate with micro-mobility protocols or networks' build-in intra-system mobility functions.
- Mobility functionality must be independent and unaffected from changes occurring at the data routing mechanisms.
- During inter-RAN handovers all application level sessions should be remain active. Also, the transition duration and phenomena (i.e. delay fluctuations and packet loss) should be minimized (optimized performance).
- Programmable handovers should be supported, through triggering events, originated either from the network management system (network-initiated handovers) or user side (terminal-initiated handovers). For accomplish this requirement the mobility mechanism should retrieve information from all protocol layers (e.g. radio or application level statistics), from both terminal and network side. Also, the concept of programmable handovers includes the mobility mechanism interaction ability, with both service and network providers, achieving this way, dynamic content adaptation or core network adjustment.

Concluding, part of the above mentioned requirements were fulfilled in the following described experimental B3G platform.

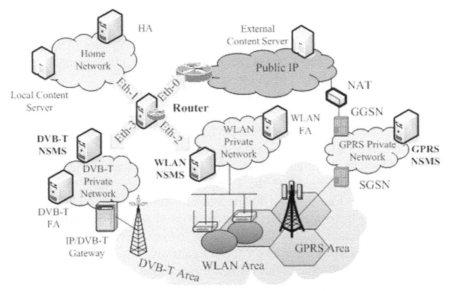

Figure 5.2 B3G prototype heterogeneous wireless networks environment.

5.3 Performance Evaluation

5.3.1 Experimental Setup

The prototype heterogeneous wireless networks environment, used for the mobility performance evaluation, consists of three different radio access technologies, GSM/GPRS, IEEE 802.11b WLAN and DVB-T. This section demonstrates the general network architecture for the exploitation of all these wireless systems, operating in the B3G context. As depicted in Figure 5.2, the private networks of WLAN and DVB-T radio access technologies are interconnected through the usage of hybrid core, while the interconnection with the GPRS network is achieved with the help of Internet (see Figure 5.2, for the high level architecture schematic).

The functionality of the platform includes the following features:

- Management systems for each radio access technology. These systems, called Network and Service Management Systems (NSMSs) are located in the relative subnets, but they can inter-communicate and cooperate [15].
- Appropriate terminals, capable of communicating over different wireless technologies. These multimode terminals are equipped with the

required intelligence for taking decisions, performing measurements and interacting with the local NSMS [16]. The management system of the terminals is called Terminal Station Management System (TSMS) and the Network Access Co-ordination Protocol [17] implemented for the interaction with the NSMS is presented below.

- Content servers for retrieving information relative to the applications and services provided.
- IPv4 backbone solution, selected for reasons explained below. Consequently, a Mobile IPv4 infrastructure is employed for the mobility management, especially during inter-system handovers. The home network (Figure 5.2), hosts the Home Agent (HA), while the Foreign Agents (FAs) are located in the corresponding subnets (WLAN and DVB-T). Moreover, the HA has been properly modified (with advanced tunneling functionality), thus enabling it to cooperate with the GPRS Network Address Translation (NAT) gateway. Also, proper modifications to the software of the DVB-T FA for enabling the establishment of the return channel were realized. The return channel is required due to the unidirectional nature of the DVB-T functionality. In the specific case, the wireless medium that acts as the missing uplink is the GPRS or the WLAN network.

It is obvious that IPv4 is selected everywhere, although IPv6 would be more convenient for the whole architecture because there is no need for including foreign agents and the NAT gateway is not necessary. The reasons for choosing IPv4 are the following:

- IPv4 is much more widely deployed and multiple commercial products and networks are based on Mobile IPv4. On the other hand, IPv6 networks are still in development and Mobile IPv6 is not standard yet.
- The commercial GPRS segment and the commercial DVB-T products used do not support IPv6.
- The applications' clients and servers used are also IPv4 based.

The TSMS-NSMS interactions, governed by the Network Access Co-ordination Protocol (NACP), include the following messages:

- *Service Request and Reply.* Through these messages, the terminal reports to the NSMS its current status (serving network, available networks, services used, request for a new service, etc) and the NSMS indicates by its response the list of the preferred networks, towards guiding the terminal in network selection. The messages are sent periodically (acting also as keep-alive probes), but also whenever a change in the current terminal status occurs (either in the network availability or in the services used).

- *Quality Report Request and Reply.* The terminal uses the request message in order to report to the NSMS quality degradation observed at the utilized services (e.g. a major traffic load alteration sensed). The NSMS after processing all the relative data instructs the terminal which is the best action suggested in this case, by sending the reply message.
- *Handover Required Notification.* This message is sent by the NSMS and forces the terminal to switch to another network. A handover indication could also be included in the service reply message, but this is sent only after the service request from the terminal. The handover required notification does not require any trigger from the terminal and covers cases where the handover is necessary, without waiting for the next service request.

The NACP, acts supplementary to the MIPv4 infrastructure, satisfying many of the mobility mechanism requirements, presented in Section 5.2.2. It is proved [18, 19] that the signalling overhead, caused by the NACP adoption, is inside acceptable bounds.

Before proceeding with the results, concerning the handover performance for various cases of source and target network, we briefly present the metric tools, used during the experiments.

5.3.2 Metric Tools

In order to evaluate the behaviour of the radio segments, it is necessary to perform an efficient set of measurements. Some of these measurements are quite simple to perform while others require some specific tools to be involved.

One of the specific purpose tools being used is a custom IP Traffic Generator [18, 21]. The TG-IP tool supports up to 10 independent source modules simultaneously. The IP destination of the traffic produced by each module can be selected prior actual traffic generation. A sequence number is also placed at a specific field inside the IP payload of each packet produced by a source module.

The TG-IP tool is used to produce:

- Test traffic of low rate. This traffic is injected into the real network and through comparison of the series of test packets being captured at source and destination points, the behaviour of the network mechanisms is revealed.

- Background traffic. In this case one or more source modules of the TG-IP are used to produce traffic similar to the one produced by actual sources. This method reduces the computational resources required for traffic generation and provides better traffic load control during the test scenarios.

The role of the TG-IP tool becomes more apparent after the introduction of a traffic analysis tool called TA-Post [18]. The TA-Post is a post-processing tool that takes as input files containing packet-capturing data provided by the well-known tcpdump utility [22]. Each line of these files contains the capturing time and size of the packet and, if necessary, its corresponding sequence number (placed prior packet generation by the TG-IP).The TA-Post provides as output graphs of one way delay and/or delay variation and of packet losses.

The general inter-working schema according to which measurements are performed using both TG-IP and TA-Post tools is depicted in Figure 5.3. In general, this mechanism provides a simple and very effective method for evaluating performance metrics in packet networks [21, 23].

According to Figure 5.3, the TG generates Test Traffic that is addressed towards the Radio Segment and is finally captured by the TA-Post tool. Application Traffic may be produced by real applications or by the TG tool itself. The OWD quantity per each packet (D_i) is extracted through the substruction $\text{TA}_i - \text{TG}_i$, where the generation time stamp (TG_i) is measured at the source point and the arrival time stamp (TA_i) is measured at the destination point.

Furthermore, according to methods and definitions described in ITU [24] and IETF [25, 26], if the inter-packet distance at the source between the i-th and the $(i + 1)$-th packet (T_i), is constant for all i and equal to T, then the computation of OWD and especially of packet delay variation ($D_{i+1} - D_i$) becomes easier and requires packet capturing only at destination point [21].

The method depicted in Figure 5.3 also incorporates techniques for anticipating common synchronization issues caused by the variable, relative time offset among source and destination computer clocks. The above engagement of TG-IP and TA-Post tools is very helpful especially in cases of UDP traffic when feedback at the opposite direction from the destination to the source is not of great importance. In cases of TCP traffic the characteristics of this feedback traffic are very important and thus metrics such as the Round Trip Time (RTT) metric cannot be omitted. A simple method for measuring RTT quantities is using the ping application.

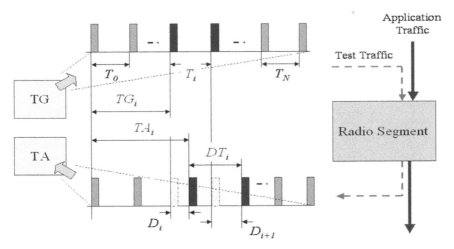

Figure 5.3 The general inter-working schema during measurements that involve the TG-IP and TA-Post tools.

5.4 Results

This section presents some indicative results describing the behaviour of the three radio segments. Furthermore it focuses on the impact of the transition among the radio segments, related to packet delay and/or packet loss issues.

Such behaviour is very important as it directly affects the quality of real-time applications where most of the traffic is carried out using the UDP protocol (MPEG streaming applications).

In order to investigate the behaviour of the radio segment, a test UDP traffic stream is used. The stream is produced by the TG-IP tool which is parameterized accordingly allowing the proper calculation of delay, delay variation and/or packet losses. The TA-Post tool carried out the later calculation task. The overall process follows the method described in Section 5.3.2.

Several indicative cases of handover occurrence have been studied. The results consist of one-way delay (OWD) traces and/or packet loss traces at the period that a handover takes place. The test UDP traffic stream is of a "regular profile" with all consecutive packets having a length of 200 bytes and an inter-packet distance of 100 or 50 msec. The OWD related results, for different handover cases, are listed below:

- When handover from the WLAN to DVB-T is considered, the delay is shifted from 1–2 up to about 40 msec, with the delay variation also increasing (the STD value goes from about 1 up to 15 msec).

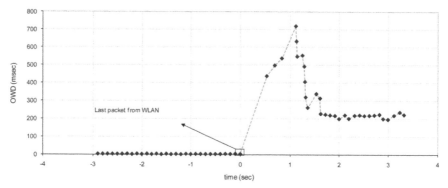

Figure 5.4 Handover from WLAN to GPRS.

- When handover from the DVB-T to GPRS is considered, the delay is shifted from about 40 up to 220 msec, with the delay variation increasing too (the STD value goes from about 15 up to 40 msec).
- When handover from the WLAN to GPRS is considered, the delay is shifted from about 1–2 up to 220 msec, with the delay variation increasing too (the STD value goes from about 1 up to 40 msec).

It can be observed that GPRS network segment exhibits the larger delays, WLAN the shorter, while DVB-T is efficient but not as good as the WLAN radio segment.

A more detailed description of the behaviour during transitions among the three radio segments follows. In all graphs the assumption that transition starts at time 0 has been made, in order the results to be more easily comparable.

During handover from WLAN to GPRS (model 1 of the loose-coupling architecture-case of transition from high to low performance network) the OWD value increases from the level of 1–2 msec up to a peak of 710 msec and falls again down to the level of 220 msec. This peak in OWD is explained by the fact that the GPRS system is of slow response, and thus several incoming packets are just queued before the GPRS system starts servicing them. Nevertheless, no packet losses are experienced. The above results indicate that this handover case is quite "safe" for applications carrying out critical data (e.g. FTP) while it may cause instant quality degradation to real-time applications (e.g. voice streaming).

During handover from GPRS to WLAN (model 1 of the loose-coupling architecture-case of transition from low to high performance network) the

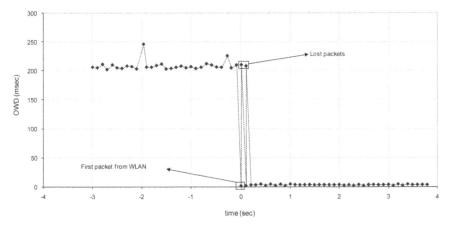

Figure 5.5 Handover from GPRS to WLAN.

OWD value decreases immediately from the level of 220 msec down to the level of 1–2 msec. The direct switching from GPRS to WLAN segment causes some of the packets already being in the GPRS queue to be lost. Thus while inter-packet distance variation becomes better some packets are lost. The above results indicate that this handover case is rather sufficient for real-time applications (e.g. voice streaming) that usually decide to sacrifice some packets in order no to exceed their tight temporal constraints. On the other hand the behaviour exhibited during this handover case may cause a data-critical application to crash.

Results as the above become more apparent as the traffic load increases. The system behaviour remains similar but its scale increases (basic condition not pushing the system beyond overloading line). A good example is given in Figure 5.6, where two graphs are depicted. Both of these graphs represent a similar system behaviour, more specifically, the peak in OWD just after the transition from WLAN to GPRS segment. The first graph (black line) corresponds to traffic load consisting of consecutive packets having a length of 200 bytes and an inter-packet distance of 100 msec. The second graph (grey line) corresponds to traffic load consisting of consecutive packets having a length of 200 bytes and an inter-packet distance of 50 msec (instead of 100 msec). Although permanent states before and after the transition are almost the same (system not overloaded) and the shape of the two peaks almost identical, the second peak is twice as big as the first one. This is explained by the fact that in the second case the queueing into the GPRS system is more intense.

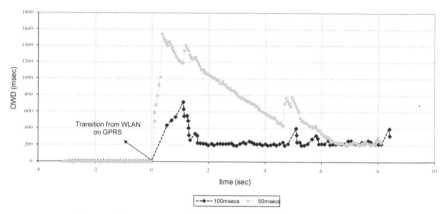

Figure 5.6 Behaviour under increased traffic load conditions.

Similar remarks can be made while increasing traffic load during handover from GPRS to WLAN. The OWD becomes much better very suddenly at the cost of increased packet losses (involving the remaining packets into the GPRS queue).

All the above results are referring to handovers between GPRS and WLAN systems. This case has intentionally been selected for thorough presentation, as the difference in terms of performance is more apparent among GPRS and WLAN (model 1 of the loose-coupling architecture) than among GPRS and DVB-T (model 1 of the loose-coupling architecture) or WLAN and DVB-T (model 2 of the loose-coupling architecture).

Nevertheless the results are quite similar during handovers among GPRS and DVB-T while a thorough study of handover behaviour among WLAN and DVB-T would require much more intense traffic load, thus leaving GPRS segment out of competition.

Figure 5.7 depicts the behaviour of the system (in OWD terms) during transition from DVB-T to GPRS segment (model 1 of the loose-coupling architecture-case of transition from high to low performance network). This behaviour is similar to the one during transition from WLAN to GPRS. The OWD value increases from the level of 40 msec up to a peak of 600 msec and falls again down to the level of 220 msec, which characterizes the GPRS performance.

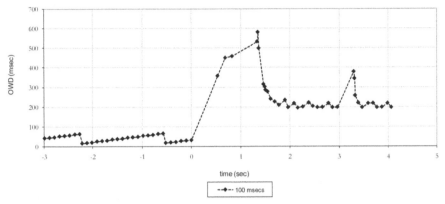

Figure 5.7 Handover from DVB-T to GPRS.

5.5 Conclusions

This paper tries to identify the typical system and user oriented handover scenarios, faced during the operation of B3G systems. At the same time, it addresses the main requirements for implementing advance mobility mechanisms. In this direction, we introduced a prototype B3G environment, comprising three different wireless segments, interconnected with MIP and enhanced management systems and multi-mode terminals, interacting through the Network Access Co-ordination Protocol. The results presented, confirm the implications that appear, during handovers from high to low performance networks and via-versa, or between networks operating inside different administrative domains. It is obvious that mobility mechanisms should be extended with the capability to inform application level entities, about the imminent change on the underlying network conditions, in an abstractive manner. Finally, more advanced mobility and transport level protocols must be introduced, confronting better with transition phenomena, appearing during inter- RAN handovers.

Acknowledgements

This work has been performed in the framework of the IST project IST-2001-33093 CREDO, which is partly funded by the European Union. The authors would like to acknowledge the contributions of their colleagues from Motorola Labs Paris, NCSR "Demokritos", IBM, ICCS/NTUA, Motorola GSG Italy, Thales Broadcast & Multimedia, Vodafone Greece.

References

[1] W. Mohr and W. Konhauser, W., Access network evolution beyond third generation mobile communications, *IEEE Communications Magazine*, vol. 38, no. 12, pp. 122–133, December 2000.

[2] P. Kenington, Emerging technologies for software radio, *IEE Elect. and Commun. Eng. J.*, vol. 11, no. 2, pp. 69–83, April 1999.

[3] P. Demestichas, N. Koutsouris, G. Koundourakis, K. Tsagkaris, A. Oikonomou, V. Stavroulaki, L. Papadopoulou, M. E. Theologou, G. Vivier and K. El Khazen, Management of networks and services in a composite radio context, *IEEE Wireless Communications*, vol. 10, no. 4, pp. 44–51, 2003.

[4] IEEE 802.11f-2003, Trial-use recommended practice for multi-vendor access point interoperability via an inter-access point protocol across distribution systems supporting.

[5] A. T. Campel, J. Gomez and S. Kim, Design implementation and evaluation of cellular IP, *IEEE Personal Communications*, vol. 7, no. 4, pp. 42–49, August 2000.

[6] R. Ramjee, et al., IP-based access network infrastructure for next-generation wireless data networks, *IEEE Personal Communications*, vol. 7, no. 4, pp. 34–41, August 2000.

[7] C. Perkins, Mobile IP, *IEEE Communications Magazine*, vol. 35, no. 5, pp. 66–82, May 1997.

[8] N. Montavont and T. Noël, Handover management for mobile nodes in IPv6 networks, *IEEE Communications Magazine*, vol. 40, no. 8, pp. 38–43, August 2002.

[9] C. Perkins, Mobile IP update version, *IEEE Communications Magazine*, 50th Anniversary Issue, pp. 66–82, May 2002.

[10] B. Ghribi and L. Logrippo, Understanding GPRS: The GSM packet radio service, *Computer Networks*, vol. 34, no. 5, pp. 763–779, November 2000.

[11] IEEE 802.11b-1999 Supplement to 802.11-1999, Wireless LAN MAC and PHY specifications: Higher Speed Physical Layer (PHY) extension in the 2.4 GHz band.

[12] ETSI, Digital Video Broadcasting (DVB), Framing structure, channel coding and modulation for digital terrestrial television, EN300-744, January 2001.

[13] M. Inoue, K. Mahmud, H. Murakami, M. Hasegawa and H. Morikawa, Novel out-of-band signalling for seamless interworking between heterogeneous networks, *IEEE Wireless Communications*, vol. 11, no. 2, pp. 54–63, April 2004.

[14] P. Demestichas, N. Koutsouris, G. Koundourakis, K. Tsagkaris, A. Oikonomou, V. Stavroulaki, L. Papadopoulou, M. E. Theologou, G. Vivier and K. El Khazen, Management of networks and services in a composite radio context, *IEEE Wireless Communications*, vol. 10, no. 4, pp. 44–51, 2003.

[15] G. Koundourakis, N. Koutsouris, V. Stavroulaki, L. Papadopoulou, V. Tountopoulos, D. Kouis, P. Demestichas and N. Mitrou, Network and service management system for optimising service delivery and traffic distribution in composite radio environments, in *Proceedings of IST Mobile & Wireless Telecommunications Summit*, Portugal, pp. 391–395, 2003.

[16] M. Catalina and P. Stathopoulos, Terminal management system for optimized service delivery in composite radio environments, in *Proceedings of IST Mobile & Wireless Telecommunications Summit*, Portugal, pp. 307–311, 2003.

[17] M. Catallina Gallego and P. Roux, Candidate network selection in composite radio environments, in *Proceedings of the 1st International Working Conference on Performance*

Modelling and Evaluation of Heterogeneous Networks (HET-NETs03), D. Kouvatsos (Ed.), Ilkley, pp. 17/1–17/10, 2003.

[18] K. Kontovasilis, C. Skianis and G. Kormentzas, Estimating signalling efficiency in a composite radio environment, in *Proceedings of the 1st International Working Conference on Performance Modelling and Evaluation of Heterogeneous Networks (HET-NETs03)*, D. Kouvatsos (Ed.), Ilkley, pp. 65/1–65/12, 2003.

[19] C. Skianis, K. Kontovasilis, G. Kormentzas and G. Lisa, Simulation study of a signalling protocol efficiency in a composite radio environment, in *Proceedings of the 2004 High Performance Computing & Simulation (HPC&S) Conference and 18th European Simulation Multiconference (ESM 2004)*, Magdeburg, Germany, 13–16 June 2004, pp. 17–22, 2004.

[20] D. Loukatos, Traffic generation and analysis, emphasizing on ATM and IP network technologies, Ph.D. Thesis, National Technical University of Athens, Greece, 2002.

[21] D. Loukatos, L. Sarakis, K. Kontovasilis and N. Mitrou, Efficient real-time traffic analysis tools for monitoring performance on packet networks, in *Proceedings of the 1st International Working Conference on Performance Modelling and Evaluation of Heterogeneous Networks (HET-NETs03)*, D. Kouvatsos (Ed.), Ilkley, pp. 52/1–52/11, 2003.

[22] V. Jacobson, C. Leres and S. McCanne, tcpdump tool, available via anonymous ftp to ftp.ee.lbl.gov, 1989.

[23] V. Paxson, G. Almes, J. Mahdavi and M. Mathis, RFC 2330 – Framework for IP Performance Metrics http://www.ietf.org, 1998.

[24] ITU-T Recommendation Y.1540 (previously numbered I.380) Internet Protocol Data Communication Service – IP Packet Transfer and Availability Performance Parameters, February 1999.

[25] C. Demichelis and P. Chimento, RFC 3077 – 3393 IP Packet Delay Variation Metric for IPPM, http://www.ietf.org, 2002.

[26] C. Demichelis, Improvement of the instantaneous packet delay variation (IPDV) concept and applications, in *Proceedings of the World Telecommunications Congress 2000*, May 2000.

6

Adaptation of Multimedia Flows in a Seamless Mobility Context Using Overlay Networks

D. Pichon[1,2], K. Guillouard[1], P. Seite[1] and J.-M. Bonnin[2]

[1] *France Télécom, Orange Labs, 35510 Cesson Sévigné, France;*
e-mail: {dominique.pichon, karine.guillouard}@orange-ftgroup.com
[2] *Institut TELECOM, Telecom Bretagne, RSM, 2 Rue de la Châtaigneraie CS 17607, 35576 Cesson Sévigné,France; e-mail: jm.bonnin@enst-bretagne.fr*

Abstract

The streaming of multimedia flows becomes an arduous task in a complex network environment where terminals may be connected to the Internet using different access networks, such as Ethernet, ADSL, GPRS, UMTS or WiMax. Indeed, the change of networks potentially entails large repercussions on the service delivery, which has a severe impact on the user-perceived quality of service of the video streaming. We propose a framework with a service provider closely working with the mobility manager in order to react to mobility events. According to the runtime service execution environment, the service manager constructs an adequate overlay network, for example by adding a proxy which can later be controlled to match the network capacities. An application scenario with Scalable Video Coding multimedia flows in a mobile context is also described.

Keywords: Overlay network, mobility, adaptivity, multimedia streaming, heterogeneous networks.

D. D. Kouvatsos (ed.), Mobility Management and Quality-of-Service for Heterogeneous Networks, 129–152.

Acronyms

3GPP	3rd Generation Partnership Project
ACS	Ambient Control Space
ADSL	Asymmetric Digital Subscriber Line
AVC	Advanced Video Coding
BL	Base Layer
CN	Correspondent Node
CSC	Call State Control
EL	Enhancement Layer
ETSI	European Telecommunications Standards Institute
FE	Functional Entity
GPRS	General Packet Radio Service
HO	HandOver
IETF	Internet Engineering Task Force
IMS	IP Multimedia Subsystem
IPTV	Internet Protocol Television
ITU	International Telecommunication Union
MG	Media Gateway
MGC	Media Gateway Controller
MGF	Media Gateway Function
MIP	Mobile IP
MIPv6	Mobile IP version 6
MM	Mobility Manager
MN	Mobile Node
MRC	Media Resource Control
MRB	Media Resource Brokering
NASS	Network Attachment Subsystem
NGN	Next Generation of Networks
QoS	Quality of Service
RACS	Resource and Admission Control Subsystem
RO	Route Optimization
RTCP	RTP Control Protocol
RTP	Real-Time Protocol
RTSP	Real Time Streaming Protocol
SCTP	Stream Control Transmission Protocol
SDP	Session Description Protocol
SIP	Session Initiation Protocol
SM	Service Manager

SSON   Service-Specific Overlay Network
SVC   Scalable Video Coding
TISPAN   Telecommunications and Internet converged Services and
          Protocols for Advanced Networking
UMTS   Universal Mobile Telecommunications System

6.1 Introduction

Diversity is a common characteristic to access networks, terminals and services. Several access network technologies exist and allow users to gain access to their preferred services. However, a strong vertical integration between terminals, access networks and services restricts the use of services to specific network infrastructures and dedicated devices.

In the past few years, a huge effort of convergence has been carried out in order to define a common network infrastructure, also known as Next Generation Network (NGN) [1]. This infrastructure is made up of a transport part and a separate control part that are available to all the different access networks, giving users access to a larger range of services.

This network convergence paves the way for a new concept called the seamless mobility, that allows users to remain connected and use their services, whatever the current access networks and devices. Services may then be accessed from any device while roaming through different access networks, which results in a need for services that may adapt themselves according to their execution environments, the so-called context-aware services.

Context-aware services may adapt themselves by exploiting the Overlay Network concept that has recently emerged in the scientific literature. Indeed, overlay networks offer a flexible way to accommodate heterogeneity by creating a virtual network dedicated to the service, e.g., by including specific required proxies. Overlay networks can also dynamically react, which is a major asset of this concept and makes it an interesting approach for the streaming media distribution.

Mobility is a major source of modifications of the service execution environment. For example, an inter-technology handover (or vertical handover) inevitably results in a variation of resources (available throughput, jitter, round trip time, etc.), which may have a deep impact on the quality of the service. The specific problem addressed by this research is how to ensure the best possible provision of services when their execution environments strongly vary. This paper will specifically focus on the adaptation of multimedia streams when a mobile user roams across heterogeneous access networks by using mobility-aware overlay networks.

The remainder of the paper is structured as follows. Section 6.2 presents the next generation network architecture and some mechanisms that can be implemented to allow a seamless mobility through the different access networks. Section 6.3 describes the potential impact of the mobility on multimedia services and the possible adaptation solutions. Related work concerning coordination between mobility and service management are then discussed in Section 6.4. Section 6.5 then shows the architecture we propose while conclusions and future work are given in Section 6.6.

6.2 Seamless Mobility through Next Generation Networks

6.2.1 NGN Overview

The Next Generation Network concept is an effort carried out by several major international standardization organizations. The International Telecommunication Union (ITU) defines it as [1]:

> a packet-based network able to provide Telecommunication Services to users and able to make use of multiple broadband, QoS-enabled transport technologies and in which service-related functions are independent of the underlying transport-related technologies. It enables unfettered access for users to networks and to competing service providers and services of their choice. It supports generalised mobility which will allow consistent and ubiquitous provision of services to users.

The first major step was the introduction of the IP Multimedia Subsystem (IMS) [2] by the 3rd Generation Partnership Project (3GPP) in the fifth release of UMTS specifications. The IMS specifies a functional architecture dedicated to the provision of multimedia services. It defines an overlay network on top of the packet-based core network. This overlay network forms the *control layer* and deals with signalling messages in order to control the service provision. The IMS also defines interfaces to the *service layer* that service providers can use to offer their services. These interfaces are conform as far as possible to the IETF "internet standards", e.g., the Session Initiation Protocol (SIP) [3].

The European Telecommunication Standards Institute (ETSI), as part of its work on Next Generation Network, kept the IMS concept as a subsystem

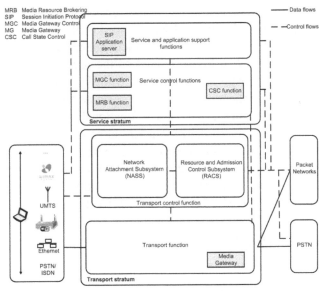

Figure 6.1 NGN functional architecture.

in its first NGN specification [4], proposed by its standardization body, Tele-communications and Internet converged Services and Protocols for Advanced Networking (TISPAN). It further specifies the transport layer defining two subsystems: the Network Attachment Subsystem (NASS) and the Resource and Admission Control Subsystem (RACS).

An overview of the NGN architecture is depicted in Figure 6.1. The two main strata are highlighted: the transport stratum and the service stratum. The former stratum allows the transport of data using any access network. This transport is controlled by transport control functions and service control functions. Transport control functions may be divided into the two subsys-tems relative to the control of new terminal attachments (NASS) and to the control of the resource allocation inside the network (RACS). NASS performs usual tasks such as the dynamic provision of network addresses, the location management, etc. As for the RACS, it allows services to be guaranteed a certain level of Quality of Service (QoS) by controlling the resource alloca-tion and verifying its correct use. The transport layer also includes transfer functions entities used by the other subsystems to perform specific tasks on data. For example, it may be a Media Gateway Function (MGF) entity used by the service stratum to provide, e.g., transcoding functions between an IP-transport domain and a switched circuit network.

Service control functions are part of the service stratum. This stratum is directly inspired by the IMS, originally developed by the 3GPP. Its main components are the Call State Control (CSC) functional entities (FE) that manage signalling messages, such as those related to the registration process. CSC FE interact with the transport stratum, e.g., with the RACS so as to ask for bandwidth, and with applications to provide value-added services. Figure 6.1 also shows other important service control functional entities in the context of multimedia transport, namely the Media Resource Control (MRC) FE (also called Media Gateway Controller) and the Media Resource Brokering (MRB) FE. The former may control a Media Resource Processing (MRP) FE (also called Media Gateway) in the transport stratum and the latter is crucial as it allows the right MRC-FE and the associated MRP-FE to be selected.

6.2.2 Support for Mobility

6.2.2.1 Handover Management

The network convergence is a major asset of the NGN architecture. Any device may then access any service by using any access network technology. However, the mobility between different access networks remains an open issue in the first release of NGN and is still an active topic of research even if many solutions have already been designed by the IETF.

In the mobility terminology, network mobility between different technologies is called *vertical mobility* or *vertical handover*. Specific mobility protocols have been implemented to allow users to stay connected even after layer-3 mobility events. From the correspondant point of view, different approaches are possible:

- hiding the layer-3 addressing change;
- communicating the layer-3 addressing change.

Let us consider a session between a mobile node (MN) and a correspondent node (CN) as depicted in Figure 6.2.

In the first case layer-3 modifications remain invisible for the correspondent node thanks to a mobility anchor situated in a specific place. In the other case the correspondent node is informed about the layer-3 change. With regard to the former we can cite a few implementations. The most well-known is MIP [5] or its counterpart MIPv6 [6] in the IPv6 world which defines protocols to provide for registering the temporary address, designated as the *care-of address* in the MIP terminology, with the help of a mobility anchor (also called home agent) located in the home network. As the home agent has

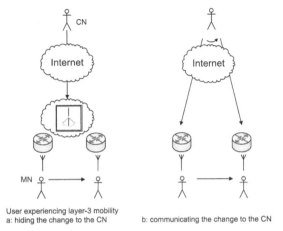

User experiencing layer-3 mobility
a: hiding the change to the CN   b: communicating the change to the CN

Figure 6.2 Communicating or not the layer-3 change.

a global knowledge of the mobile node network address we call this a global mobility management.

The mobile user can also inform correspondent nodes about its current addresses. This may be performed at the network layer by use of MIPv6 with Route Optimization (RO). In that mode, the mobile node directly informs the correspondent node via a binding update.

The IP stack can also be modified to include a shim layer that manages the translation between an upper layer identifier, namely an initially available address which remains constant throughout the session and the locator addresses, actually used to forward packets, as explained in [7].

At the transport layer, the transport protocol can provide ways for adding new IP addresses to the association between the two nodes, e.g., SCTP and its mobile extension (mobile SCTP). A thorough survey on mobility at the transport layer is available in [8]. Eventually, the handover function may reside at the application layer by means of signalling messages, such as a SIP ReInvite message, as explained in [9].

6.2.2.2 A Need for a Decision

The possibility for a terminal to be sequentially connected to different networks or in the same time (i.e., being multihomed) causes new problems. The choice of the best access network and the identity of the entity responsible for this choice are two major problems, which are correlated. This correlation is due to the fact that the decision-making process depends on the input information, which may be different according to the decision entities involved, that can be located either at the terminal side and/or the network side.

The decision-making process tries to optimize the choice of the best access network according to information such as the quality of the available access networks, e.g., the maximal throughput, the application QoS needs, any requirements expressed by the user or the network operator. However, the terminal and the network have different views, i.e., *information*. The terminal decision-maker has a full knowledge of the running environment of its terminal while it has poor information concerning the running state of the access networks and vice versa for its counterpart in the network. Information may not be exchanged for any reasons, e.g., for privacy concerns, to limit the load due to the exchange of information. Therefore, as they have different views and different goals, this will result in different mobility decisions.

In the terminal the decision-making process may be performed following two approaches: the terminal-centric view and the network-assisted view. The latter differs from the former by the fact that the network helps the terminal to choose. As for the terminal-centric view, we may cite the Ubique implementation [10]. Ubique is an advanced middleware that deals with information contained in profiles in order to select and subsequently configure the adequate interface.

Implementations of the network-located decision-making process also exist. For instance, Bonjour et al. [11] design a network entity, called *mobility manager*, which collects information such as the available access networks and their quality in order to instruct the mobile terminal which interface to select.

6.3 Mobility Impacts the Service Provision

6.3.1 Mobility Requires an Adaptation Phase

The nature of the handover and its consequences may be the source of severe disruptions in the service delivery as it may impact its running environment. Indeed, the handover may entail an interruption delay, which may be critical for the service as the mobile user is no longer able to send or receive data in the meantime. The handover has also long-term consequences on the service as it results in a modification of its runtime execution environment. These consequences are due to the fact that this environment highly depends on the quality of the access network. The access network sets the available throughput, the QoS facilities, the security management, the bit costs, which

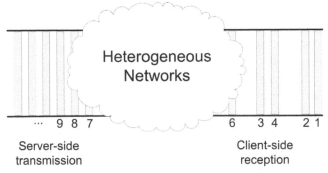

Figure 6.3 A basic streaming system.

may be crucial in the design of the service. Unlike unpredictable events such as congestion, mobility and its consequences may be foreseen and therefore anticipated in order to trigger an adaptation process.

An adaptation phase is therefore required in order to tailor services to the runtime service execution environment. The service adaptation process may be defined as a data processing phase taking into account a set of inner and outer constraints. Inner constraints are due to the nature of the service itself. For example, the service may require a limited jitter or round-trip time, a minimal throughput, etc. Outer constraints are requirements expressed by any elements taking part in the service delivery. It may be preferences stated by the service provider, the network operators and the final user, for example in the subscription phase. Constraints may also come from the limited features of the terminals, the limited capacities of the flow source, the access network characteristics, etc.

6.3.2 Focus on the Adaptation of Multimedia Flows

Video streaming is a common service with strong requirements to preserve the user-perceived QoS, such as a limited jitter, and so on. As such, it could strongly benefit from an adaptation platform.

A streaming system may be illustrated as in Figure 6.3. Basically, it is made up of a server that transmits time-constrained packets towards various clients located on heterogenous networks. Some intermediary entities, i.e., proxies, may help to perform adaptation.

As depicted in this figure, some difficulties may arise during the transmission that may considerably lower the user-perceived QoS. Packets may undergo different transmission delays, which implies that packets are no longer ordered at the reception. Packets may also be corrupted or even lost during the transmission. Usually, clients implement buffering mechanisms so that the decoder can postpone the display until enough data is received, reordered and decoded.

On the other hand, the streaming experience can also be improved by means of more complex adaptation techniques. As highlighted in [12], the following three points need to be addressed:

- the selection of the appropriate datapath;
- the use of adaptive mechanisms to transmit data;
- the use of data adapted to the crossed networks.

The selection of the path is of uppermost importance. Indeed, this datapath has inherent properties, which will limit its use to a certain kind of data and at some amounts. Therefore, a service-aware datapath has to be determined, taking into account the different available access networks. The mobility management has to be considered in this step as it determines the access networks to be used. This path from the source to the terminal may also include any entities required to the service enforcement, such as admission control entities, proxies, etc.

The second point refers to the fact that adaptive mechanisms should be used to control the transmission of data. A feedback between the client and the server allows the latter to make its transmission rate match the receiver needs. For instance, Real-Time Control Protocol (RTCP) [13] information enables a server to control the packet scheduling; the server may limit its rate to avoid congestion, or it may schedule the most important data units, such as reference picture when data is hierarchically encoded. The feedback may also be used to alter the encoding parameters related to the error correction. More redundancy may be added so that the receiver recovers more easily from losses.

Finally, data may facilitate adaptation. File switching is an ordinary technique used to adapt data to the receiver side, as detailed in [14]. Data is encoded in multiple versions, according to the whole set of receiver client possibilities. Then, switching is performed between the different versions as and when required. File switching is a simple technique; however, it implies a huge storage space and entails use of complex mechanisms to switch between versions, while keeping the temporal synchronization between flows. A possible alternative is designated as *simulcast* and allows the source to deliver the different versions simultaneously. Nevertheless, it is often an unattractive strategy as it has important network resource costs.

These problems are alleviated by a recent amendment of the Advanced Video Coding (AVC) standard, called SVC [15], i.e., Scalable Video Coding, which allows encoders to split data into several layers. Among them, the base layer can be independently decoded and reconstruct the initial sequence at a minimum quality level. The addition of enhancement layers improves the quality at a certain granularity level.

Scalability has two major assets. Firstly, it decreases the operating costs and simplifies the content management. Indeed, decoding and streaming are decoupled. A single version is made available to accommodate the whole set of requirements. Then, intermediate proxies select the adequate subparts. This greatly simplifies the content management and reduces the need for storage space. The second important asset is the possibility to make universal and seamless availability a reality. A single file is sufficient to make a stream available to heterogeneous terminals. Moreover, medium-grain scalability allows video streams to be rendered smoothly in order to get the maximum of the network capacities. The graceful degradation is a major advantage in the context of our studies as resources may sharply vary after a handover.

6.3.3 A Need for a Decision

According to the runtime service execution environment, an appropriate service-aware datapath has to be selected, which highlights the need for a decision-maker. In a similar way to the location of the mobility manager, the decision-maker role played by the service manager can be performed by a network entity (network-controlled) or by the terminal without any coordination with the network (terminal-controlled) or with its help (network-assisted).

This datapath is made up of a core network part and an access network part, which are distinct. The definition of a service-aware datapath involves the use of dedicated mechanisms in each part. In the core network part, traffic engineering mechanisms are used to optimize the performance of operational networks. Traffic flows are aggregated and classified into trunks according to criteria, e.g., the QoS requirements. Then, specific routes are defined, either manually by the operator or automatically computed, e.g., by use of a path computation element [16]. As for the access network part, the decision-maker has to select the best access network by taking into account all the inner and outer constraints already detailed in Section 6.3.1. Some implementations exist which partially perform this task, such as the Ubique framework.

The definition of the service-aware datapath may also be more complex as proxies might be required in the service delivery chain. The decision-maker

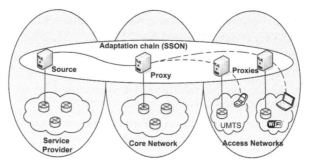

Figure 6.4 Service-aware overlay network.

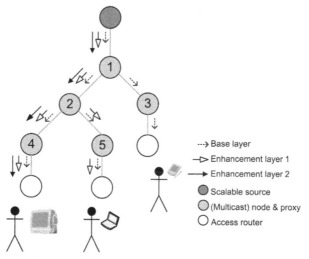

Figure 6.5 A chain of proxies to adapt SVC flows.

(in the service stratum of the NGN model) may design a specific overlay network dedicated to the service delivery, also called a service-specific overlay network (SSON), as described in [17].

As depicted in Figure 6.4, a specific chain from the source to the end terminal (in the transport stratum of the NGN model), including any necessary proxies, may be determined in order to adapt services to the runtime service execution environment. For example, it may be a chain of multimedia proxies performing packets discrimination, e.g., selecting the adequate layers of a SVC flow, as shown in Figure 6.5.

As for the optimal location, it depends on the target goals of the optimization process. If we only consider network resource costs and a unicast transmission, then the optimal place to adapt data is at the source. However, if we consider other goals such as the reduction of the adaptation phase delay, the use of proxies may be necessary.

Finally, the decision-maker needs also to command the different elements of the adaptation chain as and when required. It collects information about the service environment and decides whether or not to trigger an adaptation phase. To this end, it needs to be able to control proxies (MG), by inter-acting with the associated controllers (MGC). Several protocols have been developed by the IETF for controlling multimedia sessions. For example, we can cite the Media Gateway Control Protocol (MGCP) [18], which is dedicated to voice over IP services and allows a call agent (or MGC) to control media gateways.

6.4 Mobility and Service Coordination: Related Works

6.4.1 Service Adaptation to the Link

Some cross-layer approaches already exist to coordinate the service manage-ment and the network capacities in order to take into account information related to the network. As a first example, Van der Schaar et al. [19] exploits the cross-layer joint APPLICATION-MAC-PHY optimization by use of an adaptive application which given the channel quality indicator, can choose the PHY and MAC configuration and modify its behavior. In this work a scalable codec can divide its flows into subflows, each undergoing different treatment at the MAC and PHY layers. They use this principle to transmit an SVC flow subdivided in layers on a 802.11e network. Each layer is given preferential treatment via a per-flow QoS negotiation.

3GPP in [20] specifies methods for adapting a multimedia stream to the wireless link. To this end, it defines attributes, e.g., 3GPP-Link-Char, that are carried in usual RTSP packets and allow server to adapt its streaming to the link capacities. It also defines necessary extensions to protocols such as RTSP to report client buffer feedback.

However, such works does not study problems related to the layer-3 mo-bility management, i.e., the change of access network. Such repercussions on the service delivery are sometimes mentioned by studies focusing on the mobility management and are presented below.

6.4.2 Service Adaptation to Networks

In [21], a SIP end-to-end architecture is introduced wherein a SIP entity manages both service and mobility management. To this end, it acts as a proxy and receives data on behalf of the media player and selects the right interface triggering handovers by use of SIP ReInvite messages.

In [22], the researchers present an architecture where the mobility management is located in the end-terminal. A Mobility Manager receives information such as the availability of networks in the environment, the current interface status and then computes this data to command the layer-3 or lower layers accordingly. It also notifies applications to make them mobility-aware.

The mobility and service management may also be located inside the network. As in [11] where the mobility manager informs the source of the stream of the imminence of the handover, its execution and the new characteristics of the access networks.

These approaches exhibit some drawbacks. Firstly, applications have to be context-aware and be able to react to complex information issued from the mobility manager. Secondly, when the service management is carried out by end entities, their capacities are limited to adapt without the help of third parties. To face these problems the mobility-aware overlay network concept seems a useful solution.

6.4.3 Mobility-Aware Overlay Networks

An overlay network is a virtual network built on top of a physical network and represents a convenient way to add functionalities in a network without impacting on its components. Overlay networks offer a flexible way to improve media content streaming by adding easily required entities, such as proxies, in the virtual network between the source and the final terminal. This virtual network can be designed to accommodate any needs and is therefore considered as a useful approach to deal with heterogeneous networks. The content adaptation by use of overlay networks has become a major research challenge in the past few years.

This concept has been exploited by the european project *Ambient Networks* (Sixth Framework Program). It defines the *composition* concept which allows heterogeneous networks to collaborate in an automatic way by negotiating agreements on their mutual interfaces. The Ambient Control Space of each Ambient Network is responsible for the agreement negotiation and its enforcement. That space controls all aspects of the network management, by use of dedicated functional entities (FE), such as the mobility FE or the overlay management FE, which interact in order to improve the whole behaviour of an Ambient Network. Thus, the overlay management can be aware of mobility events.

The article [17] presents the overlay network management in an ambient network wherein the authors propose the use of service-specific overlay network (SSON) for every service. The SSON defines an appropriate high-level path including all the necessary entities to compose the service. In the ambient network terminology, these entities are named *ONodes*, and can take on the roles of Media Servers, Media Clients and Media Ports. The SSON definition, generation and monitoring is performed by the Ambient Control Space. The forwarding behaviour in the SSON is accomplished by use of tunneling. To this end, they define a header that specifies the ONode behaviour (transcoding, caching, etc.) and the next hop address.

The mobility-aware overlay network concept shows significant advantages; services are adapted while minimally involving applications. However, the ambient network solution implies a header management, that potentially entails large overhead in the network. Moreover, the adaptation decision-process is performed by the ambient control space. It does not mention in details the interface in the ACS between the mobility-dedicated entity and the overlay management entity.

We therefore propose an NGN-compliant architecture with an environment-aware service manager that manages the overlay network and which is closely working with a mobility manager, in order to accurately define the service.

6.5 Proposition of a Generic Architecture

In the previous sections we highlighted the need for a tight coordination between the service provision and the mobility management. In the following lines an identification of the requirements is performed so as to deduce a generic architecture able to adapt any service of a mobile user in an heterogeneous environment.

6.5.1 Requirements Analysis

Firstly we need to select the best access network(s) and inform services to let them adapt to the current allocated ressources. This entity is called the *Mobility Manager* (or MM).

The MM has to discover all the available access networks and their current capacities in order to be able to offer them to services. That discovery may be facilitated by use of protocols such as 802.21 [23], currently being standardized by the IEEE organization and that can be used to discover access networks or obtain current information related to them, such as the load.

We also need the MM to know the QoS needs of the different services in order to allow the MM to select the access networks accordingly. Among the different QoS requirements, we may cite:

- class;
- minimal, mean, maximal bit rate;
- maximum transmission unit;
- typical packet error rate;
- maximum delay tolerated, (i.e., round trip time);
- maximum delay variation tolerated (i.e., jitter);
- security level required;
- battery level required.

The MM registers all these requirements expressed by any services and then selects the best access network(s).

As running services are competing for the same resources, we also need to control their admission to ensure the quality of services. We can either let every service ask for its own admission or let the MM do this task on its behalf. The second choice is more appropriate because of the central role played by the MM. Indeed, thanks to its global knowledge of all the different services, the MM can ask for resources in an optimized way, limiting the number of QoS requests.

Moreover, we need an entity to route packets among the different available paths and as the MM selects the access network for each service, it has an inherent knowledge of the routing, which avoids every service to exchange routing table.

The admission control performed by the MM has also to deal with priorities, which means that a service may be stopped to allow a more urgent service to be executed.

The last requirement implies a dialog between each service and the MM. This dialog should allow services to be informed when resources at their disposal strongly vary, which may happen when a more urgent service appears or when a running service is routed to a new access network. This dialog shoud occur prior to the handover in a down mobility scenario characterized by a reduction in the available resources, which make it unable to support the service as such, e.g., a handover from a 802.11g network to a GPRS network for a video service.

As for the MM location, it may be located inside the terminal as in the Ubique framework [10] or distributed between the different elements that take part in the mobility management as in the HDHO framework [24].

As far as the service is concerned we need the service to be able to react to information provided by the MM in order to adapt itself to the current resources. The adaptation process may result in the modification of the service parameters, such as the source encoding parameters, but also in the choice of the adequate source, proxy, etc.. The adaptation process is a complete task that should not be a burden for the applications, whether in the source or target terminal.

Therefore, we define an entity specific to each service that is responsible for the management of the whole service delivery and which takes care of the adaptation process. This entity is the *Service Manager* (or SM). It acts as an application-level proxy and negotiates with both the MM and the different service delivery elements to establish the service session on behalf of the application on the terminal.

Specific to each service, the SM reads content-related information about the service and subsequently defines different states, each requiring a certain QoS. For example, we may consider a multimedia IPTV service made up of SVC layers, such as:

- a Base Layer (BL), 150 kbits/s, QoS_{BL};
- a first Enhancement Layer (EL1), 400 kbits/s, QoS_{EL1};
- a second Enhancement Layer (EL2), 1500 kbits/s, QoS_{EL2}.

Five states are specified:

1. degraded mode, corresponding to the base layer;
2. medium-level quality mode, corresponding to the base layer plus the enhancement layer 1;
3. high-level quality mode, corresponding to the whole layers;
4. on hold mode;
5. terminated mode.

These states are communicated to the MM which stores them as profiles and uses them when selecting the right access network(s) for each service.

At last but not least, the SM obviously has to react to MM state selection by adapting the overlay network. The SM learns the better state and strives for adapting the network session.

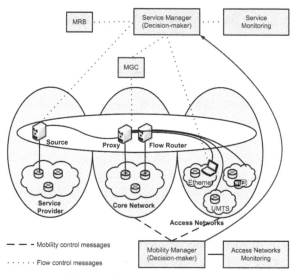

Figure 6.6 Proposed architecture.

The SM may also be informed about the current network localization(s), e.g., the IP address, and choose a better source, or proxy, according to this information.

6.5.2 The Proposed Architecture

The proposed architecture is made up of entities specific to the service and entities dedicated to the mobility, which are shared by all services. The global architecture is depicted in Figure 6.6.

Among the service entities we first define the Service Manager as an entity specific to each service. To define the service, the service manager firstly learns the content-related information about the runtime service execution environment. To this end, it detects the type of service, for example a multimedia stream made up of SVC layers and its description (localization of the source, bitrate, spatial, temporal and quality resolution, etc.). It also learns context-related data, such as requirements or preferences expressed by all the stakeholders of the service delivery. Then, it defines several adaptation states accordingly, which are communicated to the Mobility Manager. When triggerred by the MM, the SM reacts and updates the service.

Applications (in the source and the terminal) initiate, accept and terminate a network session, i.e., a service. An adaptive application also integrates a module to modify an ongoing session. This adaptive module can also be placed on the path in an intermediate node, such as a proxy, in which case, it may be controlled by a controller (MGC), as depicted in Figure 6.6.

As for the service monitoring, it is carried out by means of protocols, such as RTCP [13]. RTCP provides a useful way to report a feedback from the client to the source, e.g., to inform the source about data packet losses.

We should also mention the Service Provider Portal, which registers the whole available content. It further makes them available by giving accurate descriptions, by means of protocols, such as Session Description Protocol (SDP) [25].

Finally, the service manager also interacts with a directory, known as the Media Resource Broker (MRB), which is able to provide to the SM useful information to connect to the right SSON. The MRB determines the most appropriate SSON, e.g., a list of proxies, if several are required, according to criteria, such as the terminal localization, the availability and the current proxy load, etc.

The service manager, portal, MRB, MGC are all located in the service stratum while proxies are part of the transport stratum. This is a generic architecture not only devoted to the multimedia services. Indeed, the service manager can deal with any type of service, by triggering adaptation phases apppropriate to the service type.

As for the mobility management, we firsly specify the Mobility Manager, as explained before. The MM is responsible for:

- registering the different available paths and their characteristics in real time;
- activating/desactivating paths;
- collecting the service requirements, such as their different states;
- instructing services which state to use;
- deciding the routing of the services.

We also use network monitoring entities in order to get real-time information about the running state of the network. The MM subscribes to these entities so as to be notified of an event occurence. Among the different events that may trigger a notification, we can find:

- new network detection;
- establishment/lost of an association with the access network;
- signal power under a configured threshold;
- signal to noise ratio too low;
- quality of the path has increased/decreased;
- congestion in the access network.

Eventually the routing of the service among the different paths requires the presence of a dedicated module whose location depends on the mobility architecture implemented. For end-to-end mobility architectures, such as with SIP or shim, this module is located in both ends in order to correctly route packets in both directions. For mobility architectures with a mobility anchor, the routing module is obviously located in the anchor and in the end terminal. The adding of a mobility anchor allows the source, e.g., an IPTV server, to be released of the routing burden. The distribution of the flows among the different available paths may be performed using policies, such as those described in [26].

6.5.3 Application Scenario

We consider the scenario depicted in Figure 6.7. In this scenario the application is running on a terminal that may be potentially connected to the Internet using a UMTS interface and/or a WiFi interface. Firstly, it registers to the network using UMTS. It then asks for a video streaming service, conform to the description provided in Section 6.5.1 and adapted to its terminal. To this end, it uses the service manager as an application proxy (message 3). The service manager acts on behalf of the application and configures the service as needed. It informs the MM and waits for its admission decision (message 4). In the fifth message the SM is informed that it should use the service in its basic configuration. The service manager then asks an MRB for an SSON with the runtime service execution environment, i.e., the current localization of the terminal, the description of the service, etc. (message 6). The MRB then answers with the adequate SSON, in that case, the address of a proxy to connect to (message 7). The service manager sets up the proxy, via the associated controller (not shown, for the sake of simplicity), so that it delivers only the base layer (messages 8 and 9). Eventually, it acknowledges the MM request (message 10) and forwards the *RTSP 200 OK* (message 11) so that the application can start the service (messages 12 and 13).

Later, the terminal becomes multi-homed (UMTS and Wifi both active) (message 14) and the mobility manager makes up his mind to increase the service quality by selecting a higher state (message 15). The service manager analyses this request and triggers a new adaptation phase. In that example, it keeps the same SSON and only modifies the proxy configuration so that a new layer be added to the video streaming (messages 16 and 17). It then acknowledges the mobility manager (message 18). The quality of the service is considerably improved without involving the application, nor the source.

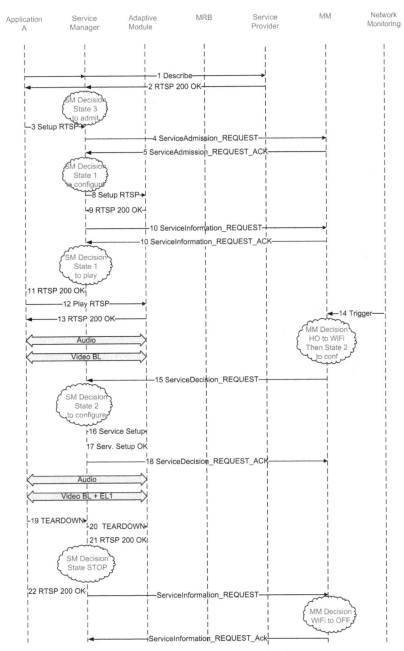

Figure 6.7 Service manager as an application proxy.

6.6 Conclusions and Future Works

In this paper we have introduced a generic architecture that enables a tight collaboration between the mobility management and the service provision. By use of the interface to the mobility manager any service is able to react to modifications of the runtime service execution environment, improving this way the quality of the service provision. The advantages of our approach are its compliance to the NGN concept, the use of overlay networks to accomodate the diversity and the maximal reuse of existing protocols. A service management framework is currently in developement using OPNET on top of an existing mobility management framework in order to accurately specify the interface between the service management and the mobility management and optimize their coordination. The scenario is also being implemented in order to show the control of multimedia streamings in a mobility context, by using SVC proxies already implemented at France Télécom.

Acknowledgements

The authors wish to acknowledge all the comments and suggestions made by the reviewers, which clearly improved the content as well as the quality of this presentation.

References

[1] ITU-T, Recommendation y.2001 general overview of NGN, 2004.
[2] 3GPP, IP Multimedia Subsystem (IMS); Stage 2, TS 23.228 v.7.5.0, 3rd Generation Partnership Project (3GPP), 2006.
[3] J. Rosenberg, H. Schulzrinne, G. Camarillo, A. Johnston, J. Peterson, R. Sparks, M. Handley and E. Schooler, SIP: Session Initiation Protocol, RFC 3261, June 2002 (Updated by RFCs 3265, 3853, 4320).
[4] ETSI/TISPAN, Es 282 001 v1.1.1 NGN functional architecture release 1, 2005-08.
[5] C. Perkins, IP Mobility support for IPv4, RFC 3344, August 2002.
[6] D. Johnson, C. Perkins, and J. Arkko, Mobility support in IPv6, RFC 3775, June 2004.
[7] E. Nordmark and M. Bagnulo, Shim6: Level 3 multihoming shim protocol for ipv6, internet-draft, draft-ietf-shim6-proto-09.txt. Work in progress, October 2007.
[8] M. Atiquzzaman and A. S. Reaz, Survey and classification of transport layer mobility management schemes, in *PIMRC'05: Proceedings of the IEEE 16th International Symposium on Personal, Indoor and Mobile Radio Communications*, 11–14 September, pp. 2109–2115, 2005.

[9] H. Schulzrinne and E. Wedlund, Application-layer mobility using SIP, in *Service Portability and Virtual Customer Environments*, pp. 29–36, 2000.

[10] L. Suciu, J-M. Bonnin, K. Guillouard and B. Stévant, Achieving "always best connected" through extensive profile management, in *PWC'04: Proceedings of the 9th International Personal Wireless Communications Conference*, Lecture Notes in Computer Science, vol. 3260, Springer, pp. 421–430, 2004.

[11] S. Bonjour, S. Athéo, E. Njedjou, S. Venant-Valéry, and N. Wiltshire, Network-controlled MIPv6 based inter-technology handover, in *Proceedings of the IST Summit 2003*, 2003.

[12] J. Chakareski and P. Frossard, Adaptive systems for improved media streaming experience, *IEEE Communications Magazine*, vol. 45, pp. 77–83, January 2007.

[13] H. Schulzrinne, S. Casner, R. Frederick and V. Jacobson, RTP: A transport protocol for real-time applications, RFC 3550 (Standard), July 2003.

[14] I. Curcio and D. Leon, Application rate adaptation for mobile streaming, in *WOWMOM '05: Proceedings of the Sixth IEEE International Symposium on World of Wireless Mobile and Multimedia Networks*, Washington, DC, IEEE Computer Society, pp. 66–71, 2005.

[15] ISO/IEC, 14496-10:2005 scalable video coding, 2007, Final Draft Amendment 3.

[16] A. Farrel, J.-P. Vasseur and J. Ash, A path computation element (PCE)-based architecture, RFC 4655 (Informational), August 2006.

[17] J. Rey, B. Mathieu, D. Lozano, S. Herbom, K. Ahmed, S. Schmid, S Goebbels, F. Hartung and M. Kampmann, Media aware overlay routing in ambient networks, in *PIMRC'05: Proceedings of the IEEE 16th International Symposium on Personal, Indoor and Mobile Radio Communications*, September, vol. 2, pp. 952–957, 2005.

[18] F. Andreasen and B. Foster, Media Gateway Control Protocol (MGCP) version 1.0, RFC 3435 (Informational), January 2003 (Updated by RFC 3661).

[19] M. van der Schaar, Y. Andreopoulos and Z. Hu, Optimized scalable video streaming over IEEE 802.11A/E HCCA wireless networks under delay constraints, *IEEE Transactions on Mobile Computing*, vol. 5, no. 6, pp. 755–768, 2006.

[20] 3GPP ETSI, TS 26 234 transparent end-to-end packet-switched streaming service (pss); protocols and codecs, version 7.4.0 Release 7, October 2007.

[21] C. A. Szabó, S. Szabó and L. Bokor, Design considerations of a novel media streaming architecture for heterogeneous access environment, in *BWAN'06: Proceedings of the 2006 Workshop on Broadband Wireless Access for Ubiquitous Networking*, New York, ACM Press, p. 3, 2006.

[22] F. Cacace and L. Vollero, Managing mobility and adaptation in upcoming 802.21 enabled devices, in *WMASH'06: Proceedings of the 4th International Workshop on Wireless Mobile Applications and Services on WLAN Hotspots*, September, ACM, New York, pp. 1–10, 2006.

[23] 802.21, Draft IEEE standard for loacl and metropolitan area networks: Media independant handover services, Technical Report, IEEE, March 2006.

[24] L. Suciu and K. Guillouard, A hierarchical and distributed handover management approach for heterogeneous networking environments, in *ICNS'07: Proceedings of the Third International Conference on Networking and Services*, Washington, DC, IEEE Computer Society, p. 77, 2007.

[25] M. Handley, V. Jacobson and C. Perkins, SDP: Session Description Protocol, RFC 4566 (Proposed Standard), July 2006.
[26] K. Mitsuya, R. Kuntz, K. Tasaka and R. Wakikawa, A policy data set for flow distribution, draft-mitsuya-monami6-flow-distribution-policy-04.txt, Internet-Draft, August 2007.

PART TWO
OPTIMAL ADMISSION CONTROL

PART TWO

7

Efficient Algorithm to Determine the Optimal Configuration of Admission Control Policies in Multiservice Mobile Wireless Networks

David Garcia-Roger, Jorge Martinez-Bauset and Vicent Pla

Departamento de Comunicaciones, Universidad Politécnica de Valencia, UPV, ETSIT, Camino de Vera s/n, 46022 Valencia, Spain; e-mail: dgroger@it.uc3m.es, {jmartinez, vpla}@dcom.upv.es

Abstract

We propose a new methodology and associated algorithms for computing the optimal configuration of the Multiple Fractional Guard Channel (MFGC) admission control policy in multiservice mobile wireless networks. Our approach is based on the solution space concept which discloses a novel insight into the problem of determining the optimal configuration parameter values of the MFGC policy and provides an heuristic evidence that the algorithm finds the optimal solution and converges in all scenarios, an evidence that was not provided in previous proposals. Besides, our algorithm is shown to be more efficient than previous algorithms appeared in the literature.

Keywords: Gradient methods, land mobile radio cellular systems, Markov processes, modeling, multimedia systems, optimal control.

7.1 Introduction

The enormous growth of mobile telecommunication services, together with the scarcity of radio spectrum has led to a reduction of the cell size in cellular

D. D. Kouvatsos (ed.), Mobility Management and Quality-of-Service for Heterogeneous Networks, 155–172.

systems. Smaller cell size entails a higher handover rate having an important impact on the radio resource management and the QoS perceived by customers. Moreover, 3G networks establish a new paradigm with a variety of services having different QoS needs and traffic characteristics. In these scenarios Admission Control (AC) is a key aspect in the design and operation of multiservice mobile networks.

In this paper we propose a new algorithm for computing the optimal configuration of a trunk reservation policy named the *Multiple Fractional Guard Channel* (MFGC) [1, 2]. The configuration of the MFGC policy specifies the average amount of resources that each service has access to. The optimal configuration maximizes the offered session rate that the system can handle while meeting certain QoS requirements, which we call the system capacity. The QoS requirements are defined as upper bounds for the blocking probabilities of both new setup and handover requests. In a wireless scenario this distinction is required because a session being forced to terminate due to a handover failure is considered more harmful than the rejection of a new session setup request. One of the important features of the MFCG policy is that it can achieve a system capacity that is very close to the optimal [3].

To the best of our knowledge only two algorithms for computing the system capacity of the MFGC policy have been proposed in the literature [2, 4]. We refer to those algorithms as HCO and PMC respectively, after their authors' initials. Our work is motivated by the fact that previous algorithms did not provide any evidence supporting that they where finding the optimal solution nor that they converged in all scenarios. Our approach provides a novel insight into the problem, which we believe that by itself it is a significant contribution, but in addition the algorithm we have developed, based on the insight provided by our study, offers computational advantages better than those provided by previous proposals.

The HCO algorithm requires the optimal *prioritization order* as input, i.e. a list of session types sorted by their relative priorities. For a system with N services, new session and handover request arrivals are considered, making a total of $2N$ arrival streams. Therefore, the MFGC policy configuration is defined by the $2N$-tuple $t = (t_1, \ldots, t_{2N})$, where the configuration parameter $t_i \in \mathbb{R}$ represents the average amount of resources that stream i can dispose of. If t_{opt} is the policy setting for which the maximum capacity is achieved, the optimal prioritization order is the permutation $\sigma^* \in \Sigma$, $\Sigma := \{(\sigma_i, \ldots, \sigma_{2N}) : \sigma_i \in \mathbb{N}, 1 \leq \sigma_i \leq 2N\}$, such that $t(\sigma_1^*) \leq t(\sigma_2^*) \leq \ldots \leq t(\sigma_{2N}^*) = C$, where $t(\sigma_i^*)$ is the σ_i^* element of t_{opt} and C is the total number of resource units of the system. Selecting the optimal prioritization order is a complicated task as it depends on both QoS constraints and system characteristics as pointed out in [2]. In general there are a total of $(2N)!$ different prioritization orders. In [2] the authors give some guidelines to construct a partially sorted list of prioritization orders according to their likelihood of being the optimal ones. Then a trial and error process is followed using successive elements of the list until the optimal prioritization order is found. For each element the HCO algorithm is run and if after a large number of iterations it did not converged, another prioritization order is tried.

The PMC algorithm does not require any a priori knowledge. Indeed, after obtaining the optimal policy configuration t_{opt} for which the maximum capacity is achieved, the optimal prioritization order is automatically determined as a by-product of the algorithm. Moreover, through numerical examples it is shown in [4] that the PMC algorithm is more efficient than the HCO algorithm even when the latter is provided with the optimal prioritization order. In [4] the optimization problem is formulated as a non-linear programming problem, which attempts to determine the MFGC policy configuration parameters in such a way to maximize the session arrival rates while keeping the blocking probabilities under specified bounds, and an algorithm for solving the non-linear programming problem is provided. Given that, in general, the blocking probabilities are non-monotonic functions both of the offered load and the thresholds that specify the policy configuration, finding the optimal solution is not an easy task and no evidence was provided supporting that the algorithm converged in all scenarios.

Our new algorithm is based on the *solution space* concept. If for each possible configuration of the MFGC policy we determine the maximum session rate that can be offered to the system while satisfying the QoS constrains, then the result of this study is called the solution space. As with the HCO and PMC algorithms, the convergence of our algorithm is based on the assumption that the solution space has a single (and thus, global) maximum. Even though a formal verification of this assumption is out of the scope of the paper, we have obtained the solution space for multiple policies and multiple scenarios and found that a single peak can always be found in the solution space, and that this peak is the system capacity [3]. Besides, the shape of the solution spaces tend to be steeper for policies that achieve higher system capacity, like the MFGC policy. These evidences suggest that a simple *hill climbing* algorithm could be deployed, and might shed light on a more formal characterization of solution spaces.

The remaining of the paper is structured as follows. In Section 7.2 the system model is described and its mathematical analysis is outlined in Section 7.3. Section 7.4 justifies the applicability of a gradient method for the determination of the optimal configuration of the MFGC policy. Section 7.5 describes in detail the new proposed algorithm. Computational complexity of the algorithm is comparatively evaluated in Section 7.6. Finally, Section 7.7 concludes the paper.

7.2 Model Description

The system has a total of C resource units, being the physical meaning of a unit of resource dependent on the specific technological implementation of the radio interface. The system offers N different classes of service. For each service new and handover session request arrivals are distinguished so that there are N types of services and $2N$ types of arrival streams. Arrivals are numbered in such a manner that for service i new session arrivals are referred to as arrival type i, whereas handover arrivals are referred to as arrival type $N + i$.

For the sake of mathematical tractability we make the common assumptions of Poisson arrival processes and exponentially distributed random variables for cell residence time and session duration.

The arrival rate for new (handover) sessions of service i is λ_i^n (λ_i^h). A request of service i consumes b_i resource units, $b_i \in \mathbb{N}$. We denote by f_i the percentage of service i new session requests and assume that its value is known. Therefore, the aggregated rate of new session requests is expressed as $\lambda^T = \sum_{i=1}^{N} \lambda_i^n$, $\lambda_i^n = f_i \lambda^T$. This is a common simplification in the literature [5].

The duration of service i sessions is exponentially distributed with rate μ_i^c. The cell residence time of a service i customer is exponentially distributed with rate μ_i^r. Hence, the resource holding time in a cell for service i is exponentially distributed with rate $\mu_i = \mu_i^c + \mu_i^r$. The exponential assumption for the cell residence time represents a good performance approximation and indicates general performance trends [6]. The exponential assumption can also be considered a good approximation for the time in the handover area [7] and for the interarrival time of handover requests [8].

Let $p = (P_1, \ldots, P_{2N})$ be the blocking probabilities, where new session blocking probabilities are $P_i^n = P_i$ and the handover ones are $P_i^h = P_{N+i}$. The forced termination probability of accepted sessions under the assumption of homogeneous cell [9] is

$$P_i^{ft} = \frac{P_i^h}{\mu_i^c/\mu_i^r + P_i^h}.$$

The system state is described by an N-tuple $x = (x_1, \ldots, x_N)$, where x_i represents the number of type i sessions in the system, regardless they were initiated as new or handover sessions. This approximation is irrelevant when considering exponential distributions due to their memoryless property. Let $b(x)$ represent the amount of occupied resources at state x, $b(x) = \sum_{i=1}^{N} x_i b_i$.

A generic definition of the MFGC and Complete-Sharing policies are now provided. For the MFGC policy, when a service i request finds the system in state x, the following decisions can be taken:

$$b(x) + b_i \begin{cases} \leq \lfloor t_i \rfloor & \text{accept} \\ = \lfloor t_i \rfloor + 1 & \text{accept with probability} \quad t_i - \lfloor t_i \rfloor \\ > \lfloor t_i \rfloor + 1 & \text{reject} \end{cases}$$

where parameters t_i are the policy configuration parameters that are set to achieve a given QoS objective.

The Complete-Sharing (CS) policy is equivalent to the absence of policy, i.e. a request is admitted provided there are enough free resource units available in the system.

7.3 Mathematical Analysis

The model of the system is a multidimensional birth-and-death process, which state space is denoted by S. Let r_{xy} be the transition rate from x to y and let e_i denote a vector whose entries are all 0 except the i-th one, which is 1.

$$r_{xy} = \begin{cases} a_i^n(x)\lambda_i^n + a_i^h(x)\lambda_i^h & \text{if } y = x + e_i \\ x_i \mu_i & \text{if } y = x - e_i \\ 0 & \text{otherwise} \end{cases}$$

The coefficients $a_i^n(x)$ and $a_i^h(x)$ denote the probabilities of accepting a new and handover session of service i respectively. Given a policy configuration

$(t_1, \ldots, t_{2N})$ these coefficients can be determined as follows

$$a_i^n(x) = \begin{cases} 1 & \text{if } b(x) + b_i \leq \lfloor t_i \rfloor \\ t_i - \lfloor t_i \rfloor & \text{if } b(x) + b_i = \lfloor t_i \rfloor + 1 \\ 0 & \text{if } b(x) + b_i > \lfloor t_i \rfloor + 1 \end{cases}$$

and

$$a_i^h(x) = \begin{cases} 1 & \text{if } b(x) + b_i \leq \lfloor t_i \rfloor \\ t_{N+i} - \lfloor t_{N+i} \rfloor & \text{if } b(x) + b_i = \lfloor t_{N+i} \rfloor + 1 \\ 0 & \text{if } b(x) + b_i > \lfloor t_{N+i} \rfloor + 1 \end{cases}$$

From the above, the global balance equations can be written as

$$p(x) \sum_{y \in S} r_{xy} = \sum_{y \in S} r_{yx} p(y) \qquad \forall x \in S \qquad (7.1)$$

where $p(x)$ is the state x stationary probability. The values of $p(x)$ are obtained from (7.1) and the normalization equation. From the values of $p(x)$ the blocking probabilities are obtained as

$$P_i = P_i^n = \sum_{x \in S} \left(1 - a_i^n(x)\right) p(x) \qquad P_{N+i} = P_i^h = \sum_{x \in S} \left(1 - a_i^h(x)\right) p(x)$$

If the system is in statistical equilibrium the handover arrival rates are related to the new session arrival rates and the blocking probabilities (P_i) through the expression [10]

$$\lambda_i^h = \lambda_i^n \frac{1 - P_i^n}{\mu_i^c / \mu_i^r + P_i^h} \qquad (7.2)$$

The blocking probabilities do in turn depend on the handover arrival rates yielding a system of non-linear equations which can be solved using a fixed point iteration method as described in [9, 10].

7.4 Determination of the Optimal Policy Configuration

We pursue the goal of computing the system capacity, i.e. the maximum offered session rate that the network can handle while meeting certain QoS requirements. These QoS requirements are given in terms of upper-bounds for the new session blocking probabilities (B_i^n) and the forced termination probabilities (B_i^{ft}). The common approach to carry out this AC synthesis process in multiservice systems is by iteratively executing an analysis process. The synthesis process is a routine that having as inputs the values of the system parameters (λ_i^n, λ_i^h, μ_i, b_i and C) and the QoS requirements (B_i^n and B_i^{ft}), produces as output the optimal configuration (the thresholds t_i). In contrast the analysis process is a routine that having as inputs the value of the system parameters and the configuration of the AC policy produces as output the blocking probabilities for the different arrival streams.

Given that, in general, the blocking probabilities are non-monotonic functions both of the offered load and the thresholds that specify most policy configurations; the common approach is to carry out a multidimensional search using, for example, meta-heuristics like genetic algorithms which are able to find a *good* configuration in a reasonable amount of time. It should be pointed out that each execution of the analysis process requires solving the associated continuous-time Markov chain.

Additional insight can be gained by determining the maximum offered session rate for each possible policy configuration. The result of this study is called the *solution space*, and its peak value is the system capacity of the AC policy, i.e. the maximum aggregated session arrival rate ($\lambda^T = \sum_{i=1}^{N} \lambda_i^n$, $\lambda_i^n = f_i \lambda^T$) that can be offered to the system in order to satisfy the QoS requirements. The surface that defines the solution space is obtained as follows. For each configuration of the thresholds t_i, $\lambda_{\max}^T$ is computed by a binary search process which has as input the value of the system parameters μ_i, b_i, C and the thresholds t_i, and produces as output the blocking probabilities (P_i^n and P_i^h). The binary search process stops when it finds the $\lambda_{\max}^T$ that meets the QoS requirements (B_i^n and B_i^{ft}), $i = 1, \ldots, N$.

In order to illustrate our algorithm we have chosen a simple example with only two services but without their associated handover streams. This allows us to represent the solution space in only three dimensions. Figure 7.1 show the solution space when MFGC policy is deployed in a scenario with $C = 10$ resource units, $\boldsymbol{b} = (1, 2)$, $\boldsymbol{f} = (0.8, 0.2)$, $\boldsymbol{\mu} = (1, 3)$, $\boldsymbol{B}^n = (0.05, 0.01)$. The configuration of the policy is defined by two parameters t_1 and t_2. It should be noted that the system capacity is expressed as a relative value to the capacity obtained for the CS policy.

The form of the solution space shown in Figure 7.1, which displays a unique maximum, suggests that a hill climbing algorithm could be an efficient approach to obtain the optimum configuration for MFGC policy in this scenario. Other optimization approaches like genetic algorithms that are more appropriate for scenarios with multiple local maxima will not be as efficient. We have obtained the solution space for multiple policies and multiple scenarios and found that a single peak can always be found in the solution space, and that this peak is the system capacity [3]. The policies that meet this condition are defined below. The solution space has been obtained for the scenarios defined in Table 7.1. To simplify the description of the admission policies we define the vector $x' = (x_1^n, \ldots, x_N^n, x_1^h, \ldots, x_N^h)$, where x_i^n (x_i^h) are the sessions in progress of service i initiated as new (handover). It is clear that $x_i = x_i^n + x_i^h$.

Integer Limit (IL) [11]. A threshold $(t; t \in \mathbb{N}, 1 \le t \le C)$ is defined for each request type. Thus, there will be a set of $2N$ thresholds $(t_1^n, \ldots, t_N^n, t_1^h, \ldots, t_N^h)$. A request of type x_i^n (x_i^h) that arrives in state x' is accepted if $x_i^n + 1 \le t_i^n$ $(x_i^h + 1 \le t_i^h)$ and blocked otherwise.

Upper Limit and Guaranteed Minimum (ULGM) [12]. Service i requests have access to two sets of resources: a private set and a shared set. The number of resource units in the private set available for new setup requests is denoted as $s_i^n \cdot b_i$ and for handover requests as $s_i^h \cdot b_i$, where $s_i^n, s_i^h \in \mathbb{N}$. Therefore the size of the shared set is $C - \sum_{i=1}^N (s_i^n + s_i^h) \cdot b_i$. A new (handover) service i request is accepted if $(x_i^n + 1) \le s_i^n$ $((x_i^h + 1) \le s_i^h)$ or if there are enough free resource units in the shared set, otherwise it is blocked.

Multiple Guard Channel (MGC) [13]. A threshold $(t; t \in \mathbb{N}, 1 \le t \le C)$ is defined for each request type. Thus, there will be a set of $2N$ thresholds $(t_1^n, \ldots, t_N^n, t_1^h, \ldots, t_N^h)$. A request of type x_i^n (x_i^h) that arrives in state x is accepted if the number of busy resource units is less than the corresponding threshold, i.e. $b(x) + b_i \le t_i^n$ $(b(x) + b_i \le t_i^h)$, and blocked otherwise.

Multiple Fractional Guard Channel (MFGC). It was defined in Section 7.2.

Figure 7.1 shows how the hill climbing algorithm works. (i) Given a starting point in a $2N$-dimensional search space (for example, point **0**), the hill climbing algorithm begins by computing the value of the function (the system capacity λ_{max}^T), and the blocking probabilities for the different arrival streams (P_i^n and P_i^h); (ii) the steepest dimension is selected as described below (in this case t_2^n); (iii) the algorithm searches for a maximum point along that dimension (in this case **1**). Note that what we have here is actually a maximization problem along a line; and (iv) return to (i) until the local

Table 7.1 Definition of the scenarios studied by Garcia et al. [3].

	b_1	b_2	f_1	f_2	$B_1^n(\%)$	$B_2^n(\%)$	$B_i^h(\%)$	λ_i^n	λ_i^h	μ_1	μ_2
A	1	2	0.8	0.2	5	1					
B	1	4	0.8	0.2	5	1					
C	1	2	0.2	0.8	5	1	$0.1B_i^n$	$f_i\lambda$	$0.5\lambda_i^n$	1	3
D	1	2	0.8	0.2	1	2					
E	1	2	0.8	0.2	1	1					

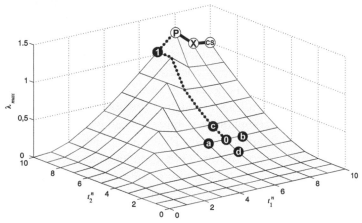

Figure 7.1 Example of the use of a hill climbing algorithm to determine the optimum configuration for the MFGC policy.

maximum **P** is found within the desired precision (the algorithm progression is shown with a dotted line). For the hill climbing algorithm explained, two important questions that arise are: (a) how the steepest dimension is selected, and (b) how the search for a maximum point is performed.

When applying gradient based methods to our problem, the function must be evaluated for the two adjacent neighbors in each of the 2N dimensions (points **a, b, c** and **d**), selecting then the steepest dimension as the one for which the function value is largest (**c**). However, if the binary search process to compute the system capacity is performed with low accuracy, or neighbors are sufficiently close to the considered point, then this method is impracticable, mainly because the function values for the neighbors are identical to the considered point, giving no information at all.

As this is the case, another approach is required to determine the steepest dimension. We define the relative distance to the QoS objective B_i of an arrival stream i with blocking probability P_i (assuming that meets its objective, i.e. $P_i < B_i$) by the quotient $(B_i - P_i)/B_i$. Since the relative distance to the QoS objective of an arrival stream is usually an indication that requests from that arrival stream may be blocked further while still meeting the QoS objective (and thus, providing an additional capacity for the remaining arrival streams), we choose as the steepest dimension the configuration parameter t_i associated with the arrival stream i which maximizes the relative distance to the QoS objective. Besides, this approach has an additional benefit in terms of computational complexity: it removes the need for computing the system capacity in the neighboring points in order to find out the steepest dimension.

In its search for the maximum along this steepest dimension our algorithm makes a number of successive unitary steps and it stops when it reaches the peak. When the solution space is continuous, as with the MFGC policy, a gradual refinement process is needed to reduce the size of the step once a promising region has been found, which is possibly close to the optimal. A further reduction of the computation complexity can be obtained by observing that the optimum configuration (point **P**) for any policy is near the CS configuration (point **CS**), and therefore it is a good idea to select it as the starting point. Figure 7.1 illustrates a typical progression (solid line) of the proposed algorithm starting from the CS configuration (point **CS**) towards point **X** and ending at the peak (point **P**).

7.5 Hill Climbing Algorithm

The capacity optimization problem can be formally stated as follows

Given: $C, b_i, f_i, \mu_i^c, \mu_i^r, B_i^n, B_i^{ft}$; $i = 1, \ldots, N$

Maximize: $\lambda^T = \sum_{1 \leq i \leq N} \lambda_i^n, \lambda_i^n = f_i \lambda^T$
    by finding the appropriate MFGC parameters t_i; $i = 1, \ldots, 2N$

Subject to: $P_i^n \leq B_i^n, P_i^{ft} \leq B_i^{ft}$; $i = 1, \ldots, N$

We propose an algorithm to work out this capacity optimization problem. Our algorithm has a main part (Algorithm 1 solveMFGC) from which the procedure capacity (see Algorithm 2) is called. The procedure capacity does, in turn, call another procedure (MFGC) that calculates the blocking probabilities. For the sake of notation simplicity we introduce the 2N-tuple

$p_{\mathbf{max}} = (B_1^n, \ldots, B_N^n, B_1^h, \ldots, B_N^h)$ as the upper-bounds vector for the blocking probabilities, where the value of B_i^h is given by

$$B_i^h = \frac{\mu_i^c}{\mu_i^r} \frac{B_i^{ft}}{1 - B_i^{ft}} \tag{7.3}$$

Following the common convention we used bold-faced font to represent array variables in the pseudo-code of algorithms.

Algorithm 1 ($\lambda_{\max}^T$, t_{opt}) =solveMFGC(C, $p_{\mathbf{max}}$, b, μ_c, μ_r) (calculates MFGC parameters)

1:   $\varepsilon_2 :=<$ desired precision $>$
2:   current$\varepsilon_2 := 1$
3:   point $:= C$
4:   direction $:= -1$
5:   $\boldsymbol{step} := (1, 1, \ldots, 1)$ $<$size $2N>$
6:   steepest $:= 0$
7:   changeOfDirection:=FALSE
8:   $t_{\text{opt}} := (C, C, \ldots, C)$; $t := t_{\text{opt}}$
9:   $\lambda_{\max}^T := \mathbf{0}$; $\lambda^T := \mathbf{0}$;
10: $p_{\text{opt}} := \mathbf{0}$; $p := \mathbf{0}$;
11: $d p_{\text{opt}} := \mathbf{0}$; $d p := \mathbf{0}$;
12:
13: $(\lambda_{\max}^T, p_{\text{opt}})$:=capacity($p_{\mathbf{max}}, t_{\text{opt}}, \mu_c, \mu_r, b, C$)
14: $d p_{\text{opt}} := (p_{\mathbf{max}} - p_{\text{opt}})/p_{\mathbf{max}}$; $d p := d p_{\text{opt}}$
15: current$\varepsilon_2 := max(d p_{\text{opt}})$
16: steepest $:= <$ the stream i which maximizes $d p_{\text{opt}}(i)$ $>$
17:
18: **while** current$\varepsilon_2 > \varepsilon_2$ **do**
19:     point:= t_{opt}(steepest)
20:     direction:= -1
21:     **if** $\boldsymbol{step}$(steepest)$<>$1 **then**
22:        $\boldsymbol{step}$(steepest)=0.5
23:     **end if**
24:     changeOfDirection:= FALSE
25:
26:     **repeat**
27:        **if** direction$= -1$ **then**
28:           point=point$-\boldsymbol{step}$(steepest)
29:        **else**

30:          point=point+$step$(steepest)
31:       **end if**
32:
33:       $t := t_{\mathrm{opt}}$; t(steepest):= point;
34:       $(\lambda^T, p) :=$ capacity$(p_{\mathbf{max}}, t, \mu_c, \mu_r, b, C)$
35:       $dp := (p_{\mathbf{max}} - p)/p_{\mathbf{max}}$;
36:
37:       **if** $\lambda^T >= \lambda^T_{\mathrm{max}}$ **then**
38:          t_{opt}(steepest):= point; $\lambda^T_{\mathrm{max}} := \lambda^T$;
39:          $p_{\mathrm{opt}} := p$; $dp_{\mathrm{opt}} := dp$;
40:       **end if**
41:
42:       **if** dp(steepest)$> \varepsilon_2$ **then**
43:          **if** direction$= -1$ **then**
44:             **if** changeOfDirection **then**
45:                $step$(steepest) $:= step$(steepest)/2
46:             **end if**
47:          **else**
48:             $step$(steepest) $:= step$(steepest)/2
49:             direction:$= -1$
50:             changeOfDirection:= TRUE
51:          **end if**
52:       **else**
53:          **if** $\lambda^T < \lambda^T_{\mathrm{max}}$ **then**
54:             **if** direction:$= +1$ **then**
55:                **if** changeOfDirection **then**
56:                   $step$(steepest) $:= step$(steepest)/2
57:                **end if**
58:             **else**
59:                $step$(steepest) $:= step$(steepest)/2
60:                direction:$= +1$
61:                changeOfDirection:= TRUE
62:             **end if**
63:          **end if**
64:       **end if**
65:    **until** $(dp$(steepest)$< \varepsilon_2)$ AND $(\lambda^T >= \lambda^T_{\mathrm{max}})$
66:
67:    steepest:= < the stream i which maximizes $dp_{\mathrm{opt}}(i)$ >
68:    current$\varepsilon_2 := max(dp_{\mathrm{opt}})$
69: **end while**

Algorithm 2 $(\lambda^T_{\max}, \boldsymbol{p})$=capacity$(\boldsymbol{p}_{\mathbf{max}}, \boldsymbol{t}, \boldsymbol{\mu_c}, \boldsymbol{\mu_r}, \boldsymbol{b}, \boldsymbol{C})$

INPUTS: $\boldsymbol{p}_{\mathbf{max}}, \boldsymbol{t}, \boldsymbol{\mu_c}, \boldsymbol{\mu_r}, \boldsymbol{b}, \boldsymbol{C}$
OUTPUTS: $\lambda^T_{\max}, \boldsymbol{p}$
1:    $\varepsilon_1 :=<$ desired precision $>$
2:    current$\varepsilon_1 := 1$
3:    $L := 0$
4:    $U :=<$ high value $>$
5:    meetQoSrequirements:=FALSE
6:
7:    **while** (current$\varepsilon_1 > \varepsilon_1$) OR NOT(meetQoSrequirements) **do**
8:      $\lambda^T_{\max} := (U + L)/2$
9:      $p := \text{MFGC}(t, \lambda_n, \mu_c, \mu_r, b, C)$
10:     current$\varepsilon_1 := min((\boldsymbol{p}_{\mathbf{max}} - p)/\boldsymbol{p}_{\mathbf{max}})$
11:     **if** current$\varepsilon_1 < 0$ **then**
12:       $U := \lambda^T_{\max}$
13:       meetQoSrequirements:= FALSE
14:     **else**
15:       $L := \lambda^T_{\max}$
16:       meetQoSrequirements:= TRUE
17:     **end if**
18: **end while**

The algorithm `solveMFGC` begins computing the CS configuration as the starting point (lines 13–14), and selects the arrival stream i for which $(B_i - P_i)/B_i$ is the highest as the steepest dimension, (line 16). The whole hill climbing loop starts in line 18, and the maximization loop along a dimension starts in line 26. Note (line 21) that the first time a dimension is chosen as the steepest, the hill climbing step equals to 1, however if a dimension has been chosen previously, initial hill climbing steps will be reduced to 0.5 because a certain locality of the optimal configuration is assumed. Lines (37–64) perform the hill climbing algorithm along the steepest dimension. This subroutine performs tasks like changing the direction of the successive steps and its refinement once a promising configuration has been found out. The algorithm `capacity` is basically a binary search of $\lambda^T_{\max}$ that calls procedure (MFGC) in each iteration in order to calculate the blocking probabilities.

7.5.1 On the Procedure MFGC

The procedure MFGC, which is invoked in the inner-most loop of our algorithm, is used to obtain the blocking probabilities, p := MFGC($t, \lambda_n, \mu_c, \mu_r, b, C$), by using the Gauss–Seidel method to solve the continuous-time Markov chain (CTMC) that models the system. The major part of the computational complexity of the algorithms described in this paper comes from solving many different times the CTMC, therefore the difference among different algorithms basically yields on how many times a CTMC has to be solved. Note that in Section 7.6 only two types of services will be considered. For scenarios with a higher number of services, the Markov chain would have $2N$ dimensions (new and handover requests). To obtain valuable results the number of resource units of the system would have to be dimensioned appropriately, which will cause an explosion of the state space and thus rendering a numerical evaluation of any algorithms simply unfeasible. In order to compare the algorithms in scenarios with a high number of dimensions it will be required to tackle more efficiently the curse of dimensionality inherent to these scenarios using an approximate method to solve the associated CTMC. Nevertheless, solving the Markov chain with a lower precision will have an impact on the behaviour of the algorithms, which may vary from one to the other. A detailed study of the new behaviour of the algorithms in the presence of inaccurate solutions of the CTMC is outside the scope of this paper.

Furthermore, for the computation of the blocking probabilities, an additional fixed point iteration procedure is also required in order to obtain the value of the handover request rates (see the end of Section 7.3). At each iteration a multidimensional birth-and-death process must be solved. Solving this process, that in general will have a large number of states, constitutes the most computationally expensive part of the algorithm.

We make use of the same enhancement explained in [4] to eliminate the fixed point iteration to compute the handover arrival rates. Each run of capacity finds a $\lambda_{\max}^T$ (within tolerance limit) so that $p \leq p_{\mathbf{max}}$. Thus, instead of using (7.2) to compute λ_i^h we use the expression

$$\lambda_i^h = \lambda_i^n \frac{1 - B_i^n}{\mu_i^c / \mu_i^r + B_i^h} \qquad (7.4)$$

Although (7.2) and (7.4) look very similar there is a substantial difference between the two. In Eq. (7.4), λ_i^h is explicitly defined whereas in (7.2) it is not as P_i^n and P_i^h depend on λ_i^h. Note that $p = p_{\mathbf{max}}$ only when $\lambda_{\max}^T$ is the system capacity (within tolerance limit), but using (7.4) reduces considerably the computation cost and therefore speeds-up the convergence rate of the algorithm.

We use expression (7.4) because by properly setting the configuration parameters of the MFGC policy it is possible to meet the QoS objectives with high precision (provided that there is a feasible solution). It is clear that when the aggregated arrival rate equals the system capacity then the value of the configuration parameters are such that the blocking probabilities perceived by the different streams are very close to their objectives. Therefore, even if the handover request rates computed in the beginning are imprecise, their precision improve as the algorithm progresses towards the maximum.

7.6 Numerical Evaluation

In this section we evaluate the computational complexity of our algorithm and compare it to the complexity of the HCO and PMC algorithms.

For the numerical examples we considered a system with two services ($N = 2$), and to assess the impact of mobility on computational complexity, five different scenarios (A, B, C, D, and E) were considered with varying mobility factors (μ_i^r / μ_i^c). The set of parameters that define scenario A are: $b = (1, 2)$, $f = (0.8, 0.2)$, $\mu_c = (1/180, 1/300)$, $\mu_r = (1/900, 1/1000)$, $B^n = (0.02, 0.02)$, $B^{ft} = (0.002, 0.002)$; all tolerances have been set to $\epsilon = 10^{-2}$. By (7.3), $B^h \approx (0.01002, 0.00668)$ and then $p_{\mathrm{max}} \approx (0.02, 0.02, 0.01002, 0.00668)$.

For the rest of scenarios the parameters have the same values as the ones used in scenario A except μ_i^r, which is varied to obtain four different mobility factor combinations: (B) $\mu_1^r = 0.2\mu_1^c$, $\mu_2^r = 0.2\mu_2^c$; (C) $\mu_1^r = 0.2\mu_1^c$, $\mu_2^r = 1\mu_2^c$; (D) $\mu_1^r = 1\mu_1^c$, $\mu_2^r = 0.2\mu_2^c$; (E) $\mu_1^r = 1\mu_1^c$, $\mu_2^r = 1\mu_2^c$.

A comparison of the number of floating point operations (flops) required by the HCO and PMC algorithms and our algorithm is shown in Table 7.2. The three algorithms were tested with the speed-up technique (see Section 7.5.1). It is worth noting that, as expected, the values obtained for the optimal capacity computed using the different methods were within tolerance in all tested cases.

Note that the HCO algorithm is provided with the prioritization order as input and therefore it does not need to search for it as their authors propose, which is a substantial advantage in terms of computation cost. Additionally, in its original version it does not implement the speed-up technique introduced in [4], without which the flops count is much higher than the one shown

Table 7.2 Comparison of the HCO (with known prioritization order) and PMC algorithms with our algorithm for different mobility factors (in Mflops).

	HCO				PMC				Our algorithm			
C	5	10	20	**TOT**	5	10	20	**TOT**	5	10	20	**TOT**
A	2.00	20.00	156.00	178.00	0.39	4.53	46.60	51.52	0.21	1.17	9.28	10.66
B	2.08	17.54	74.33	93.95	0.35	4.42	53.64	58.41	0.27	1.92	8.70	10.89
C	2.67	14.06	147.13	163.86	0.34	3.87	43.01	47.22	0.26	1.28	12.90	14.44
D	1.12	24.54	110.41	136.07	0.38	3.93	47.95	52.26	0.23	1.74	12.36	14.33
E	2.24	16.86	121.39	140.49	0.31	3.93	45.92	50.16	0.26	2.39	13.56	16.21
TOT	28.11	93.00	609.26	730.37	1.77	20.68	237.12	259.57	1.23	8.50	56.80	66.53

in Table 7.2. For example, for scenario A, the HCO algorithm with speed-up technique requires 2, 20 and 156 Mflops for $C = 5$, 10 and 20 respectively, while without the speed-up technique it needs 5.7, 60.2 and 438 Mflops, i.e. the speed-up technique divides the flop count by a factor of about three.

Our algorithm performs better than the other algorithms. The gain factor ranges from 4.9 to 17 with respect to the HCO algorithm and from 1.2 to 6.3 with respect to the PMC algorithm, with an average gain of 10.3 and 2.81 for HCO and PMC algorithms, respectively.

7.7 Conclusions

We proposed a new algorithm for computing the optimal setting of the configuration parameters for the Multiple Fractional Guard Channel (MFGC) admission policy in multiservice mobile wireless networks. The optimal configuration maximizes the offered traffic that the system can handle while meeting certain QoS requirements.

Compared to two recently published algorithms (HCO and PMC) ours, which is based on a simple and intuitive hill climbing approach, is less computationally expensive in all scenarios studied. Besides, the solution space concept discloses a novel insight into the problem of determining the optimal configuration parameter values of the MFGC policy, providing an heuristic evidence that the algorithm finds the optimal solution and converges in all scenarios.

Acknowledgements

This work has been supported by the Spanish Ministry of Education and Science (30%) and by the European Union (FEDER 70%) under projects TIN2008-06739-C04-02 and TSI2007-66869-C02-02. It is also supported by the "Catedra Telefonica" of the Universidad Politecnica de Valencia and Vodafone Spain.

D. Garcia-Roger was supported by the Generalitat Valenciana under contract CTB/PRB/2002/267.

References

[1] H. Heredia-Ureta, F. A. Cruz-Pérez and L. Ortigoza-Guerrero, Multiple fractional channel reservation for optimum system capcity in multi-service cellular networks, *Electronics Letters*, vol. 39, no. 1, pp. 133–134, January 2003.

[2] H. Heredia-Ureta, F. A. Cruz-Pérez and L. Ortigoza-Guerrero, Capacity optimization in multiservice mobile wireless networks with multiple fractional channel reservation, *IEEE Transactions on Vehicular Technology*, vol. 52, no. 6, pp. 1519–1539, November 2003.

[3] D. García, J. Martínez and V. Pla, Admission control policies in multiservice cellular networks: Optimum configuration and sensitivity, in *Mobile and Wireless Systems*, Lecture Notes in Computer Science (LCNS), vol. 3427, Springer-Verlag, Berlin/Heidelberg, pp. 121–135, 2005.

[4] V. Pla and V. Casares-Giner, Analysis of priority channel assignment schemes in mobile cellular communication systems: A spectral theory approach, *Performance Evaluation*, vol. 59, nos. 2-3, pp. 199–224, February 2005.

[5] S. Biswas and B. Sengupta, Call admissibility for multirate traffic in wireless ATM networks, in *Proceedings of IEEE INFOCOM*, vol. 2, pp. 649–657, 1997.

[6] F. Khan and D. Zeghlache, Effect of cell residence time distribution on the performance of cellular mobile networks, in *Proceedings of VTC'97*, IEEE, New York, pp. 949–953, 1997.

[7] V. Pla and V. Casares-Giner, Effect of the handoff area sojourn time distribution on the performance of cellular networks, in *Proceedings of IEEE MWCN*, September, pp. 401–405, 2002.

[8] P. V. Orlik and S. S. Rappaport, On the handoff arrival process in cellular communications, *Wireless Networks Journal (WINET)*, vol. 7, no. 2, pp. 147–157, March/April 2001.

[9] D. Hong and S. S. Rappaport, Traffic model and performance analysis for cellular mobile radio telephone systems with prioritized and nonprioritized handoff procedures, *IEEE Transactions on Vehicular Technology*, vol. VT-35, no. 3, pp. 77–92, August 1986. (See also: CEAS Technical Report No. 773, College of Engineering and Applied Sciences, State University of New York, Stony Brook, June 1999.)

[10] Y.-B. Lin, S. Mohan and A. Noerpel, Queueing priority channel assignment strategies for PCS hand-off and initial access, *IEEE Transactions on Vehicular Technology*, vol. 43, no. 3, pp. 704–712, August 1994.

[11] V. B. Iversen, The exact evaluation of multi-service loss systems with access control, in *Proceedings of the Teleteknik and Seventh Nordic Teletraffic Seminar (NTS-7)*, vol. 31, Lund (Sweden), pp. 56–61, August 1987.

[12] C.-T. Lea and A. Alyatama, Bandwidth quantization and states reduction in the broadband ISDN, *IEEE/ACM Transactions on Networking*, vol. 3, no. 3, pp. 352–360, June 1995.

[13] B. Li, C. Lin and S. T. Chanson, Analysis of a hybrid cutoff priority scheme for multiple classes of traffic in multimedia wireless networks, *Wireless Networks Journal (WINET)*, vol. 4, no. 4, pp. 279–290, 1998.

8

Optimal Admission Control Using Handover Prediction in Mobile Cellular Networks

Vicent Pla[1], Jose Manuel Gimenez-Guzman[2],
Jorge Martinez-Bauset[1] and Vicente Casares-Giner[1]

[1]*Department of Communications, Universidad Politécnica de Valencia (UPV),
ETSIT Camí de Vera s/n, 46022 Valencia, Spain;
e-mail: {vpla, jmartinez, vcasares}@dcom.upv.es*
[2] *Departamento de Automática, Universidad de Alcalá, 28871 Alcalá de Henares,
Madrid, Spain; e-mail: josem.gimenez@uah.es*

Abstract

In this paper we study the impact of incorporating handover prediction information into the session admission control process in mobile cellular networks. The objective is to compare the performance of optimal policies obtained with and without the predictive information. A prediction agent classifies mobile users in the neighborhood of a cell into two classes, those that will probably be handed over into the cell and those that probably will not. We consider the classification error by modeling the false-positive and non-detection probabilities. Two different approaches to compute the optimal admission policy were studied: *dynamic programming* and *reinforcement learning*. Results show significant performance gains when the predictive information is used in the admission process.

Keywords: Mobile cellular networks, quality of service, admission control, optimization, handover prediction, reinforcement learning.

D. D. Kouvatsos (ed.), Mobility Management and Quality-of-Service for Hetero-geneous Networks, 173–194.

8.1 Introduction

Future mobile communication systems are expected to support broadband multimedia services with diverse Quality of Service (QoS) requirements. The cellular architecture is used in wireless networks to utilize the radio spectrum efficiently. Since mobile users may change cells a number of times during the lifetime of their sessions, availability of wireless network resources at the session setup time does not necessarily guarantee that will be available throughout the lifetime of a session. Thus users may experience a performance degradation due to their mobility. This problem is magnified by the current trend to reduce the cell size to accommodate more mobile users in a given area as handover events will occur at a much higher rate [1].

Session Admission Control (AC) is a key aspect in the design and operation of multiservice cellular networks that provide QoS guarantees. The design of the AC system must take into account not only packet level issues (like delay, jitter or losses) but also session level issues (like blocking probabilities of both session setup and handover requests) [2]. This paper explores the second type of issues from a novel optimization perspective.

AC in single service cellular systems has been thoroughly studied, see for instance the seminal work by Hong and Rappaport [3] or more recent papers like [4–6] and references therein. While most of these papers provide intuitive reservation schemes for AC a more insightful approach is adopted in [7] and [8], where AC in single service scenarios is regarded as an optimization problem. Admission control in the presence of mobility and multiple services is not that well studied although some contributions in this direction can be found in the literature [9–12].

Most of the proposed AC policies take the admission decision using only state information local to the cell, such as the number of active sessions per service. However, mobile cellular networks permit to have some anticipated knowledge about forthcoming requests, and more importantly, this predictive information concerns the most sensitive requests, namely, the handover attempts. Following that observation, several mobility prediction schemes and associated AC policies have appeared for single service scenarios, see for example [2,13–16] and references therein. A common feature of these studies is the proposal of heuristic AC policies which exploit the specific information provided by each mobility prediction scheme.

In this paper we study AC policies that make use of predictive information from an optimization perspective, in both single service and multiservice scenarios. Our goal is to obtain the optimal policy for a given amount of information provided by the mobility prediction scheme. We consider that such approach has not been sufficiently explored. The deployment of classical optimization techniques applied to this type of scenario provide results that help to define theoretical limits for the gain that can be expected, which could not be set by simply deploying heuristic approaches.

For a single service scenario, in [17] the authors determine a near-optimal policy by means of a genetic algorithm that takes into account not only the cell state but also the state of neighboring cells. However, results in [17] show that performance gain obtained when using such additional information is rather insignificant. We reached the same conclusion using a different optimization method. These disappointing results suggest that the prediction of possible forthcoming handovers obtained from the occupancy state of the neighboring cells is not sufficiently specific.

Here we go a step further and evaluate the performance gain that can be obtained when the AC process is provided with more specific information. In our study the total population of active terminals in the surroundings of a cell is divided into two classes: those that will handover into the cell with high probability and those that will not. Our model of the prediction agent (PA) does not provide information about the time instant at which the handover will occur. We postpone the study of this scenario for a future work. Obviously, the more information is provided by the PA the better the performance of the AC policy will be. Unfortunately, the complexity of the PA and the optimization process increases as more information is provided.

The rest of the paper is structured as follows. In Section 8.2 we describe the model of the system and of the PA. The optimization approaches, both in single service and multiservice scenarios, are presented in Section 8.3. The numerical evaluation of the proposed model is introduced in Section 8.4. Finally, a summary of the paper and some concluding remarks are given in Section 8.5.

8.2 Model Description

We consider a single cell system and its neighborhood, where the cell has a total of C resource units, being the physical meaning of a unit of resource dependent on the specific technological implementation of the radio interface. A total of N different services are offered by the system. For each service new and handover session arrivals are distinguished so that there are N services and $2N$ types of arrivals.

For the sake of mathematical tractability we make the common assumptions of Poisson arrival processes and exponentially distributed random variables for cell residence time and session duration. The arrival rate for new (handover) sessions of service i is λ_i^n (λ_i^h) and a request of service i consumes b_i resource units, $b_i \in \mathbb{N}$. For service i, the session duration and cell residence rates are μ_i^s and μ_i^r respectively. The resource holding time in a cell for service i is also exponentially distributed with rate $\mu_i = \mu_i^s + \mu_i^r$.

8.2.1 Prediction Agent

Two main types of prediction systems have been studied in the literature [18]: history-based and positioning-based. Schemes of the first group compute movement patterns to determine movement predictions statistically, like for example estimation of the handover arrival rate to a cell. Given that mobile terminals (MTs) having a similar movement history are more likely to have common movement patterns, measurement data can be aggregated into groups, improving in this way the performance of the prediction system [2]. For schemes of the second group [19], the probability of reserving resources for a handover session increases as the MT approaches the cell. A further enhancement can be achieved by estimating the direction and speed of the MT and extrapolate this information to determine future movements [16]. It is clear that both methods can be combined to improve performance even further .

Given that the focus of our study was not the design of the PA we used a generic model of it instead. The PA informs the AC system about the number of active terminals in the neighborhood that are forecasted to produce a handover into the cell. The amount of time elapsed since an active MT is deemed as "probably producing a handover" until the handover actually occurs is not predicted by the PA and we model it by an exponential random variable, which is an approximation widely used in the literature. The exponential assumption has been considered a good approximation for the cell dwell times [20], where in the worst case indicates general performance trends, for the time in the handover area [21] and for the inter-arrival time of handover requests [22].

An active MT entering the cell neighborhood is labeled by the PA as "probably producing a handover" (H) or the opposite (NH), according to some of its characteristics (position, trajectory, velocity, historic profile, etc.) and/or some other information (road map, hour of the day, etc.). After an exponentially distributed time, the actual destiny of the MT becomes defin-

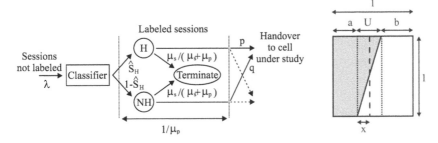

(a) Basic operation of the prediction scheme   (b) Proposed model of the PA

Figure 8.1 Model of the PA.

itive and either a handover into the cell occurs or not (for instance because the session ends or the MT moves to another cell). The AC system is aware at any time of the number of MTs labeled as H. In general, the classification into H or NH is not completely accurate and therefore our model incorporates the probabilities of non-detection and false-positive. We assume that only one session is active per MT.

Our model of the PA is characterized by three parameters: the average sojourn time of the MT in the predicted stage μ_p^{-1}, the probability p of producing a handover if labeled as H and the probability q of producing a handover if labeled as NH. Note that in general $q \neq 1 - p$. The basic operation of the prediction model is shown in Figure 8.1a.[1] The values of p and q relate to each other through the specific model of the PA, which is shown in Figure 8.1b. In the figure there is a square (with a surface equal to one) representing the population of MTs that is going to be classified by the PA. The shaded area represents the fraction of active MTs that will ultimately move into the cell, while the white area represents the rest of active MTs. It should be pointed out that those active MTs that will ultimately move into the cell might do so after the session has terminated. The classifier sets a threshold, which is represented by a vertical dashed line, to discriminate between those MTs that will likely produce a handover and those that will not. MTs falling on the left side of the threshold are labeled as H and those on the right side as NH. There exists an uncertainty zone, which is represented by the slope of the line separating the shaded and white areas. Parameter x represents the

---

[1] For the sake of clarity of notation, throughout this section we omit the subscript that refers to the type of service. Thus, $\lambda_n = \lambda_i^n$, $\lambda_h = \lambda_i^h$, $\mu_s = \mu_i^s$, $\mu_r = \mu_i^r$, $\mu = \mu_i$, $P_n = P_i^n$, $P_h = P_i^h$ and $P_{ft} = P_i^{ft}$.

relative position of the classifier threshold. This uncertainty produces classification errors: the white area on the left of the threshold and the shaded area on the right of the threshold. Although for simplicity we use a linear model for the uncertainty zone it would be rather straightforward to consider a different model. Let us introduce the following notation referring to the areas in Figure 8.1b: S_H denotes the shaded area and represents the fraction of active MTs in the cell neighborhood that will ultimately move into the cell, while the white area represents the rest of active MTs; $\hat{S}_H$ denotes the surface on the left of the threshold and represents the fraction of MTs labeled as H; $\hat{S}_H^e$ denotes the white surface on the left of the threshold and represents the fraction of MTs labeled as H that will not produce a handover; $\hat{S}_{NH}^e$ denotes the shaded surface on the right of the threshold and represents the fraction of MTs labeled as NH that will produce a handover. From Figure 8.1b it follows that

$$1 - p = \frac{\hat{S}_H^e}{\hat{S}_H} = \frac{x^2}{2U(a+x)} \; ; \qquad q = \frac{\hat{S}_{NH}^e}{1 - \hat{S}_H} = \frac{(U-x)^2}{2U(1-a-x)} \, .$$

Parameters a and b can be expressed in terms of the fraction of MTs moving into the target cell S_H and the degree of uncertainty in the prediction U,

$$a = S_H - U/2 \; ; \qquad b = 1 - S_H - U/2,$$

and then

$$1 - p = \frac{\hat{S}_H^e}{\hat{S}_H} = \frac{x^2}{U(2S_H - U + 2x)} \, ,$$

$$q = \frac{\hat{S}_{NH}^e}{1 - \hat{S}_H} = \frac{(U-x)^2}{U(2 - 2S_H + U - 2x)} \, .$$

Referring to Figure 8.1a, the value of the session rate entering the classifier λ is chosen so that the system is in statistical equilibrium, i.e. the rate at which handover sessions enter a cell (λ_h^{in}) is equal to the rate at which handover sessions exit the cell (λ_h^{out}). We can write

$$\lambda_h^{in} = \lambda S_H \frac{\mu_p}{\mu_p + \mu_s} \, ,$$

$$\lambda_h^{out} = \frac{\mu_r}{\mu_r + \mu_s} [(1 - P_n)\lambda_n + (1 - P_h)\lambda_h^{in}], \qquad (8.1)$$

where P_n (P_h) is the blocking probability of new (handover) requests.

Making $\lambda_h^{\text{in}} = \lambda_h^{\text{out}}$, substituting P_h by

$$P_h = \frac{P_{ft}}{1 - P_{ft}} \cdot \frac{\mu_s}{\mu_r},$$

where P_{ft} is the probability of forced termination of a successfully initiated session, and after some algebra we get

$$\lambda = (1 - P_n)(1 - P_{ft})\lambda_n \left(\frac{\mu_r}{\mu_s} + \frac{\mu_r}{\mu_p}\right)\left(\frac{1}{S_H}\right).$$

8.3 Optimization of the Admission Policy

The information provided by the PA and the state of the cell (number of active sessions) is used to find the optimal admission policy and its performance. The generic definition of the system state space is

$$\mathcal{S} := \left\{x = (x_1, \ldots, x_N, x_{N+1}, \ldots, x_{2N})\right\},$$

where x_i is the number of ongoing sessions of service i, $1 \leq i \leq N$, in the cell under study and x_{i+N} is the number of ongoing sessions of service i in the cell neighborhood which are labeled as H.

We make use of the theory of *Markov decision processes* (MDPs) [23] to find a policy that minimizes the average expected cost rate. A MDP can be viewed as a stochastic automaton in which an agent's actions influence the transitions between states, and costs are imputed depending on the states visited by an agent. Formally, a MDP can be defined as a tuple $\{\mathcal{S}, \mathcal{A}, \mathcal{P}, \mathcal{C}\}$, where $\mathcal{S}$ is a finite set of states, $\mathcal{A}$ is a finite set of actions, $\mathcal{P}$ is a state transition function and $\mathcal{C}$ is a cost function. The agent can control the state of the system by choosing actions a from $\mathcal{A}$, influencing in this way the state transitions. The results of an action are stochastic in that the actual transition cannot be predicted with certainty. The transition function $\mathcal{P} : \mathcal{S} \times \mathcal{A} \rightarrow \mathcal{S}$ specifies the effect of taking an action at a given state. We denote by $p_{xy}(a)$ the transition probability from state x to state y when action a is taken at state x, and require that $0 \leq p_{xy}(a) \leq 1 \, \forall x, y \in \mathcal{S}, \forall a \in \mathcal{A}$ and $\sum_{y \in S} p_{xy}(a) = 1$.

The agent knows the state of the system x at any time and it chooses actions based only on the current state. We consider deterministic stationary Markovian policies, $\pi : \mathcal{S} \rightarrow \mathcal{A}$, which define the next action of the agent based only on the current state x, i.e. an agent adopting this policy performs action $\pi(x)$ in state x. For the problems we consider, optimal stationary Markovian policies always exist.

We assume a bounded, integer-valued cost function $\mathcal{C} : \mathcal{S} \rightarrow \mathbb{N}$, and denote by $c(x, a)$ the finite cost for executing action a in state x. Different optimality criteria can be adopted to measure the cost of a policy π, all measuring in some way the cost accumulated by the agent as it follows policy π. In this work we focus on the average cost criterion because is more appropriate for the problem under study than other discounted cost approaches [24].

When the agent minimizes a discounted cumulative sum of costs, and we suppose that starting from state x_0 and using policy π the system evolves through states $\{x_0, x_1, \ldots, x_t\}$ in interval $[0, t]$, then the discounted cost of policy π is defined as

$$v^{\pi}(x_0) = \lim_{t \to \infty} E\left(\sum_{m=0}^{t} \gamma^m c(x_m, \pi(x_m)) \right),$$

where $\gamma \leq 1$ is the discount factor. This allows simpler computational methods to be used, as discounted total reward will be finite. Note that an agent minimizing $v^{\pi}(x)$ will prefer actions that generate an immediate cost reduction instead of those ones generating the same cost reduction some steps into the future, due to the discounted factor. In many situations, discounted methods can be justified by the nature of the problem, like in economics. Notwithstanding, a more natural measure of optimality exists for infinite-horizon tasks like the one studied in this work, based on minimizing the average cost per action. We define the total cost accumulated in the interval $[0, t]$ as

$$w^{\pi}(x_0, t) = \sum_{m=0}^{t} c(x_m, \pi(x_m)).$$

If the environment is stochastic then $w^{\pi}(x_0, t)$ is a random variable. Under the average cost criterion we seek to minimize the average expected cost rate over time t, as $t \to \infty$. When the system starts at state x and follows policy π, the average expected cost rate is denoted by $g^{\pi}(x)$ and is defined as

$$g^{\pi}(x) = \lim_{t \to \infty} \frac{1}{t} E[w^{\pi}(x, t)].$$

In this work we minimize a weighted sum of loss rates and therefore the average cost criterion is more appropriate for the problem under study than other discounted cost approaches. In a system like ours, it is not difficult to see that for every deterministic stationary policy the embedded Markov chain has a unichain transition probability matrix, and therefore the average expected cost rate does not vary with the initial state [25]. We call it the "cost rate" of the policy π, denote it by g^π and consider the problem of finding the policy π^* that minimizes g^π, which we name the optimal policy.

It can be shown that for our system

$$g^\pi = \sum_{i=1}^{N} (\beta_i^n P_i^n \lambda_i^n + \beta_i^h P_i^h \lambda_i^h),$$

where β_i^n (β_i^h) is the relative weight associated to the blocking of a new (handover) request and P_i^n (P_i^h) is the blocking probability of new (handover) requests, both of service i. In general, $\beta_i^n < \beta_i^h$ to account for the fact that the blocking of a handover request is less desirable than the blocking of a new session request.

Two different optimization approaches have been used to find the optimal policy. The first approach is based on *dynamic programming* (DP) [23], specifically we used a policy improvement method [25]. This approach is applied to a single service scenario. The second is an automatic learning approach based on the theory of *reinforcement learning* (RL) [26], more specifically we used the average reward reinforcement learning algorithm proposed in [27]. This approach is applied to a multiservice scenario. DP gives an exact solution and allows to evaluate the theoretical limits of incorporating movement prediction in the AC problem, whereas RL tackles more efficiently the curse of dimensionality and offers the important advantage of being a model-free method, i.e. transition probabilities and average costs are not needed by the method. As a consequence, in the RL approach, neither the numerical values of the PA parameters (p, q, μ_p) nor the arrival rates and holding times need to be known beforehand. Moreover, the learning algorithm can adapt to variations of those parameters.

8.3.1 Single Service

In this section we describe the optimization approach based on DP. Since there is only one service type, throughout this section we simplify notation by omitting the subscript thar refers to it, i.e. $\lambda_n = \lambda_1^n$, $\lambda_h = \lambda_1^h$, $\mu_s = \mu_1^s$, $\mu_r = \mu_1^r$, $\mu = \mu_1$, $P_n = P_1^n$, $P_h = P_1^h$ and $P_{ft} = P_1^{ft}$. Without loss of generality, we assume that $b_1 = 1$.

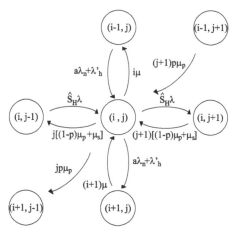

Figure 8.2 Transition rates.

Let us represent the system state by (i, j) where i is the number of active sessions in the cell and j is the number of MTs in the cell neighborhood labeled as H. The set of possible states of the system is

$$\mathcal{S} := \big\{ x = (i, j) : 0 \leq i \leq C; \ 0 \leq j \leq C_p \big\},$$

where C_p represents the maximum number of MT that can be labeled as H at a given time. We use a large value for C_p so it has no practical impact on our results. At each state $(i, j), i < C$, the set of possible actions is defined by $\mathcal{A} := \{ a : a = 0, 1 \}$, being $a = 0$ the action that rejects an incoming new session and $a = 1$ the action that accepts an incoming new session. Handover sessions have priority over new sessions and they are accepted as long as resources are available $(i < C)$. At state (C, j) only the action $a = 0$ is possible.

Figure 8.2 shows the transition rates from and to state (i, j). Note that some of the transition rates depend on the action $a = 0, 1$. In the figure we introduced λ_h' which is the average arrival rate of handovers that have not been predicted, and it is given by

$$\lambda_h' = (1 - \hat{S}_H) \frac{\mu_p}{\mu_p + \mu_s} q \lambda,$$

where λ is the input rate to the PA.

The model described is a continuous-time Markov chain, which we convert to a *discrete time Markov chain* (DTMC) by applying uniformization (see [28, section 4.7]). It can be shown that

$$\Gamma = C_p(\mu_p + \mu_s) + C(\mu_r + \mu_s) + \lambda + \lambda_n$$

is an uniform upper-bound for the outgoing rate of all the states. If $r_{xy}(a)$ denotes the transition rate from state x to state y when action a is taken at state x, then the transition probabilities of the resulting DTMC are given by

$$p_{xy}(a) = \frac{r_{xy}(a)}{\Gamma} \quad \text{if } y \neq x \qquad \text{and} \qquad p_{xx}(a) = 1 - \sum_{y \in S} p_{xy}(a).$$

We define the incurred cost rate at state x when action a is taken by

$$c(x, a) = \begin{cases} 0, & i < C, \ a = 1, \\ \lambda_n, & i < C, \ a = 0, \\ \lambda_n + \beta(\lambda_h' + jp\mu_p), & i = C, \ a = 0. \end{cases}$$

The weighting factor β (typically) accounts for the fact that blocking a handover request is less desirable than blocking of a new session request. Costs are defined so that the average expected cost rate g^π equals a weighted sum of the average loss rate of new sessions ($P_n\lambda_n$) and handover attempts ($P_h\lambda_h$), i.e.

$$g^\pi = \lim_{n \to \infty} E\left[\frac{1}{n+1} \sum_{t=0}^{n} c(x(t), \pi(x(t)))\right] = P_n\lambda_n + \beta P_h\lambda_h,$$

where $x(t)$ is the state visited at time t when policy π is deployed. Loss rates are given by

$$P_n\lambda_n = \sum_{x:\ \pi(x)=0} \lambda_n p(x); \qquad P_h\lambda_h = \sum_{\substack{x=(C,j) \\ 0 \leq j \leq C_p}} (\lambda_h' + jp\mu_p) p(x),$$

where $p(x)$ is the stationary probability of state x. Thus, the optimization problem pursues to find the policy π^* that minimizes g^π. Since state $(0, 0)$ can be reached from any other state regardless of the policy deployed, by virtue of the *Corollary 6.20* and the subsequent remark, both in [23], we know that an optimal stationary policy exists.

If we denote by $h(x)$ the relative cost rate of state x under policy π, then we can write

$$h(x) = c(x, \pi(x)) - g^\pi + \sum_y p_{xy}(\pi(x)) h(y) \qquad \forall x \qquad (8.2)$$

from which we can obtain the average cost and the relative costs $h(x)$ up to an undetermined constant. Thus we arbitrarily set $h(0, 0) = 0$ and then solve the linear system of equations (8.2) to obtain g^π and $h(x)$, $\forall x$. Having obtained the average and relative costs under policy π an improved policy π' can be calculated as

$$\pi'(x) = \arg\min_{a=0,1} \left\{ c(x, a) - g^\pi + \sum_y p_{xy}(a)h(y) \right\}$$

so that the following relation holds $g^{\pi'} \le g^\pi$. Moreover, if the equality holds then $\pi' = \pi = \pi^*$, where π^* denotes the optimal policy, i.e. $g^{\pi^*} \le g^\pi$, $\forall \pi$.

The solution of the linear system in (8.2) followed by the policy improvement is repeated iteratively until a policy which does not change after improvement is obtained. This process is called *Policy Iteration* [25, section 8.6] and it leads to the average optimal policy in a finite – and typically small – number of iterations.

8.3.2 Multiservice

In this section we study a scenario in which the information available to the AC system is also the state of the cell and the state of the cell neighborhood. The state could be conceptually represented by a vector with $2N$ elements, each of them being the number of ongoing sessions initiated either as new or handover requests. However, we adopt a more compact representation of the state space, including only the number of resource units occupied in the cell and in its neighborhood. This is motivated by the fact that reducing the state space helps the RL algorithm to find better solutions and by the conclusions of previous studies [12], which show that the performance of policies which base their decisions only on the number of resource units occupied (trunk reservation policies), are close to the performance of the optimum policy.

We formulate the optimization problem as an infinite-horizon finite-state semi-Markov decision process (SMDP) under the average cost criterion. It is evident that we search for policies that minimize g^π. Decision epochs correspond to time instants at which arrivals occur. Given that no actions are taken at session departures, then only the arrival events are relevant to the optimization process. At each decision epoch the system has to select one action from the set of possible actions $\mathcal{A} := \{0 = \text{reject}, 1 = \text{admit}\}$.

The state space is defined as

$$\mathcal{S} := \{x = (x_0, x_1, k) : x_0, x_1, k \in \mathbb{N}; x_0 \le C; x_1 \le C_p, 1 \le k \le (2N - 1)\},$$

where x_0 is the number of resource units occupied in the cell under study, x_1 is the number of resource units occupied in the cell neighborhood by ongoing sessions labeled as H by the PA and k, $1 \leq k \leq (2N - 1)$, is the arrival type. We select one of the $2N$ arrival types as the highest priority one, being its requests always admitted while free resources are available, and therefore no decisions are taken for them.

The cost structure is defined as follows. At any decision epoch, the cost incurred by accepting any arrival type is zero and by rejecting a new (handover) request of service i is β_i^n (β_i^h). With this framework, further accrual of cost occurs when the system has to reject requests of the highest priority arrival type between two decision epochs.

We denote by $h(x)$ the relative cost rate of state x, which can be interpreted as the expected long-term advantage in total cost for starting in state x in addition to $t \cdot g^\pi$, the expected total cost at time t on the average. The Bellman optimality recurrence equations for an SMDP under the average cost criterion have the form

$$h^*(x) = \min_{a \in A_x} \left\{ c(x, a) - g^* \tau(x, a) + \sum_{x \in S} p_{xy}(a) h^*(y) \right\},$$

where $h^*(x)$ is an optimal state dependent relative value function and $c(x, a)$ and $\tau(x, a)$ are the average cost and the average sojourn time when taking decision a in state x. The greedy policy π^* defined by selecting actions that minimize the right-hand side of the above equation is gain-optimal [27].

If the parameters of the model can be derived, then the solution to the Bellman equations can be obtained through dynamic or linear programming techniques. In multiservice scenarios, where the number of states can be large, the derivation of the model parameters can be complex and make the problem intractable (curse of dimensionality). We propose an alternative approach based on a reinforcement learning algorithm named Semi-Markov Average Reward Technique (SMART) [27].

The Bellman equations can be rewritten as

$$h^*(x, a) = \min_{a \in A_x} \left\{ c(x, a) - g^* \tau(x, a) + \sum_{x \in S} p_{xy}(a) \min_{a' \in A_y} h^*(y, a') \right\},$$

where $h^*(x, a)$ is the average expected relative value of taking the optimal action a in state x and then continuing indefinitely by choosing actions optimally. Then, the optimal policy is

$$\pi^*(x) = \arg \min_{a \in A_x} h^*(x, a).$$

The SMART algorithm estimates $h^*(x, a)$ by simulation, using a temporal difference method (TD(0)). If at the $(m - 1)$-th decision epoch the system is in state x, action a is taken and the system is found in state y at the m-th decision epoch, then we update the relative state-action values as follows:

$$h_{\text{new}}(x, a) = (1 - \alpha_m)h_{\text{old}}(x, a)$$

$$+ \alpha_m \left\{ c_m(x, a, y) - g_m \tau_m(x, a, y) + \min_{a' \in A_y} h_{\text{old}}(y, a') \right\},$$

where $c_m(x, a, y)$ is the actual cumulative cost incurred between the two successive decision epochs, $\tau_m(x, a, y)$ is the actual sojourn time between the decision epochs, α_m is the learning rate parameter at the mth decision epoch, and g_m is the average cost rate estimated as

$$g_m = \frac{\sum_{k=1}^{m} c_k\left(x_{(k)}, a_{(k)}, y_{(k)}\right)}{\sum_{k=1}^{m} \tau_k\left(x_{(k)}, a_{(k)}, y_{(k)}\right)}.$$

8.4 Numerical Evaluation

We evaluated the performance gain when introducing prediction by the ratio g_{wp}^{π}/g_p^{π}, where g_p^{π} (g_{wp}^{π}) is the expected average cost rate of a policy that is optimal in a system with (without) prediction. We assume a circular-shaped cell of radio r and a holed-disk-shaped neighborhood with inner (outer) radio $1.0r$ ($1.5r$).

The values of the parameters that define the scenario are: $C = 10$ and $C_p = 60$ resource units, $N_h = \mu_i^r/\mu_i^s = 1$, $\mu_i^r/\mu_i^p = 0.5$, $S_H = 0.4$, $x = U/2$. For the single service scenario ($N = 1$) we use $b_1 = 1$, $\lambda_1^n = 1$, $\mu_1 = \mu_1^s + \mu_1^r = 1$, $\beta_1^n = 1$, and $\beta_1^h = \beta = 20$. As mentioned in Section 8.2.1, the value of λ is chosen so that the system is in statistical equilibrium, i.e. the rate at which handover sessions enter a cell equals the rate at which handover sessions exit the cell. For small values of P_1^n ($\approx 10^{-2}$) and P_1^{ft} ($\approx 10^{-3}$), we make the approximation $\lambda = 0.989\lambda_1^n(N_h + \mu_1^r/\mu_1^p)(1/S_H)$. For the multiservice scenario we use $N = 2$ services, $b_1 = 1$ and $b_2 = 2$ resource units. The arrival rates of new sessions to the cell are $\lambda_1^{nc} = 0.8\lambda_T$, $\lambda_2^{nc} = 0.2\lambda_T$, where $\lambda_T = 2$. The ratio of arrival rates of new sessions to the cell neighborhood (ng) and to the cell (nc) is made equal to the ratio of their surfaces, $\lambda_i^{ng} = 1.25\lambda_i^{nc}$. The ratio of handover arrival rates to the cell neighborhood from the outside of the system (ho) and from the cell (hc) is made equal to the ratio of their perimeters, $\lambda_i^{ho} = 1.5\lambda_i^{hc}$. From equation (8.1) it follows that $\lambda_i^{hc} = (1 - P_i^n)(1 - P_i^{ft})N_h\lambda_i^{nc}$, which we approximate by $\lambda_i^{hc} = 0.989N_h\lambda_i^{nc}$. We also set $\mu_1 = \mu_1^s + \mu_1^r = 1$, $\mu_2 = \mu_2^s + \mu_2^r = 3$, $\beta_1^n = 1$, $\beta_2^n = 20$, $\beta_1^h = 10$ and $\beta_2^h = 200$.

With regard to the reinforcement learning algorithm, we use a constant learning rate $\alpha_m = 0.01$ but the exploration rate p_m is decayed to zero by using the following rule $p_m = p_0/(1+u)$, where $u = m^2/(\gamma + m)$. We used $\gamma = 1.0 \cdot 10^{11}$ and the algorithm starts with an exploration rate $p_0 = 0.1$.

8.4.1 Single Service

When no predictive information is used in the single service scenario, the optimization is carried without considering the second component of the system state, i.e. the number of MT labeled as H, and the optimal policy results to be of the *guard channel* type [7].

The curves in Figure 8.3 represent the quotient between the average expected cost rate of the optimal policy when no prediction is deployed and the optimal policy deploying prediction. As expected, using prediction induces a gain in all cases and that gain decreases as prediction uncertainty (U) increases. In Figure 8.3a we varied the average number of handovers per session. In Figure 8.3b we varied the weighting factor β which quantifies the priority of handover requests over new sessions: the higher the value of β the lower the blocking probability of handover sessions compared to the blocking probability of new sessions. It is observed that higher values of β lead to higher performance gains. The position of the decision threshold within the uncertainty zone is evaluated in Figure 8.3c, the curves indicate that a threshold in the middle of the uncertainty zone is the best choice. Finally, Figure 8.3d shows the effect of the elapse time since an MT is classified as H until it is handed over into the target cell or it moves to another cell. Both, short and long prediction periods, have a negative effect on the performance gain.

In all the cases that we examined the optimal policy when prediction is deployed had a *dynamic guard channel* structure, in which the number of reserved channels increases with the number of MTs labeled as H. More formally, let $p(i, j)$ be the probability of accepting a new session when the

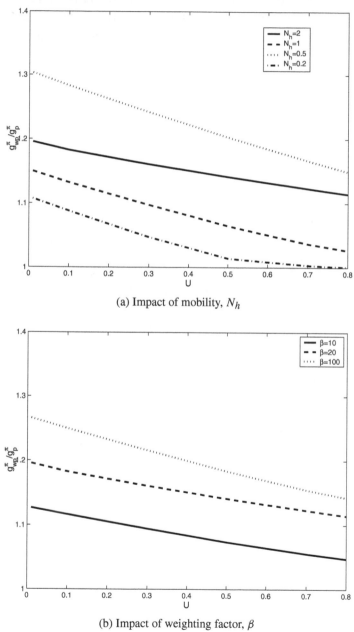

(a) Impact of mobility, N_h

(b) Impact of weighting factor, β

Figure 8.3 Performance comparison in the single service scenario.

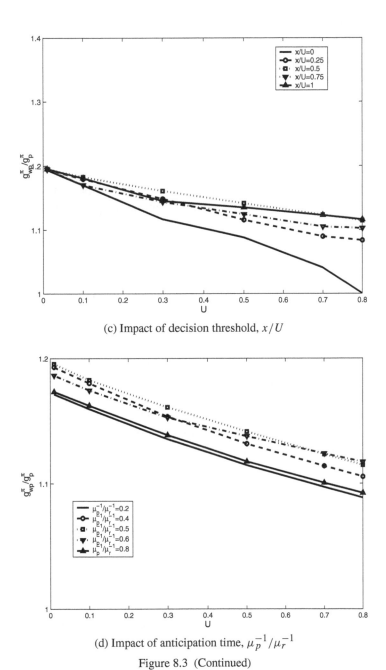

(c) Impact of decision threshold, x/U

(d) Impact of anticipation time, μ_p^{-1}/μ_r^{-1}

Figure 8.3 (Continued)

system is at state (i, j), then

$$p(i, j) = \begin{cases} 1, & \text{if } i \leq i_{\text{th}}(j) \\ 0, & \text{if } i > i_{\text{th}}(j) \end{cases} \quad \text{and} \quad i_{\text{th}}(j) \leq i_{\text{th}}(j') \quad \text{if} \quad j > j',$$

where $i_{\text{th}}(j)$ is the threshold for a given j.

8.4.2 Multiservice

When no predictive information is used in the multiservice scenario, the optimization is carried without considering the second component of the system state, i.e. the number of resources occupied in the neighborhood by sessions labeled as H.

Figure 8.4a displays the variation of the ratio g_{wp}^{π}/g_p^{π} with different values of the uncertainty U in multiservice scenarios. As a reference, the same figure also shows the variation of the ratio g_{wp}^{π}/g_p^{π} for single service scenarios. In the multiservice scenario, for each value of U we run 10 simulations with different seeds and we display the averages. As expected, using prediction induces a gain in all cases and that gain decreases as the prediction uncertainty (U) increases.

Finally it is worth noting that the main challenge in the design of efficient bandwidth reservation techniques for mobile cellular networks is to balance two conflicting requirements: reserving enough resources to achieve a low forced termination probability and keeping the resource utilization high by not blocking too many new setup requests. Figure 8.4b, which shows the variation of the utilization gain, i.e. the ratio utilization$_{wp}$/utilization$_p$, for different values of U, justifies the efficiency of our optimization approach.

8.5 Conclusion

In this paper we analyzed the performance gain that can be obtained when handover prediction information is considered in order to optimize the admission control policy in a mobile cellular network. Predictive information is provided by a prediction agent that labels the active mobile terminals in the neighborhood of the cell which will probably produce a handover into the cell. The policy optimization has been performed in a Markov or semi-Markov decision process framework and two optimization methods have been applied: policy iteration and a model free reinforcement learning methods. Our numerical results show that typical performance gains are around 10%

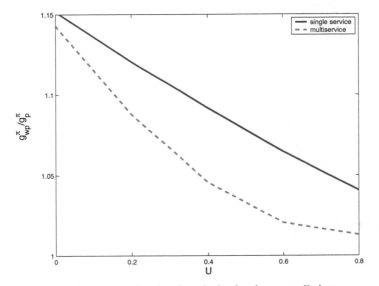

(a) Performance gain when introducing hand-over prediction

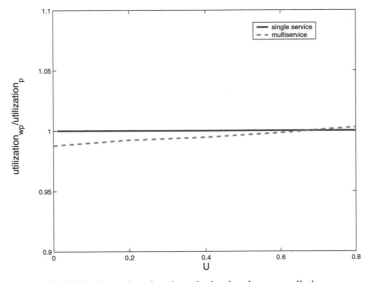

(b) Utilization gain when introducing hand-over prediction

Figure 8.4 Performance comparison in the multiservice scenario.

although improvement ratios up to 30% have also been observed in some specific scenarios. In future work we will consider a more sophisticated model of the prediction agent including, for instance, a more precise estimation of the time instant at which a handover will occur.

Acknowledgements

This work has been supported by the Spanish Ministry of Education and Science (30%) and by the European Commission (FEDER 70%) through projects TSI2007-66869-C02-02 and TIN 2008-06739-C04-02. It is also supported by the "Catedra Telefonica" of the Universidad Politecnica de Valencia.

References

[1] W. Y. Lee, Smaller cells for greater performance, *IEEE Communications Magazine*, vol. 29, no. 11, pp. 19–23, November 1991.

[2] S. Choi and K. G. Shin, Adaptive bandwidth reservation and admission control in QoS-sensitive cellular networks, *IEEE Transactions on Parallel and Distributed Systems*, vol. 13, no. 9, pp. 882–897, September 2002.

[3] D. Hong and S. S. Rappaport, Traffic model and performance analysis for cellular mobile radio telephone systems with prioritized and nonprioritized handoff procedures, *IEEE Transactions on Vehicular Technology*, vol. VT-35, no. 3, pp. 77–92, August 1986. (See also: CEAS Technical Report No. 773, College of Engineering and Applied Sciences, State University of New York, Stony Brook, June 1999.)

[4] F. Barceló, Performance analysis of handoff resource allocation strategies through the state-dependent rejection scheme, *IEEE Transactions on Wireless Communications*, vol. 3, no. 3, pp. 900–909, May 2004.

[5] D. J. Daley and L. D. Servi, Loss probabilities of hand-in traffic under various protocols: II. Model comparisons, *Performance Evaluation*, vol. 55, nos. 3–4, pp. 231–249, February 2004.

[6] V. Pla and V. Casares-Giner, Analysis of priority channel assignment schemes in mobile cellular communication systems: A spectral theory approach, *Performance Evaluation*, vol. 59, nos. 2–3, pp. 199–224, February 2005.

[7] R. Ramjee, R. Nagarajan and D. Towsley, On optimal call admission control in cellular networks, *Wireless Networks Journal (WINET)*, vol. 3, no. 1, pp. 29–41, 1997.

[8] N. Bartolini, Handoff and optimal channel assignment in wireless networks, *Mobile Networks and Applications (MONET)*, vol. 6, no. 6, pp. 511–524, 2001.

[9] N. Bartolini and I. Chlamtac, Call admission control in wireless multimedia networks, in *Proceedings of IEEE PIMRC*, vol. 1, pp. 285–289, 2002.

[10] H. Heredia-Ureta, F. A. Cruz-Pérez and L. Ortigoza-Guerrero, Capacity optimization in multiservice mobile wireless networks with multiple fractional channel reservation,

IEEE Transactions on Vehicular Technology, vol. 52, no. 6, pp. 1519 – 1539, November 2003.

[11] V. Pla and V. Casares-Giner, Optimal admission control policies in multiservice cellular networks, in *Proceedings of the International Network Optimization Conference (INOC)*, October, pp. 466–471, 2003.

[12] D. García, J. Martínez and V. Pla, Admission control policies in multiservice cellular networks: Optimum configuration and sensitivity, in *Wireless Systems and Mobility in Next Generation Internet*, G. Kotsis and O. Spaniol (Eds.), Lecture Notes in Computer Science, vol. 3427, Springer, pp. 121–135, 2005.

[13] D. Levine, I. Akyildiz and M. Naghshineh, A resource estimation and call admission algorithm for wireless multimedia networks using the shadow cluster concept, *IEEE/ACM Transactions on Networking*, vol. 5, no. 1, pp. 1–12, February 1997.

[14] J. Hou and Y. Fang, Mobility-based call admission control schemes for wireless mobile networks, *Wireless Communications and Mobile Computing*, vol. 1, no. 3, pp. 269–282, 2001.

[15] F. Yu and V. Leung, Mobility-based predictive call admission control and bandwidth reservation in wireless cellular networks, *Computer Networks*, vol. 38, no. 5, pp. 577–589, 2002.

[16] W.-S. Soh and H. S. Kim, A predictive bandwidth reservation scheme using using mobile positioning and road topology information, *IEEE/ACM Transactions on Networking*, vol. 14, no. 5, pp. 1078–1091, October 2006.

[17] A. Yener and C. Rose, Genetic algorithms applied to cellular call admission: local policies, *IEEE Transactions on Vehicular Technology*, vol. 46, no. 1, pp. 72–79, 1997.

[18] R. Zander and J. M. Karlsson, Predictive and adaptive resource reservation (PARR) for cellular networks, *International Journal of Wireless Information Networks*, vol. 11, no. 3, pp. 161–171, July 2004.

[19] Y. Zhao, Standardization of mobile phone positioning for 3G systems, *IEEE Communications Magazine*, vol. 40, no. 7, pp. 108 –116, July 2002.

[20] F. Khan and D. Zeghlache, Effect of cell residence time distribution on the performance of cellular mobile networks, in *Proceedings of VTC'97*, IEEE, pp. 949–953, 1997.

[21] V. Pla and V. Casares-Giner, Effect of the handoff area sojourn time distribution on the performance of cellular networks, in *Proceedings of IEEE MWCN*, September, pp. 401–405, 2002.

[22] P. V. Orlik and S. S. Rappaport, On the handoff arrival process in cellular communications, *Wireless Networks Journal (WINET)*, vol. 7, no. 2, pp. 147–157, March/April 2001.

[23] S. M. Ross, *Applied Probability Models with Optimization Applications*, Holden-Day, 1970.

[24] S. Mahadevan, Average reward reinforcement learning: Foundations, algorithms, and empirical results, *Machine Learning*, Special Issue on Reinforcement Learning (L. Kaebling, Ed.), vol. 22, no. 1–3, pp. 159–196, 1996.

[25] M. L. Puterman, *Markov Decision Processes: Discrete Stochastic Dynamic Programming*, John Wiley & Sons, 1994.

[26] R. Sutton and A. G. Barto, *Reinforcement Learning*, The MIT Press, Cambridge, MA, 1998.

[27] T. K. Das, A. Gosavi, S. Mahadevan and N. Marchalleck, Solving semi-Markov decision problems using average reward reinforcement learning, *Management Science*, vol. 45, no. 4, pp. 560–574, 1999.

[28] R. W. Wolff, *Stochastic Modeling and the Theory of Queues*, Prentice Hall, Englewood Cliffs, NJ, 1989.

PART THREE
PERFORMANCE MODELLING STUDIES

9

Queueing Delay of Stop-and-Wait ARQ over a Wireless Markovian Channel

S. De Vuyst, S. Wittevrongel and H. Bruneel

Stochastic Modeling and Analysis of Communication Systems (SMACS) Research Group, Department of Telecommunications and Information Processing, Ghent University, Sint-Pietersnieuwstraat 41, 9000 Gent, Belgium; e-mail: sdv@telin.ugent.be

Abstract

In this paper, we present the analysis of the Stop-and-Wait ARQ (Automatic Repeat reQuest) protocol with the notable complication that the transmission errors occur in a bursty, correlated manner. Fixed-length packets of data are sent from transmitter to receiver over an error-prone channel. The receiver notifies the transmitter whether a packet was received correctly or not by returning a feedback message over the backward channel. If necessary, the packet is retransmitted until it is received correctly, after which the transmission of another packet starts.

We model the transmitter side as a discrete-time queue with infinite storage capacity and independent and identically distributed (*iid*) packet arrivals. Arriving packets are stored in the queue until they are successfully transmitted over the channel. The probability of an erroneous transmission is modulated by a two-state Markov Chain, rather than assuming stationary channel errors. In previous work we have analysed the queue content distribution of the transmitter, while in the present paper, we give an intuitive derivation of the throughput of the system and the distribution of the packet delay. For the latter, we use the spectral decomposition theorem from linear algebra and give an accurate approximation for the asymptotic behaviour.

D. D. Kouvatsos (ed.), Mobility Management and Quality-of-Service for Heterogeneous Networks, 197–227.

Finally, we illustrate the importance of accounting for the error correlation by means of some numerical examples.

Keywords: Discrete-time queueing model, ARQ, correlated errors, analytic study.

9.1 Introduction

Whenever packets of data need to be transmitted from point A (the *transmitter*) to point B (the *receiver*), there is always a chance that an error occurs while they move through the medium between A and B (the *channel*): some packets may be corrupted or even lost entirely. To cope with this, ARQ (Automatic Repeat reQuest) protocols have been used to provide a more reliable way of communication between the transmitter and the receiver.

There are many types of ARQ, but all protocols have the same two prerequisites. First, there must be some way for the receiver to *check* the integrity of the data it receives. Usually, the transmitter will add some redundant bits to a packet before it is sent to enable the receiver to detect (or even correct) the most commonly occurring errors. The more redundancy bits, the more reliably errors can be detected, although there is no way to detect every possible error with certainty. Secondly, the channel must be *bi-directional*. When a packet is received, the receiver must send a message back to the transmitter to notify the transmitter of the condition of that packet, i.e. an acknowledgement (ACK) if it is intact or a negative acknowledgement (NACK) if the packet is in error. Since not all arriving packets can be transmitted immediately, one needs to implement a queue at the transmitter side. For some types of ARQ (i.e. Selective Repeat), the protocol does not ensure that the transmitted packets will be received at the other side in their original order, so an additional queue is needed at the receiver side (a resequencing buffer).

In the present paper, we study the Stop-and-Wait ARQ protocol (SW-ARQ), and more specifically the packet delay in the transmitter queue. In SW-ARQ, the transmitter sends a packet available in its queue over the channel and then simply *waits* until it receives the corresponding feedback message. If the packet was transmitted correctly (ACK), the next packet in the queue is transmitted. Otherwise, if an error occurred (NACK), the packet is retransmitted. Note that the transmitter is inactive during the whole time a packet is travelling through the channel, is being processed by the receiver and while the feedback message is travelling back. We will refer to this time period as the *feedback delay* and it is clear that for long such delays, quite some time is wasted simply waiting for acknowledgements, resulting in a low throughput. However, SW-ARQ is simple to implement and ensures that packets are received in the same order as they arrived to the transmitter such that no resequencing is needed.

The model we propose distinguishes itself from previous studies in that we allow the errors occurring in the channel to be *correlated in time*. Instead of assuming that the probability of an erroneous packet is static in time, we allow this probability to depend on what *state* the channel is in when the packet is transmitted. Specifically, the channel alternates between two states which could be termed the GOOD state and the BAD state, both of which reflect different conditions with regard to the error probability. The channel state process is modelled as a two-state Markov Chain with a fixed error probability in either state, resulting in what is also known as the Gilbert–Elliott model [1, 2]. This complication is inspired by the observation that real-life communication channels rarely have the same error sensitivity during their whole time of operation. Factors such as electromagnetic interference, availability of intermediate network nodes and links, presence of data traffic with higher priority and so on, may all influence the behaviour of the channel and are mostly time-varying in nature. This holds especially when the *wireless* medium is considered where the conditions may change on an even more diverse set of timescales than in a wired medium due to user mobility, noise, reflection, scattering, shadowing, or any other physical effect that causes the radio signal to change in both amplitude and phase. Such inherent signal changes over time and space impairing the data transmission is known as *channel fading*.

In wireless radio communication, a signal usually reaches the receiver along many different reflective paths while the motion between transmitter and receiver results in propagation changes along each of these paths. Therefore, the actually received signal is a superposition of many time-variant components that occasionally interfere in a destructive way. If the multiple reflective paths are large in number and there is no line-of-sight component, this fading on a small scale (as small as half a wavelength) is called *Rayleigh fading* [3]. The conditions of flat (i.e. frequency non-selective) Rayleigh fading are statistically described by Clarke's model [4] and are widely accepted as a general-purpose description of wireless signal propagation [5–11].

During the last decade, there has been a lot of discussion whether or not the transmission errors on a Rayleigh fading channel can be adequately modelled as a Markov process [5]. Using an information theoretic metric, it was shown for the first time in [6] that the envelope of a Rayleigh fading process does indeed have a Markov character. Shortly thereafter, Zorzi et al. [7] drew the same conclusion for the successes and failures of packet transmissions over such a channel. However, according to Tan et al. [8], some serious caveats must be placed with these results in case a slowly fading channel is observed for a longer period of time. In that paper, the authors use stochastic process theory rather than the information metric of [6, 7] and compare the autocorrelation function of the Rayleigh fading process with that of some well-chosen Markov models under various conditions. Nevertheless, although they may not always be able to capture all nuances of the error statistics on the physical level, Markovian channel models for the wireless link have been used very frequently, especially for the performance analysis of the communication protocols in the upper layers. These models are certainly more accurate than independent error models, while still allowing for a reasonably tractable analysis, often resulting in closed-form intuitive results.

Several error models with an underlying Markov chain have been proposed, differing in the number of states to be used, the allowed transitions between the states and the error statistics within each state. The capacity of a general Finite-State Markov Channel (FSMC) with a specific error probability in each state was studied in [9], where the correlated Rayleigh channel is given as an example. The FSMC proposed in [10] is derived from modelling the signal-to-noise ratio at the receiver under Rayleigh fading assumptions. Here, only transitions to adjacent states are possible, which is a valid assumption in case the channel fades slowly compared to the time between packet transmissions. A classic type of FSMCs known as the Fritchman models [11] divides the N states into two classes, namely error-free states and error states, without restrictions for the transition probabilities. As to the number of states required to give a good description, that is a matter of trade-off between accuracy and tractability (see the discussion in [5]). Interestingly, Chen and Rao [12] explores the conditions under which the number of states can be reduced and shows how to stochastically upper and lower bound an N-state Markovian channel model with a 2-state model. Note, however, that in a two-state Markovian channel both BAD and GOOD periods have geometrically distributed lengths, which is not always in agreement with measurements in practical systems, for instance in the case of UDP traffic over 802.11 Wireless LAN [13, 14]. Both papers suggest to use a two-state *semi*-Markov process instead, by allowing a non-geometric distribution for the sojourn time in each state.

Nevertheless, the Gilbert–Elliott channel model with only two states certainly has its merits when used for the analysis of ARQ protocols, see e.g. [5, 18, 29] as well as the examples in [26]. As demonstrated in the current paper, some of the *qualitative* effects on the queueing performance of the protocol that are caused by the correlation in the error process are manifested in their purest form precisely when using only two states. We also note that, in principle, our analytic technique can equally well be used in the case of a model with N states, although explicit results for the queue content and delay distribution can in general only be provided for $N = 2$.

The performance of SW-ARQ has been studied before, both in terms of throughput and queueing behaviour, but almost always in case of static error probabilities. Several modifications have been proposed to enhance the performance of SW-ARQ, such as sending multiple copies of a packet during the time the transmitter is waiting for feedback [15–18] or combining ARQ with improved error correction techniques, known as hybrid ARQ [19]. In [18], the influence of decoding with memory has been studied as well. Another improvement is the SW-ARQ protocol suggested in [20], where each packet is divided in n parts and only those parts that were received in error are retransmitted. Saeki and Rubin [21] study the throughput and packet delay distribution of all three major ARQ protocols over a TDMA channel with static errors, although for Selective Repeat ARQ (SR-ARQ), only bounds for the mean packet delay are given. In [22], the transmitter queue behaviour of SW-ARQ is analysed in a continuous-time setting. Towsley [25, 26] has worked on the SW-ARQ model with correlated errors as well, but did not have results for the distribution of the packet delay as we do in Section 9.5. Also, our analysis is quite different on several accounts and we provide more explicit results in the specific case of a two-state error channel. The queueing performance of SR-ARQ with correlated errors has been studied in e.g. [23, 24], but with the assumption that the feedback is returned immediately after the transmission of a packet ('ideal' SR). The transmission and resequencing delay for SR-ARQ was studied in [27] but for a less versatile model of the channel correlation.

The organisation of the paper is as follows. The mathematical model of the SW-ARQ transmitter queue is introduced in Section 9.2, along with some specific assumptions. In Section 9.3 we identify a sufficient description for the state of the system at an arbitrary slot and provide the main equations that govern its behaviour. We obtain closed-form expressions for the joint pgf of the system state as well as the pgf of the queue content in equilibrium. In Section 9.4 we prove a simple expression for the throughput of the sys-

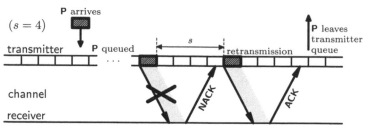

Figure 9.1 Operation of the transmitter queue under SW-ARQ, meaning of the feedback delay *s*.

tem by observing the system at departure slots rather than arbitrary slots. In Section 9.5 the concept of conditional service period is used to obtain the probability generating function (pgf) of the packet delay, hereby referring to Appendix A for a brief review of the spectral decomposition theorem and Appendix B for the asymptotic analysis of the delay distribution. Some numerical examples are discussed in Section 9.6 and finally, conclusions are drawn in Section 9.7.

9.2 Model Description

We model the transmitter of a system operating under the Stop-and-Wait (SW) ARQ protocol as a discrete-time queue. We assume that time is divided in fixed-length intervals called *slots*, whereby one slot is the time required to transmit one packet from the queue into the channel. In our analysis, the length of a certain time period is always expressed as an (integer) number of slots. Let us assume the feedback delay is a fixed number of slots too, denoted by *s*. The operation of SW-ARQ is illustrated in Figure 9.1. A packet *P* arrives at the system and is queued for some time until all preceding packets are transmitted correctly and the ACK of the previous packet is received. Then *P* is transmitted for the first time. If the transmission is erroneous, a NACK is returned and *P* is retransmitted $s+1$ slots after its previous transmission. The packet is retransmitted until finally, an ACK is returned. Then the transmitter knows *P* was transmitted correctly and there is no need to keep it in the queue any longer. Our model does not account for erroneous ACK/NACK messages, which is in fact a non-restrictive assumption [27].

Packets of information enter the system according to a *general independent arrival process*, i.e. the numbers of arrivals during consecutive slots form a sequence of independent and identically distributed random variables with

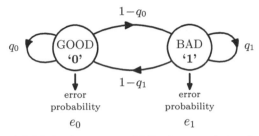

Figure 9.2 Markovian error model for the transmission channel.

common mass function $a(n) = \text{Prob}[a = n]$ $(n \geqslant 0)$ and probability generating function $A(z) = \sum_{n=0}^{\infty} a(n)z^n$. Let a_k be the number of arriving packets during slot k, then we denote their mean value by $\alpha \triangleq E[a_k] = A'(1)$. Furthermore, we assume that the arrivals a_k are not stored in the queue until the end of slot k. This way, an arriving packet can only be served (i.e. transmitted) for the first time during the *next* slot $(k + 1)$ at the very earliest.

When a packet is transmitted, its successful receipt depends on the channel state during the slot in which the feedback message is returned to the transmitter. The transitions between those states, the 0-state (GOOD) and the 1-state (BAD), are governed by a two-state Markov Chain as depicted in Figure 9.2: if the feedback of a packet is returned during a 0-slot, this means the packet was transmitted erroneously with probability e_0. Similarly, during a 1-slot a feedback message is a NACK with probability e_1 and ACK with probability $1 - e_1$. In other words, if we adopt the notation $\bar{q} \triangleq 1 - q$, then $\bar{e}_i$ is the probability of a correct transmission, $i = 0, 1$. Evidently, the designations GOOD and BAD are only meaningful if $e_0 < e_1$, although this condition is not a requirement for the analysis. At first it may seem strange that we choose not to probe the channel state during the slot in which the packet *is transmitted* but during the slot in which its feedback *is returned* to the transmitter. However, this modelling choice makes the analysis less complicated while the results regarding the equilibrium behaviour stay the same. Indeed, it is only a matter of definition to which actual slot $k - s$ we refer to by the name 'channel state in slot k'. Since the evolution of the channel state is not influenced by the rest of the system, the results of the analysis are not affected by a fixed time shift of this definition.

First of all, as a convention for the remainder of this paper, let the index i always be either 0 or 1. As indicated in Figure 9.2, the probability of remaining in state i during a slot transition is given by q_i. We denote the channel state (0 or 1) in slot k by r_k and let $\omega_{i,k} \triangleq \text{Prob}[r_k = i]$. Then we have

$$\omega_{k+1} = \omega_k \mathbf{q} \qquad \text{with } \mathbf{q} \triangleq \begin{bmatrix} q_0 & \bar{q}_0 \\ \bar{q}_1 & q_1 \end{bmatrix}, \qquad (9.1)$$

where ω_k is the row vector with elements $\omega_{0,k}$ and $\omega_{1,k}$ and $\mathbf{q}$ is the transition probability matrix of the channel state process. Rather than using q_0 and q_1, we define the parameters

$$\sigma = \frac{1 - q_0}{2 - q_0 - q_1} \quad \text{and} \quad K = \frac{1}{2 - q_0 - q_1},$$

to be understood as follows. Suppose the channel is in state 0 with probability $\bar{\sigma}$ and in state 1 with probability σ, independently from slot to slot, such that the mean sojourn times are $1/\sigma$ and $1/\bar{\sigma}$ respectively. It is clear that the overall fraction of 1-slots remains equal to σ if the mean lengths of 0- and 1-periods are both multiplied by the same factor K, i.e. if the geometric distributions are chosen such that the mean lengths are $1/(1 - q_0) = K/\sigma$ and $1/(1 - q_1) = K/\bar{\sigma}$ respectively. Therefore, the factor K can be seen as a measure for the *absolute* lengths of the 0- and 1-periods, while σ character-ises their *relative* lengths. Indeed, σ is the relative fraction of 1-slots, since we have from (9.1) in equilibrium:

$$\omega = \lim_{k \to \infty} \omega_k = \begin{bmatrix} \bar{\sigma} & \sigma \end{bmatrix}. \qquad (9.2)$$

Moreover, the correlation coefficient ϕ between the channel states in two consecutive slots is given by $1 - K^{-1}$. Note that $\phi = 0$ and $K = 1$ for uncorrelated errors, whereas for positive correlation we have $0 < \phi < 1$ and $K > 1$. The more correlation present in the channel state process $\{r_k\}$, the higher K is and the fewer the channel changes state. For example, if $\sigma = 0.5$, typical samples of $\{r_k\}$ may be:

01100011010100011011100110100100101010000110111 K small,

00000000111111111110000000011111111111000000 K large.

9.3 Preliminary Analysis: Joint Distribution of the System State

The analysis of this model with regard to the queue content was presented before in detail in [29, 30]. Since the analysis of the packet delay relies on the equilibrium distribution of the system state, we repeat here the main steps and extend them where necessary. Our study of the transmitter queue described

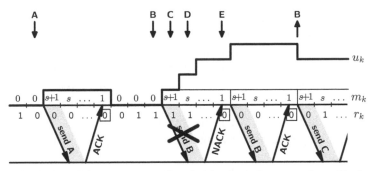

Figure 9.3 Evolution of the system state given by the channel state r_k, the residual roundtrip time m_k and the queue content u_k.

above is done by modelling the system as a (multi-dimensional) Markov chain and calculating its equilibrium distribution, assuming such equilibrium exists.

As the system state variables, we choose the set illustrated in Figure 9.3, constructed as follows. Let u_k be the queue content at the beginning of slot k, which obviously needs to be included since $\{u_k\}$ is the process we are interested in. Next, we also need to know how far a packet has progressed through the channel during slot k and when we can expect its feedback message. For this purpose we define the supplementary variable m_k, in a similar way as was done in [31]. The *residual roundtrip time* m_k indicates the remaining number of slots at the beginning of slot k, needed to complete the roundtrip of the most recently transmitted packet if $u_k \geqslant 1$, and $m_k = 0$ if and only if $u_k = 0$. So $m_k = s + 1$ when a packet is transmitted and then counts one down in each of the following slots. After s slots, when $m_k = 1$, we know that the feedback message for this packet is being returned and the packet will either leave the queue at the end of the slot (ACK) or be retransmitted in the next slot (NACK). Finally, as before, let the random variable r_k be the channel state during slot k. The channel state comes into play during slots with $m_k = 1$, where it determines the probability that either an ACK or a NACK is returned, or equivalently, that the packet departs from the queue or is to be retransmitted. Let d_k be equal to 1 if a packet departs at the end of slot k and equal to 0 otherwise. Then if $m_k = 1$, the probability of an error, and therefore of '$d_k = 0$', is e_i if $r_k = i$. Hence, the pgf of d_k is given by $d_i(z) = \bar{e}_i z + e_i$ if $r_k = i$. We also define

$$\bar{d}_i(z) \triangleq e_i z + \bar{e}_i , \qquad (9.3)$$

which is the pgf of $\bar{d}_k = 1 - d_k$.

One verifies that the triple (r_k, m_k, u_k) is an adequate Markovian description of the system state at the beginning of slot k. The transitions from slot k to slot $k + 1$ in this (three-dimensional) Markov Chain are described by the system equations given in [29] and are used to obtain separate expressions for the distribution of the system state during slots where the queue is *idle* ($m_k = 0$) and during slots where the queue is *busy* ($m_k > 0$). First, let us define $p_{i,k}$ as the probability that the queue is empty and that the channel is in state i in slot k,

$$p_{i,k} \triangleq \text{Prob}[m_k = 0, r_k = i]. \tag{9.4}$$

Secondly, let $yzH_{i,k}(y, z)$ be the joint partial pgf of the residual roundtrip time and the queue content in slot k for a busy queue and channel state i in that slot:

$$H_{i,k}(y, z) \triangleq \text{E}[y^{m_k-1}z^{u_k-1}\{m_k > 0, r_k = i\}], \tag{9.5}$$

where we use the notation $\text{E}[X\{Y\}] = \text{E}[X|Y]\text{Prob}[Y]$. Additionally, we define $zR_{i,k}(z)$ as the partial pgf of the queue content in slot k for the case it is the last slot of a roundtrip period ($m_k = 1$) and the channel state is i:

$$R_{i,k}(z) \triangleq \text{E}[z^{u_k-1}\{m_k = 1, r_k = i\}] = H_{i,k}(0, z). \tag{9.6}$$

We assume that for $k \to \infty$, the system reaches equilibrium, such that $p_{i,k}$ and the functions $R_{i,k}(z)$ and $H_{i,k}(y, z)$ converge to a limiting value which we indicate by dropping the index k. Let r, m, u and e denote the channel state, the remaining roundtrip time, the queue content and the number of arrivals respectively, in an arbitrary slot during equilibrium. We also define the row vectors $\mathbf{p}$ and $\mathbf{H}(y, z)$ as

$$\mathbf{p} = \begin{bmatrix} p_0 & p_1 \end{bmatrix}, \qquad \mathbf{H}(y, z) = \begin{bmatrix} H_0(y, z) & H_1(y, z) \end{bmatrix}. \tag{9.7}$$

In [29, 30] we then find the following relations between the probabilities p_0, p_1 and $R_0(0)$, $R_1(0)$:

$$p_0 + p_1 = \frac{A(0)}{1 - A(0)}\left(\bar{e}_0 R_0(0) + \bar{e}_1 R_1(0)\right) \quad \text{and}$$

$$\sigma p_0 - \bar{\sigma} p_1 = \frac{\phi A(0)}{1 - \phi A(0)}\left(\sigma \bar{e}_0 R_0(0) - \bar{\sigma} \bar{e}_1 R_1(0)\right), \tag{9.8}$$

and explicit expressions for $H_i(y, z)$:

$$(y - A(z))(y - \phi A(z)) H_i(y, z)$$

$$= \frac{A(z)}{z} \Big[y \bar{q}_{\bar{i}} \big(y^{s+1} \bar{d}_{\bar{i}}(z) - z \big) R_{\bar{i}}(z)$$

$$+ (y q_i - \phi A(z)) \big(y^{s+1} \bar{d}_i(z) - z \big) R_i(z) \Big]$$

$$+ \frac{y^{s+1}}{(1 - \phi) z} \Big[\bar{q}_{\bar{i}}(\phi A(z) - y)(1 - A(z))(p_0 + p_1)$$

$$+ (\phi A(z) - 1)(y - A(z))(\bar{q}_{\bar{i}} p_{\bar{i}} - \bar{q}_i p_i) \Big]. \tag{9.9}$$

Herein the functions $R_i(z)$ are determined as

$$R_i(z) = \frac{A(z)^s}{N(z)} \Big[\phi^s (\phi A(z) - 1) \big(A(z)^{s+1} \bar{d}_{\bar{i}}(z) - z \big) (\bar{q}_{\bar{i}} p_{\bar{i}} - \bar{q}_i p_i)$$

$$- \bar{q}_{\bar{i}} (A(z) - 1) \big((\phi A(z))^{s+1} \bar{d}_{\bar{i}}(z) - z \big) (p_0 + p_1) \Big], \tag{9.10}$$

with the denominator $N(z)$ being a known function of the system parameters:

$$N(z) \triangleq \sum_{i=0}^{1} \bar{q}_i \big((\phi A(z))^{s+1} \bar{d}_i(z) - z \big) \big(A(z)^{s+1} \bar{d}_{\bar{i}}(z) - z \big). \tag{9.11}$$

The only remaining unknowns are the probabilities p_0 and p_1. To determine those, we need two additional relations. A first relation can be found from the normalisation condition $1 = p_0 + p_1 + H_0(1, 1) + H_1(1, 1)$, which after taking the limit $y \to 1$ and $z \to 1$ in (9.9), turns out to be equivalent to $A'(1) = \bar{e}_0 R_0(1) + \bar{e}_1 R_1(1)$. By taking the limit $z \to 1$ in (9.10), we explicitly find

$$(1 - \phi^{s+1}) \Big[(s + 1) A'(1) + (p_0 + p_1 - 1)(\bar{\sigma} \bar{e}_0 + \sigma \bar{e}_1) \Big]$$

$$= \phi^s (s + 1)(e_0 - e_1)(\bar{q}_0 p_0 + \bar{q}_1 p_1). \tag{9.12}$$

A second relation for p_0 and p_1 can be obtained from (9.10) by exploiting the fact that $R_i(z)$ is a (partial) pgf and must be analytic and therefore bounded in the unit disc $|z| < 1$. In an appendix of [30] we prove that the denominator $N(z)$ has exactly one zero z^* inside the unit disc, i.e. $N(z^*) = 0$. This zero can be calculated numerically and a relation between p_0 and p_1 is obtained by substituting z^* into the numerator of (9.10) and let it equal zero. Now, from

(9.9) and (9.10), one finds the (unconditional) joint pgf of the system state in equilibrium as

$$P(x, y, z) \triangleq \mathrm{E}[x^r y^m z^u] = p_0 + x p_1 + y z H_0(y, z) + x y z H_1(y, z). \quad (9.13)$$

Obviously, if we know the joint distribution of the system state (r, m, u), we can also obtain the marginal distribution of the queue content u in equilibrium. Let $U(z)$ be the pgf of the queue content at the beginning of an arbitrary slot, then

$$U(z) = P(1, 1, z) = \mathrm{E}[z^u] = \sum_{i=0}^{1} \mathrm{E}[z^u \{r = i\}]$$

$$= \sum_{i=0}^{1} \mathrm{E}[z^u \{r = i, m = 0\}] + \mathrm{E}[z^u \{r = i, m > 0\}]$$

$$= \sum_{i=0}^{1} p_i + z H_i(1, z) = \frac{(1 - z) A(z)}{1 - A(z)} \left(\bar{e}_0 R_0(z) + \bar{e}_1 R_1(z) \right), \quad (9.14)$$

for which we took the limit $y \to 1$ in (9.9). After using the expressions (9.10) for $R_i(z)$ and the normalisation condition (9.12), we finally find

$$
\begin{aligned}
U(z) = {} & \frac{(1 - z) A(z)^{s+1}}{(1 - A(z)) N(z)} \Big[z(1 - \phi^{s+1})(1 - \phi A(z))(1 - A(z)^{s+1}) \\
& \times \left(A'(1) + \frac{\bar{\sigma} \bar{e}_0 + \sigma \bar{e}_1}{s + 1} (p_0 + p_1 - 1) \right) + (p_0 + p_1) \bar{\phi}(1 - A(z)) \\
& \times \left(\bar{e}_0 \bar{e}_1 (1 - z) \phi^{s+1} A(z)^{s+1} - z(\bar{\sigma} \bar{e}_0 + \sigma \bar{e}_1)(1 - \phi^{s+1} A(z)^{s+1}) \right) \Big].
\end{aligned}
$$
$$(9.15)$$

As was explained in [29, 30], various interesting measures concerning the behaviour of the queue content can be derived from this pgf, such as the mean $\mathrm{E}[u]$, variance $\mathrm{Var}[u]$ and tail distribution $\mathrm{Prob}[u = n]$ (n large).

9.4 Calculation of the Throughput

We now focus on the calculation of the throughput of the SW-ARQ system described above. In what follows we explain some concepts necessary in the

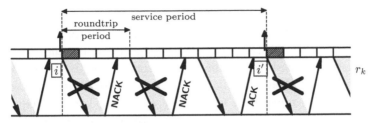

Figure 9.4 The service time of a packet of $n = 3$ roundtrip periods starting in channel state i and ending in state i'.

next section and provide a new and more intuitive proof for the throughput expression already stated in [26]. Let us call the *service period* of a packet the period it stays in the server of the queue, i.e. from the slot in which it is transmitted for the first time, up to and including the slot in which it departs from the queue. We define the *throughput* η of the system as the maximum number of packets per slot that can be correctly delivered to the receiver. Hence, η is a measure for the maximum output rate at which the system can transmit incoming packets and should therefore be compared to $A'(1)$, the mean arrival rate. Indeed, in order to have a stable queue that reaches equilibrium, we must have $A'(1) < \eta$. Obviously, the maximum output rate is only achieved if the system operates under overload conditions, i.e. if we assume there are *always* packets waiting in the queue for transmission. Under these conditions, we can also define the throughput η as the inverse of the mean service time of an arbitrary packet. Therefore, we proceed by deriving the service time distribution of the packets.

   Unfortunately, as a consequence of the correlated nature of the channel, the service times are not independent. Specifically, a packet's service time distribution will generally be different when its initial transmission happens during a 0-slot than during a 1-slot. Let us say that the service time of a packet *starts* in channel state i if the channel is in state i in the slot *before* the initial transmission of the packet, as indicated in Figure 9.4. Conversely, the service *ends* in channel state i' if the channel state is i' in the slot where the packet leaves the queue. Now, let us define $\gamma_{ii'}(n)$ $(n \geq 1)$ as the conditional probability that the service of a packet requires n roundtrip periods (or n transmission attempts) and that the channel state is i' at the end of the service *given* that the service time starts in state i $(i, i' = 0, 1)$, i.e.

$$\gamma_{ii'}(n) = \text{Prob}[\text{ packet needs } n \text{ transmissions}, r_{k+n(s+1)} = i' \mid r_{k-1} = i\,],$$
$$(9.16)$$

if the packet is first transmitted in slot k.

Since we assume that the service periods are not interrupted by idle periods, every service period starts in the same channel state that the previous one ended with, which clearly is also the channel state during the departure slot of the previous packet. The *embedded* channel state process *at departure slots* only is of particular importance to us, since we need to know the equilibrium probabilities of being in either channel state when a service period starts (or ends). Let r_k^* be the channel state during the k-th departure slot and $\pi_{i,k}$ the probability that $r_k^* = i$. Then we have for the row vector π_k with elements $\pi_{0,k}$ and $\pi_{1,k}$ respectively,

$$\pi_{k+1} = \pi_k \sum_{n=1}^{\infty} \gamma(n) \qquad \text{with } \gamma(n) = \begin{bmatrix} \gamma_{00}(n) & \gamma_{01}(n) \\ \gamma_{10}(n) & \gamma_{11}(n) \end{bmatrix}, \qquad (9.17)$$

which is to be compared with (9.1) for the channel state probabilities ω_k at consecutive slots. The row vector of equilibrium probabilities of $\{r_k^*\}$ is $\pi = \lim_{k\to\infty} \pi_k$ which is different from the probabilities ω in (9.2), as will be shown.

The probabilities $\gamma(n)$ can be found as follows. According to (9.1), the $(s+1)$-*step transition probabilities* of the channel state process $\{r_k\}$ are given by the matrix $\mathbf{q}^{s+1}$, such that $[\mathbf{q}^{s+1}]_{ii'}$ is the probability that the channel state is i' at the end of a roundtrip period given that it is i at the end of the previous roundtrip. This matrix has eigenvalues 1 (since it is stochastic) and ϕ. The spectral decomposition representation of $\mathbf{q}^h$ is given by

$$\mathbf{q}^h = \begin{bmatrix} \bar{\sigma} & \sigma \\ \bar{\sigma} & \sigma \end{bmatrix} + \phi^h \begin{bmatrix} \sigma & -\sigma \\ -\bar{\sigma} & \bar{\sigma} \end{bmatrix} \qquad (h \geqslant 0). \qquad (9.18)$$

Now, the matrix $\gamma(n)$ in (9.17) is found as

$$\gamma(n) = (\mathbf{q}^{s+1}\mathbf{e})^{n-1} \mathbf{q}^{s+1}\bar{\mathbf{e}}, \qquad n \geqslant 1, \qquad (9.19)$$

where the channel error probabilities are arranged in the matrices $\mathbf{e} \triangleq \begin{bmatrix} e_0 & 0 \\ 0 & e_1 \end{bmatrix}$ and $\bar{\mathbf{e}} \triangleq \begin{bmatrix} \bar{e}_0 & 0 \\ 0 & \bar{e}_1 \end{bmatrix}$. The geometric-like expression (9.19) can easily be interpreted as follows. If the service of a packet requires n roundtrip periods, then there are first $n - 1$ roundtrip periods, each of length $s + 1$, wherein an error occurred followed by one roundtrip without channel error.

The z-transform of the matrix $\gamma(n)$ gives us the probability generating matrix (or *pgm*) $\mathbf{g}(z)$ of the number of roundtrip periods required for the successful transmission of a packet, accounting for the channel state at the start and end of the service. From (9.19) we find

$$\mathbf{g}(z) = \sum_{n=1}^{\infty} \boldsymbol{\gamma}(n) z^n = (\mathbf{I} - z\,\mathbf{q}^{s+1}\mathbf{e})^{-1}\mathbf{q}^{s+1}\bar{\mathbf{e}}z$$

$$= \frac{z}{\nu(z)}\left(\mathbf{q}^{s+1}\bar{\mathbf{e}} - z\,\phi^{s+1}\begin{bmatrix} \bar{e}_0 e_1 & 0 \\ 0 & e_0 \bar{e}_1 \end{bmatrix}\right), \tag{9.20}$$

where $\mathbf{I}$ is the 2×2 identity matrix and where

$$\nu(z) = \det(\mathbf{I} - z\,\mathbf{q}^{s+1}\mathbf{e}) = 1 - z\big(e_0(\bar{\sigma} + \phi^{s+1}\sigma) + e_1(\sigma + \phi^{s+1}\bar{\sigma})\big) + z^2 e_0 e_1 \phi^{s+1}. \tag{9.21}$$

Note that the inverse matrix in (9.20) always exists for $|z| \leqslant 1$, because the elements of $\mathbf{g}(z)$ are (partial) pgfs and must be analytic in that region. For future purposes, we also introduce the pgm $\mathbf{S}(z) \triangleq \mathbf{g}(z^{s+1})$ which gives the distributions of the number of *slots* in a service period rather than the number of roundtrips. The nice thing about the matrix representation $\mathbf{S}(z)$ is that it allows us to handle the distribution of contiguous service times *as if they were independent*. Indeed, the conditional distribution of the length of n contiguous services (i.e. without idle periods in between) ending in state i', given that the first service starts in state i is simply given by $[\mathbf{S}(z)^n]_{ii'}$.

From (9.16) and (9.20), the transition probability matrix of the embedded process $\{r_k^*\}$ is seen to be given by $\mathbf{g}(1)$ and is easily obtained from (9.20) as

$$\mathbf{g}(1) = \begin{bmatrix} q_0^* & \bar{q}_0^* \\ \bar{q}_1^* & q_1^* \end{bmatrix} \quad \text{with } q_i^* = \frac{\bar{e}_i}{\nu(1)}\big(\bar{\sigma} + \phi^{s+1}(\sigma - e_{\bar{i}})\big).$$

Hence, the equilibrium probabilities $\boldsymbol{\pi}$ of the channel state r^* at an arbitrary departure slot must satisfy $\boldsymbol{\pi} = \boldsymbol{\pi}\mathbf{g}(1)$, from which

$$\boldsymbol{\pi} = \begin{bmatrix} \pi_0 & \pi_1 \end{bmatrix} = \left[\frac{\bar{e}_0 \bar{\sigma}}{\bar{e}_0 \bar{\sigma} + \bar{e}_1 \sigma} \quad \frac{\bar{e}_1 \sigma}{\bar{e}_0 \bar{\sigma} + \bar{e}_1 \sigma}\right]. \tag{9.22}$$

Remember that for the queue working under overload conditions we have defined the throughput η as the inverse of the mean service time. The distributions of the service times conditioned on the channel state in which the service starts are given by $\mathbf{S}(z)$ whereas $\boldsymbol{\pi}$ in (9.22) are the equilibrium probabilities of being in either state 0 or 1 at the start of a service time. Therefore, using the moment-generating property of pgfs, we find the throughput as

$$\eta^{-1} = \boldsymbol{\pi}\,\mathbf{S}'(1)\,\mathbf{1} = (s+1)\boldsymbol{\pi}\,\mathbf{g}'(1)\,\mathbf{1} \quad \text{with } \mathbf{1} \triangleq \begin{bmatrix} 1 \\ 1 \end{bmatrix}, \tag{9.23}$$

where $\mathbf{1}$ is the 2×1 column vector with 1 on both entries and where $\mathbf{S}'(1)$ indicates the matrix $\mathbf{S}(z)$ with each element differentiated to z and evaluated for $z = 1$ (and likewise for $\mathbf{g}'(1)$). After properly evaluating (9.23), we finally find the following expression for the throughput of the Stop-and-Wait ARQ protocol over the correlated error channel:

$$\eta = \bar{\sigma}\,\frac{\bar{e}_0}{s+1} + \sigma\,\frac{\bar{e}_1}{s+1}\,. \tag{9.24}$$

9.5 Analysis of the Packet Delay

In this section, we derive an expression for the pgf $D(z)$ of the total delay d experienced by an arbitrary packet traversing the transmitter queue. Of all the packets arriving to the system, consider an arbitrary packet and tag it as packet $\mathcal{P}$. Also, let us mark the arrival slot of $\mathcal{P}$ as slot I. We define the delay d as the number of slots between the end of the slot in which $\mathcal{P}$ arrives (slot I) and the end of the slot in which $\mathcal{P}$ departs from the transmitter queue. The delay of $\mathcal{P}$ foremost depends on the system state (r_I, m_I, u_I) at the beginning of slot I. Hence, we first need an expression for the system state distribution $P_I(x, y, z)$ in slot I. Now, it has been argued before (see e.g. [28]) that due to the uncorrelated (iid) nature of the packet arrival process, the system as 'seen' by an *arbitrary arriving packet* has the same distribution as the system in an *arbitrary slot*, i.e. $P_I(x, y, z) = P(x, y, z)$. Apart from the system state at the beginning of slot I, the delay of $\mathcal{P}$ also depends on the number and order of arrivals *during* that slot. Let ℓ be the number of packets arriving in the queue in slot I that will be served no later than (but including) $\mathcal{P}$, as indicated in Figure 9.5. The pgf $L(z)$ of ℓ is found as (see [28])

$$L(z) = \frac{z(1 - A(z))}{A'(1)(1 - z)}\,, \tag{9.25}$$

by considering the fact that the pgf of the number of arrivals a_I in slot I is not $A(z)$, but rather $z A'(z)/A'(1)$ and that $\mathcal{P}$ could be *any* of the a_I arrivals with equal probability. In the following, we derive the pgf $D(z)$ of the delay of $\mathcal{P}$ by conditioning on the system state at the beginning of slot I. Specifically, as in Section 9.3, we make the distinction between the cases where the system is *idle* in slot I (i.e. $m_I = 0$) or *busy* ($m_I > 0$). In both cases, we can refer to Figure 9.5 for a visual representation of the time periods that constitute the total delay of $\mathcal{P}$.

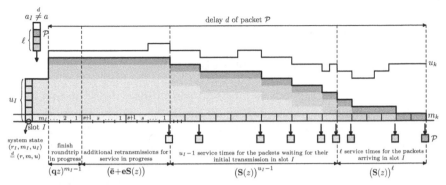

Figure 9.5 The delay d of the arbitrary packet $\mathcal{P}$ that arrives in slot I with system state (r_I, m_I, u_I).

Let us first consider the case where $\mathcal{P}$ arrives when the queue is idle. This means that the first of the ℓ packets will immediately be transmitted in the next slot. If the service period of the first packet is finished, the next of the ℓ packets is served, without interruption in between, and so forth until finally, the packet $\mathcal{P}$ is served. Therefore, we know from the previous section that the pgfs of the length of these ℓ contiguous services ending in channel state i' given that they start in state i ($i, i' = 0, 1$) are the entries of the pgm

$$\mathbf{D}_{m_I=0}(z) = \sum_{v=1}^{\infty} \mathbf{S}(z)^v \operatorname{Prob}[\ell = v] = L(\mathbf{S}(z)). \tag{9.26}$$

The probabilities that the queue is empty and in either channel state during slot I, are given by the vector $\mathbf{p}$ in (9.7). Hence,

$$\mathrm{E}[z^d \{m_I = 0\}] = \mathbf{p} \, \mathbf{D}_{m_I=0}(z) \, \mathbf{1} = \mathbf{p} \, L(\mathbf{S}(z)) \mathbf{1}. \tag{9.27}$$

In Appendix A, we derive the spectral decomposition (9.46) of the matrix $\mathbf{S}(z)$ with eigenvalue functions $\lambda_1(z)$ and $\lambda_2(z)$. Let us agree that the index j is always used to indicate one of the two eigenvalues (i.e. $j = 1, 2$) and that we may write λ_j when in fact, we mean $\lambda_j(z)$. We find

$$\mathrm{E}[z^d \{m_I = 0\}] = \sum_{j=1}^{2} L(\lambda_j) \, \mathbf{p} \, \mathbf{S}_j(z) \mathbf{1}. \tag{9.28}$$

Secondly, we consider the more complicated case when $\mathcal{P}$ arrives when the queue is busy serving another packet. Suppose the system state at the beginning of slot I is (r_I, m_I, u_I). From Figure 9.5 it is seen that the pgfs of the delay period ending in channel state i' given that it starts in state i (i.e. $r_I = i$) $(i, i' = 0, 1)$ are the entries of the pgm

$$\mathbf{D}_{m_I > 0}(m_I, u_I, z) = (\mathbf{q}z)^{m_I - 1} \cdot (\bar{\mathbf{e}} + \mathbf{e}\,\mathbf{S}(z)) \cdot \mathbf{S}(z)^{u_I - 1} \cdot L(\mathbf{S}(z)), \quad (9.29)$$

where each factor corresponds to a certain part of the delay. The first factor indicates the number of slots needed to finish the roundtrip period of the packet being served during slot I. If this roundtrip period is finished, either an ACK or NACK was returned to the transmitter. In case of an ACK, no channel error occurred (probabilities $\bar{\mathbf{e}}$) which means the service is finished and the packet departs from the queue. In case of a NACK, a channel error occurred (probabilities $\mathbf{e}$) and the packet is retransmitted such that an additional *remaining* service time must be accounted for. Note that the conditional length of this remaining service time has a distribution that is also given by $\mathbf{S}(z)$. Hence the second factor $\bar{\mathbf{e}} + \mathbf{e}\,\mathbf{S}(z)$. The third factor is due to the service times of the $u_I - 1$ packets waiting in the queue at the beginning of slot I but that were not yet transmitted then. Finally, the fourth factor accounts for the packets arriving during slot I that are transmitted *before* $\mathcal{P}$ and for $\mathcal{P}$ itself, similar to (9.26). Now, for $1 < h \leqslant s + 1$ and $n \geqslant 1$ let

$$\chi(h, n) \triangleq \left[\text{Prob}[r_I = 0, m_I = h, u_I = n] \quad \text{Prob}[r_I = 1, m_I = h, u_I = n] \right]; \tag{9.30}$$

then by using (9.29) we find for the distribution of the delay d in case the queue is busy during slot I:

$$E[z^d \{m_I > 0\}] = \sum_{h=1}^{s+1} \sum_{n=1}^{\infty} \chi(h, n)\, \mathbf{D}_{m_I > 0}(h, n, z)\mathbf{1}$$

$$= \sum_{h=1}^{s+1} \sum_{n=1}^{\infty} \chi(h, n)(\mathbf{q}z)^{h-1} \sum_{j=1}^{2} (\bar{\mathbf{e}} + \mathbf{e}\lambda_j)\lambda_j^{n-1} L(\lambda_j) \mathbf{S}_j(z)\mathbf{1}$$

$$= \sum_{j=1}^{2} \sum_{h=1}^{s+1} \sum_{n=1}^{\infty} L(\lambda_j) \left[z^{h-1}\lambda_j^{n-1} \chi(h, n) \begin{bmatrix} \bar{\sigma} & \sigma \\ \bar{\sigma} & \sigma \end{bmatrix} (\bar{\mathbf{e}} + \mathbf{e}\lambda_j) \mathbf{S}_j(z)\mathbf{1} \right.$$

$$\left. + (\phi z)^{h-1}\lambda_j^{n-1} \chi(h, n) \begin{bmatrix} \sigma & -\sigma \\ -\bar{\sigma} & \bar{\sigma} \end{bmatrix} (\bar{\mathbf{e}} + \mathbf{e}\lambda_j) \mathbf{S}_j(z)\mathbf{1} \right]. \tag{9.31}$$

Note that we have used the spectral decomposition (9.46) again applied to $\mathbf{D}_{m_I > 0}(h, n, z)$ in the second line and the representation (9.18) for $\mathbf{q}^{h-1}$ in the third. From (9.5) and (9.7) it is clear that the pgfs of the entries of $\chi(h, n)$ are given by $H_0(y, z)$ and $H_1(y, z)$ respectively, which were determined Section 9.3. Hence, we find for the delay if the packet $\mathcal{P}$ arrives in a busy slot:

$$\mathrm{E}[z^d\{m_I > 0\}] = \sum_{j=1}^{2} L(\lambda_j) \, \mathbf{C}_j(z)(\bar{\mathbf{e}} + \mathbf{e}\lambda_j)\mathbf{S}_j(z)\mathbf{1}, \qquad (9.32)$$

where we have used the row vectors $\mathbf{C}_j(z)$ defined as

$$\mathbf{C}_j(z) \triangleq \mathbf{H}(z, \lambda_j) \begin{bmatrix} \bar{\sigma} & \sigma \\ \bar{\sigma} & \sigma \end{bmatrix} + \mathbf{H}(\phi z, \lambda_j) \begin{bmatrix} \sigma & -\sigma \\ -\bar{\sigma} & \bar{\sigma} \end{bmatrix}. \qquad (9.33)$$

The entries $C_{0j}(z)$ and $C_{1j}(z)$ follow from (9.9) by applying the appropriate substitutions for z and y:

$$
\begin{aligned}
C_{ij}(z) &= \frac{A(\lambda_j)\bar{\phi}^{-1}}{\lambda_j(z - A(\lambda_j))}\big[\big((\bar{q}_{\bar{i}} + \phi^{s+1}\bar{q}_i)z^{s+1}\bar{d}_i(\lambda_j) - \bar{\phi}\lambda_j\big)R_i(\lambda_j) \\
&\quad + \bar{q}_{\bar{i}}(1 - \phi^{s+1})z^{s+1}\bar{d}_{\bar{i}}(\lambda_j)R_{\bar{i}}(\lambda_j)\big] \\
&\quad + \frac{-z^{s+1}\bar{\phi}^{-1}}{\lambda_j(z - A(\lambda_j))}\big[(1 - A(\lambda_j))\bar{q}_{\bar{i}}(p_0 + p_1) \\
&\quad + \phi^s(1 - \phi A(\lambda_j))(\bar{q}_i p_i - \bar{q}_{\bar{i}} p_{\bar{i}})\big],
\end{aligned}
\qquad (9.34)
$$

in which everything is already known from Sections 9.2 and 9.3. After substitution of expression (9.10) for the functions $R_i(z)$ into (9.34), we find the more explicit expression

$$
\begin{aligned}
C_{ij}(z) &= \frac{A^{s+1}(\lambda_j) - z^{s+1}}{\bar{\phi}\lambda_j(z - A(\lambda_j))N(\lambda_j)} \\
&\quad \times \Big[\big[N(\lambda_j) - A^{s+1}(\lambda_j)\bar{d}_i(\lambda_j)\big(\bar{d}_i(\lambda_j)\phi^{s+1}\bar{\phi}A^{s+1}(\lambda_j) - \lambda_j(\bar{q}_{\bar{i}} + \phi^{s+1}\bar{q}_i)\big)\big] \\
&\quad \times \big[\bar{q}_{\bar{i}}(1 - A(\lambda_j))(p_0 + p_1) + \phi^s(1 - \phi A(\lambda_j))(\bar{q}_i p_i - \bar{q}_{\bar{i}} p_{\bar{i}})\big] \\
&\quad + \bar{q}_{\bar{i}} A^{s+1}(\lambda_j)\bar{d}_{\bar{i}}(\lambda_j)\lambda_j(1 - \phi^{s+1}) \\
&\quad \times \big[\bar{q}_i(1 - A(\lambda_j))(p_0 + p_1) - \phi^s(1 - \phi A(\lambda_j))(\bar{q}_i p_i - \bar{q}_{\bar{i}} p_{\bar{i}})\big]\Big].
\end{aligned}
\qquad (9.35)
$$

Finally, we can bring together (9.28) and (9.32) to obtain the *unconditional* pgf $D(z)$ of the packet delay d:

$$D(z) = \mathrm{E}[z^d] = \mathrm{E}[z^d\{m_I = 0\}] + \mathrm{E}[z^d\{m_I > 0\}]$$

$$= \sum_{j=1}^{2} L(\lambda_j)[\mathbf{p} + \mathbf{C}_j(z)(\bar{\mathbf{e}} + \mathbf{e}\lambda_j)]\mathbf{S}_j(z)\mathbf{1}. \tag{9.36}$$

This can be simplified further by observing that $\bar{e}_i + e_i\lambda_j$ is in fact $\bar{d}_i(\lambda_j)$ and by using the expression (9.47) for the vector $\mathbf{S}_j(z)\mathbf{1}$. We find

$$D(z) = \sum_{j=1}^{2}\sum_{i=0}^{1}\frac{1}{2}L(\lambda_j)[p_i + \bar{d}_i(\lambda_j)C_{ij}(z)]$$

$$\times \left[1 \pm \frac{(-1)^i\bar{\phi}\,\mu(z^{s+1}) + 2\bar{e}_i\bar{q}_i(1 - \phi^{s+1})}{\bar{\phi}\sqrt{\psi(z^{s+1})}}\right], \tag{9.37}$$

with $\pm$ being $+$ if $j = 1$ and $-$ if $j = 2$. To summarise, in (9.37) one has to substitute (9.35) for $C_{ij}(z)$, (9.43) for the eigenvalue functions $\lambda_j(z)$,(9.3) for $\bar{d}_i(z)$, (9.42) for $\mu(z)$, (9.41) for $\psi(z)$ and (9.25) for $L(z)$. It is possible to derive the moments of the packet delay from (9.37) by using the moment-generating property of pgfs. For example, we have verified numerically that the mean packet delay $\mathrm{E}[d]$ found as $D'(1)$ is exactly equal to $U'(1)/A'(1)$, with $U'(1)$ the mean queue content found in [29], as required by Little's theorem [33]. We refer to Appendix B for the tail distribution of the packet delay where an elegant and accurate approximation is presented.

9.6 Numerical Examples

In order to illustrate how the equilibrium distribution of the packet delay d is influenced by the parameters of the model we now consider some practical examples. Figures 9.6 to 9.10 are plots of the mass function $d(n)$ of the packet delay. These probabilities were obtained by numerical inversion of the pgf $D(z)$ given by (9.37) using the algorithm presented in [37]. The inversion formula for z-transforms is

$$d(n) = \frac{1}{2\pi J}\oint_{C_r} D(z)z^{-n-1}dz, \tag{9.38}$$

where J is the complex imaginary unit $\sqrt{-1}$ and C_r is a circular contour around the origin with radius $0 < r < 1$. In [37], this integral is approximated

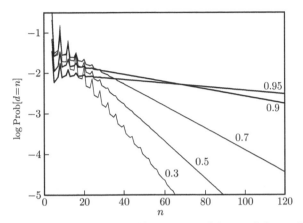

Figure 9.6 Logarithmic plot of Prob[$d = n$] in case $e_0 = 0.1$, $e_1 = 0.5$, $\sigma = 0.5$ and $K = 10$ for the channel model, feedback delay $s = 3$ and arrivals with geometric distribution and for various values of the load $\rho = 0.3, 0.5, 0.7, 0.9, 0.95$.

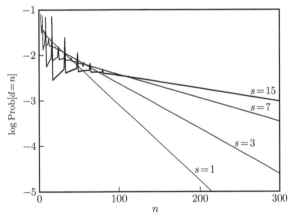

Figure 9.7 Logarithmic plot of Prob[$d = n$] in case $e_0 = 0.1$, $e_1 = 0.5$, $\sigma = 0.5$ and $K = 10$ for the channel, Poisson arrivals with load $\rho = 0.9$, for various values of the roundtrip delay $s + 1 = 2, 4, 8, 16$.

by sampling the integrand on $2n$ points of the contour C_r. Using the discrete Poisson summation formula, the error bound for this approximation is shown to be r^{2n} for large n and small r, such that any desired accuracy is guaranteed by choosing r sufficiently small.

In Figure 9.6 we show a logarithmic plot of the mass function $d(n)$ for increasing values of the load $\rho = A'(1)/\eta$. The fraction of slots in either channel state is the same ($\sigma = 0.5$), the correlation factor $K = 10$ and the

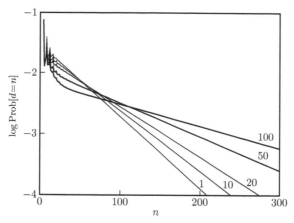

Figure 9.8 Logarithmic plot of Prob[$d = n$] in case $e_0 = 0.1$, $e_1 = 0.5$, $\sigma = 0.5$ with feedback delay $s = 3$ and Poisson arrivals with load $\rho = 0.9$, for various values of $K = 1, 10, 20, 50, 100$.

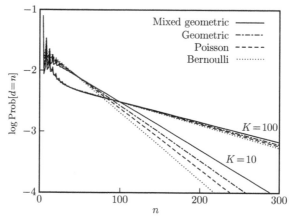

Figure 9.9 Logarithmic plot of Prob[$d = n$] in case $e_0 = 0.1$, $e_1 = 0.5$, $\sigma = 0.5$ and feedback delay $s = 3$. For $K = 10$ and $K = 100$ the distribution is plotted for four different arrival distributions $A(z)$: Bernoulli, Poisson, geometric and mixed geometric, with load $\rho = 0.9$.

error probabilities in the GOOD and BAD state are 0.1 and 0.5 respectively. The roundtrip delay is $s + 1 = 4$ slots and the numbers of arrivals per slot have a geometric distribution. As expected, we observe that the probability mass shifts towards higher values as the load increases. If more packets enter the queue in the same time period, they will have to wait longer before they can be transmitted. Secondly, as with all of the following plots, we see that

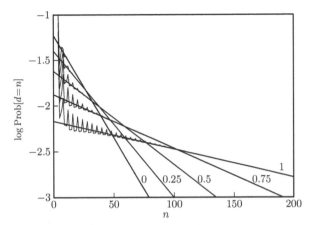

Figure 9.10 Logarithmic plot of Prob$[d = n]$ in case $e_0 = 0$, $\sigma = 0.5$, $K = 10$, $s = 3$ and Poisson arrivals with load $\rho = 0.9$. The error probability in the BAD state takes values $e_1 = 0, 0.25, 0.5, 0.75, 1$. The bold line indicates the approximated tail distribution.

$d(n) = 0$ for $n < s + 1$ since the minimal delay of a packet is one roundtrip period of $s + 1$ slots. A packet experiences this minimal delay when it is the first packet arriving during an idle slot and its first transmission is successful. Notice also the peaks in the mass function for n equal to multiples of the roundtrip delay $s + 1$. The peaks are more pronounced and last into higher values of n when the load is lower, which can be explained as follows. For low load, there is a high probability that packets arrive during a slot when the system is idle. Hence, the first of the packets will be transmitted immediately in the next slot and the delay of all those packets will be a multiple of $s + 1$. Note that the probability mass $d(n)$ for n *not* a multiple of $s+1$ is entirely due to arrivals in slots when $m_k = 2, \ldots, s + 1$, which is more likely to happen when the load increases. The same observations can be made from Figure 9.7, where various values of the roundtrip delay are considered. The arrivals have a Poisson distribution with load $\rho = 0.9$ and the other parameters are as in Figure 9.6.

To illustrate the impact of the *correlated nature* of the transmission errors occurring in the channel, we have plotted in Figure 9.8 the delay distribution for increasing values of the correlation factor $K = 1, 10, 20, 50, 100$. The arrivals are Poisson with load 0.9 for all curves and the other parameters are the same as in Figure 9.7. Observe that although the number of slots in the BAD state is the *same* for all curves, the delay increases drastically with the factor K. The simple fact that *both* the BAD and the GOOD periods *last longer* (see Section 9.2) for higher K, results in a higher delay for the arbitrary packet. This result strongly shows the importance of accounting for possible correlation in the transmission channel when estimating the packet delay (and the queue content, see [29,30]). Indeed, the delay may be severely underestimated when assuming only *static* errors ($K = 1$). In Figure 9.9 we show the influence of the arrival distribution $A(z)$. Consider the following pgfs for the number of arrivals per slot:

$$A_1(z) = \alpha z + 1 - \alpha, \qquad A_2(z) = \exp(\alpha(z - 1)),$$

$$A_3(z) = \frac{1}{1 + \alpha - \alpha z}, \qquad A_4(z) = \frac{0.5}{1 + \alpha_1 - \alpha_1 z} + \frac{0.5}{1 + \alpha_2 - \alpha_2 z},$$

i.e. a Bernoulli, Poisson, geometric and mixed geometric distribution respectively. For the latter, we choose $\alpha_1 = 0.1\alpha$ and $\alpha_2 = 1.9\alpha$, such that all distributions have mean value $\alpha = 0.9\eta$. The other system parameters are the same as in Figure 9.6. The plot shows the logarithmic mass function of the packet delay for these four types of arrivals, both in case $K = 10$ and $K = 100$. The Bernoulli, Poisson, geometric and mixed geometric distributions have variances 0.132, 0.157, 0.182 and 0.222 respectively and we observe that a higher variance of the arrival distribution results in a higher packet delay.

Finally, in Figure 9.10, we assume that there are no errors in the GOOD state ($e_0 = 0$) while we consider increasing values for the error probability in the BAD state, $e_1 = 0, 0.25, 0.5, 0.75, 1$. As expected, the delay increases if the error probability is higher. Additionally, we demonstrate the effectiveness of the dominant pole approximation for the tail distribution of the delay as discussed in Appendix B. The bold lines correspond to the geometric decay of the tail distribution calculated from (9.51), which gives very good results for high n.

9.7 Conclusion

We have analysed the transmitter queue of the SW-ARQ protocol in case of a two-state Markovian channel. Closed-form expressions are derived for the pgf of the equilibrium distribution of the system state and the queue content. By using the eigenvalues and the spectral decomposition of the matrix with the conditional lengths of the service time, we obtained the pgf of the delay experienced by an arbitrary packet. We also gave an accurate approximation for the tail distribution of the delay and found that the asymptotic contribution of the second eigenvalue function is zero. By means of some examples we have discussed the influence of the model parameters on the delay performance. The most important observation is that the delay increases drastically as the correlation in the channel is higher (i.e. if K increases), although the overall fraction of BAD slots remains the same. This result emphasises the importance of taking into account the correlation of the errors when dimensioning the buffer space or calculating the packet loss.

Appendix

A Spectral Decomposition of $\mathbf{S}\,(z)$

For the analysis of the delay in Section 9.5, we need a suitable representation for a function f of the pgm $\mathbf{S}(z)$. The following spectral decomposition theorem allows us to determine $f(\mathbf{S}(z))$ by evaluating f for *scalars* rather than matrices and can be found in any textbook on matrix algebra (such as e.g. [34, 35]).

A square $N \times N$ matrix $\mathbf{A}$ is diagonalisable if there exists a similarity transform $\mathbf{P}$ such that $\mathbf{P}^{-1}\mathbf{A}\mathbf{P} = \mathbf{D}$, with $\mathbf{D}$ a diagonal matrix with entries taken from the spectrum $\sigma(\mathbf{A})$ of $\mathbf{A}$. The spectrum $\sigma(\mathbf{A}) = \{\lambda_1, \ldots, \lambda_k\}$ $(k \leqslant N)$ is the set of eigenvalues of $\mathbf{A}$. Let $f(z)$ be a function that is defined on $\sigma(\mathbf{A})$, then there exist matrices $\mathbf{G}_j$ $(1 \leqslant j \leqslant k)$ such that

$$f(\mathbf{A}) = f(\lambda_1)\mathbf{G}_1 + f(\lambda_2)\mathbf{G}_2 + \cdots + f(\lambda_k)\mathbf{G}_k\,.$$

The matrices $\mathbf{G}_j$ are called the *constituents* or *spectral projectors* of $\mathbf{A}$ belonging to the eigenvalue λ_j and have the following properties:

- $\mathbf{G}_j$ is idempotent, i.e. $\mathbf{G}_j^2 = \mathbf{G}_j$.
- $\mathbf{G}_1 + \mathbf{G}_2 + \cdots + \mathbf{G}_k = \mathbf{I}$, with $\mathbf{I}$ the $N \times N$ identity matrix.
- $\mathbf{G}_j\mathbf{G}_{j'} = \mathbf{0}$ whenever $j \neq j'$ $(1 \leqslant j, j' \leqslant k)$.

The projectors $\mathbf{G}_i$ can be obtained by using the Lagrange interpolation formula on $\sigma(\mathbf{A})$:

$$\mathbf{G}_j = \prod_{j'=1, j'\neq j}^{k} (\mathbf{A} - \lambda_{j'}\mathbf{I}) \Bigg/ \prod_{j'=1, j'\neq j}^{k} (\lambda_j - \lambda_{j'})\,. \tag{9.39}$$

The extension of this theorem to non-diagonalisable matrices $\mathbf{A}$ also exists but is slightly more complicated.

We can use the above to obtain the spectral decomposition of the pgm $\mathbf{g}(z)$ given by (9.20). First, let us denote the two eigenvalues of $\mathbf{g}(z)$ as $\lambda_1^*(z)$ and $\lambda_2^*(z)$ respectively, which are the solutions of the characteristic equation $\det(\lambda \mathbf{I} - \mathbf{g}(z)) = 0$ equivalent to

$$\lambda^2 \nu(z) + \lambda z \left[z \phi^{s+1}(e_0 \bar{e}_1 + \bar{e}_0 e_1) - \bar{e}_0(\bar{\sigma} + \phi^{s+1}\sigma) - \bar{e}_1(\sigma + \phi^{s+1}\bar{\sigma}) \right]$$
$$+ \frac{z^2 \bar{e}_0 \bar{e}_1 \phi^{s+1}}{\nu(z)} = 0. \tag{9.40}$$

Let $z^2 \psi(z)$ be the discriminant of this ordinary square root equation in λ, i.e.

$$\psi(z) \triangleq \mu^2(z) + 4\bar{e}_0 \bar{e}_1 \sigma \bar{\sigma} (1 - \phi^{s+1})^2, \tag{9.41}$$

with

$$\mu(z) \triangleq \bar{e}_0(\bar{\sigma} + \phi^{s+1}\sigma) - \bar{e}_1(\sigma + \phi^{s+1}\bar{\sigma}) - z(e_1 - e_0)\phi^{s+1}. \tag{9.42}$$

As in Section 9.5, we agree that the index j is always either 1 or 2. Solving the characteristic equation then yields the eigenvalue functions

$$\lambda_j^*(z) = \frac{z}{2\nu(z)} \left[\bar{e}_0(\bar{\sigma} + \phi^{s+1}\sigma) + \bar{e}_1(\sigma + \phi^{s+1}\bar{\sigma}) \right.$$
$$\left. - z \phi^{s+1}(\bar{e}_0 e_1 + e_0 \bar{e}_1) \pm \sqrt{\psi(z)} \right], \tag{9.43}$$

where $\pm$ means $+$ for $j = 1$ and $-$ for $j = 2$. It can be verified that if $|z| < 1$, the eigenvalue functions $\lambda_j^*(z)$ are situated inside the unit disc too. Therefore, in case $f(z)$ is a pgf and thus analytic for $|z| < 1$, the spectral decomposition is always well defined. Also, we have that $|\lambda_1^*(z)| \geqslant |\lambda_2^*(z)|$, i.e. $\lambda_1^*(z)$ is the 'larger' of the two eigenvalue functions and yields the PF-eigenvalue $\lambda_1^*(1) = 1$ of the stochastic matrix $\mathbf{g}(1)$ with eigenvector $\boldsymbol{\pi}$ as given by (9.22). However, although $\lambda_1^*(1) = 1$, the function $\lambda_1^*(z)$ is in general *not* a pgf, since its singularities with smallest modulus are not real, as would be required by Pringsheim's theorem (a.k.a. Vivanti's theorem, see [36]). These singularities are the branch points given by the two (complex conjugate) roots of $\psi(z)$. The spectral projectors $\mathbf{G}_j$ of $\mathbf{g}(z)$ are found from (9.39) and (9.43) as

$$\mathbf{G}_j(z) = \pm \frac{1}{2\sqrt{\psi(z)}} \cdot \begin{bmatrix} \mu(z) \pm \sqrt{\psi(z)} & 2\bar{e}_1\sigma(1 - \phi^{s+1}) \\ 2\bar{e}_0\bar{\sigma}(1 - \phi^{s+1}) & -\mu(z) \pm \sqrt{\psi(z)} \end{bmatrix}, \tag{9.44}$$

with the same convention for $\pm$ as before. Since $\mathbf{S}(z) = \mathbf{g}(z^{s+1})$, we now also have the spectral decomposition of $\mathbf{S}(z)$, with eigenvalues $\lambda_j(z)$ and spectral projectors $S_j(z)$ given as

$$\lambda_j(z) = \lambda_j^*(z^{s+1}), \qquad S_j(z) = G_j(z^{s+1}). \tag{9.45}$$

Hence, we can write for the pgm $S(z)$:

$$f(S(z)) = f(\lambda_1(z)) S_1(z) + f(\lambda_2(z)) S_2(z). \tag{9.46}$$

From (9.44) and (9.45) one also finds for the projectors $S_j(z)$ postmultiplied by 1:

$$S_j(z)1 = \frac{1}{2}1 \pm \frac{1}{2\sqrt{\psi(z^{s+1})}}\left[\begin{matrix} \mu(z^{s+1}) + 2\bar{e}_1\sigma(1 - \phi^{s+1}) \\ -\mu(z^{s+1}) + 2\bar{e}_0\bar{\sigma}(1 - \phi^{s+1}) \end{matrix} \right]. \tag{9.47}$$

We prove the following properties regarding the eigenvalue functions:

Proposition 1. *The function $\lambda_1(z)$ is related to the throughput as*

$$\lambda_1'(1) = \eta^{-1}. \tag{9.48}$$

This follows directly from (9.43), (9.45) and (9.24). $\qquad\qquad\square$

Proposition 2. *For $j = 1, 2$, and the function $N(z)$ given by (9.11):*

$$z - A(\lambda_j(z)) = 0 \;\Rightarrow\; N(\lambda_j(z)) = 0. \tag{9.49}$$

The eigenvalues $\lambda_j(z)$ are functions for which the characteristic polynomial $\det(\lambda_j(z)I - S(z))$, given by (9.40) with all occurrences of z replaced by z^{s+1}, is zero. Now, substitution of $\lambda_j(z)$ into (9.11) yields an expression for $N(\lambda_j(z))$ which after some careful rearranging of the terms and substitution of $A(\lambda_j(z)) = z$ reduces exactly to the characteristic polynomial of $S(z)$ in $\lambda_j(z)$. Hence, $N(\lambda_j(z)) = 0$. $\qquad\qquad\square$

B Asymptotic Analysis of the Packet Delay Distribution

We now use the pgf $D(z)$ obtained in (9.37) to assess the asymptotic behaviour of the delay distribution, i.e. the mass function $d(n)$ when large n. We use a well-known approximation technique (see e.g. [32]) to find $d(n)$ based on the *dominant* singularity of the pgf $D(z)$. Using analytic continuation and the Cauchy theorem, it is possible to extend the contour C_r in (9.38) to contain the whole complex plane $(r \to \infty)$ *except* for the singularities of the integrand. Instead of one contour around the origin, there is now a big contour around the z-plane and a number of small contours around the singularities of $D(z)z^{-n-1}$ other than the origin. For large n, the contribution to $d(n)$ of the contour around the singularities with smallest modulus will dominate the contributions of the other contours. Note that since $D(z)$ is analytic within the unit disc, all these singularities have modulus larger than 1. In the case at hand, the dominant singularity is a single pole z_d, which must necessarily be real and positive in order to ensure a nonnegative mass function $d(n)$.

After close inspection of the factors in (9.37), we found that the dominant pole z_d is always the smallest zero larger than one of the factor $z - A(\lambda_1(z))$ in the denominator of the functions $C_{i1}(z)$ given by (9.34). It can be shown that the other poles all have a larger modulus. From (9.49) we see that z_d is also a zero of the denominator $N(\lambda_1(z))$ of the functions $R_i(\lambda_1(z))$ (see (9.10)) in $C_{i1}(z)$, which would lead us to believe that z_d is a pole of $C_{i1}(z)$ with multiplicity 2. However, after elimination of the functions $R_i(z)$ it becomes clear in (9.35) that the numerator is *also* zero for $z = z_d$ due to the factor $A^{s+1}(\lambda_1(z)) - z^{s+1}$. The zero in numerator and denominator annihilates for $z \to z_d$ (using de l'Hôpital's rule) as

$$\lim_{z \to z_d} \frac{A^{s+1}(\lambda_1(z)) - z^{s+1}}{z - A(\lambda_j(z))} = -(s+1)z_d^s, \tag{9.50}$$

and z_d proves to be a pole with single multiplicity due to the factor $N(\lambda_1(z))$ in the denominator of $C_{i1}(z)$. Now, from the partial fractions expansion of $D(z)$ around z_d, it follows that the (dominant) contribution to $d(n)$ of the contour around z_d can be expressed by the following geometric form:

$$\text{Prob}[d = n] \cong -\theta \, z_d^{-n-1}, \tag{9.51}$$

where θ is the residue of $D(z)$ in the point $z = z_d$, i.e.

$$\theta = \text{Res}_{z_d} D(z) = \lim_{z \to z_d} (z - z_d) D(z). \tag{9.52}$$

For convenience, let us define

$$y_d \triangleq \lambda_1(z_d) \quad \text{and} \quad y_d' \triangleq \lambda_1'(z_d), \tag{9.53}$$

then we have from (9.37):

$$\theta = \sum_{i=0}^{1} \frac{1}{2} L(y_d) \bar{d}_i(y_d) \text{Res}_{z_d} C_{i1}(z) \left[1 + \frac{(-1)^i \bar{\phi} \, \mu(z_d^{s+1}) + 2\bar{e}_i \bar{q}_i (1 - \phi^{s+1})}{\bar{\phi} \sqrt{\psi(z_d^{s+1})}} \right], \tag{9.54}$$

where the residues of $C_{i1}(z)$ in z_d follow from (9.35) as

$$
\begin{aligned}
\mathrm{Res}_{z_d} C_{i1}(z) \;=\; & \frac{(s+1)z_d^{2s+1}}{\bar{\phi} N'(y_d) y_d' y_d} \Big[\bar{q}_{\bar{i}} \bar{d}_{\bar{i}}(y_d) y_d (1 - \phi^{s+1}) \\[4pt]
& \times \big[-\bar{q}_i (1 - z_d)(p_0 + p_1) + \phi^s (1 - \phi z_d)(\bar{q}_i p_i - \bar{q}_{\bar{i}} p_{\bar{i}}) \big] \\[4pt]
& + \bar{d}_i(y_d)\big(\bar{d}_{\bar{i}}(y_d)\phi^{s+1}\bar{\phi} z_d^{s+1} - (\bar{q}_{\bar{i}} + \phi^{s+1}\bar{q}_i) y_d \big) \\[4pt]
& \times \big[\bar{q}_{\bar{i}}(1 - z_d)(p_0 + p_1) + \phi^s (1 - \phi z_d)(\bar{q}_i p_i - \bar{q}_{\bar{i}} p_{\bar{i}}) \big] \Big].
\end{aligned}
$$

$$(9.55)$$

Note that the second eigenvalue function $\lambda_2(z)$ is of no importance to the tail distribution of the packet delay. We conclude that to determine the tail distribution, it is enough to calculate z_d and θ, as indicated by (9.51). The value of z_d must be calculated numerically as the smallest root of $z - A(\lambda_1(z))$ larger than one after which θ follows directly from (9.55). Finally, we note that there seems to be an interesting relation between the decay rate z_d^{-1} of the delay distribution we just discussed and the corresponding decay rate z_u^{-1} of the *queue content* distribution. As we discussed in [29, 30] and as can be seen from expression (9.14), the tail distribution of the queue content is determined by the smallest (real) zero z_u larger than one of the function $N(z)$, i.e. $N(z_u) = 0$. It then is a direct consequence of property (9.49) that $z_u = \lambda_1(z_d)$.

References

[1] E. N. Gilbert, Capacity of a burst-noise channel, *Bell Systems Technical Journal*, vol. 39, no. 9, pp. 1253–1265, 1960.

[2] E. O. Elliott, Estimates of error rates for codes on burst-noise channels, *Bell Systems Technical Journal*, vol. 42, no. 9, pp. 1977–1997, 1963.

[3] B. Sklar, Rayleigh fading channels in mobile digital communications systems, Part I: Characterization, *IEEE Communications Magazine*, vol. 35, pp. 90–100, July 1997.

[4] R. H. Clarke, A statistical theory of mobile radio reception, *Bell Systems Technical Journal*, vol. 47, no. 7, pp. 957–1000, 1968.

[5] M. Zorzi and R. R. Rao, On channel modeling for delay analysis of packet communications over wireless links, in *Proceedings of the 36th Allerton Conference on Communications, Control and Computing*, Monticello, IL, 23–25 September, 1998.

[6] H. S. Wang and P. C. Chang, On verifying the first-order Markovian assumption for a Rayleigh fading channel model, *IEEE Transactions on Vehicular Technology*, vol. 45, no. 5, pp. 353–357, 1996.

[7] M. Zorzi, R. R. Rao and L. B. Milstein, ARQ error control for fading mobile radio channels, *IEEE Transactions on Vehicular Technology*, vol. 46, no. 5, pp. 445–455, 1997.

[8] C. C. Tan and N. C. Beaulieu, On first-order Markov modeling for the Rayleigh fading channel, *IEEE Transactions on Communications*, vol. 48, no. 12, 2000, pp. 2032–2040.

[9] A. J. Goldsmith and P. P. Varaiya, Capacity, mutual information, and coding for finite-state Markov channels, *IEEE Transactions on Information Theory*, vol. 42, no. 5, pp. 868–886, 1996.

[10] H. S. Wang and N. Moayeri, Finite state Markov channel – A useful model for radio communication channels, *IEEE Transactions on Vehicular Technology*, vol. 44, no. 1, pp. 163–171, 1995.

[11] B. D. Fritchman, A binary channel characterization using partitoned Markov chains, *IEEE Transactions on Information Theory*, vol. 13, no. 2, pp. 221–227, 1967.

[12] A. M. Chen and R. R. Rao, On tractable wireless channel models, in *Proceedings of the 9th IEEE Internatonal Symposium on Personal, Indoor and Mobile Radio Communications, PIMRC98*, Boston, MA, September, pp. 825–830, 1998.

[13] J. Aráuz and P. Krishnamurthy, Markov modeling of 802.11 channels, in *Proceedings of the IEEE Vehicular Technology Conference, VTC 2003*, Orlando, FL, 4–9 October, pp. 771–775, 2003.

[14] J. McDougall and S. Miller, Sensitivity of wireless network simulations to a two-state Markov model channel approximation, in *Proceedings of the IEEE Global Tele-communications Conference, GLOBECOM 2003*, San Francisco, CA, 1–5 December, pp. 697–701, 2003.

[15] M. Moeneclaey, H. Bruneel, I. Bruyland and D. Y. Chung, Throughput optimization for a generalized stop-and-wait ARQ scheme, *IEEE Transactions on Communications*, vol. 34, no. 2, pp. 205–207, February 1986.

[16] R. Fantacci, Performance evaluation of some efficient stop-and-wait techniques, *IEEE Transactions on Communications*, vol. 40, no. 11, pp. 1665–1669, November 1992.

[17] M. De Munnynck, A. Lootens, S. Wittevrongel and H. Bruneel, Transmitter buffer be-haviour of stop-and-wait ARQ schemes with repeated transmissions, *IEE Proceedings-Communications*, vol. 149, no. 1, pp. 13–17, 2002.

[18] C. Tison and H. Bruneel, Improving the throughput of stop-and-wait ARQ schemes with repeated transmissions, *AEÜ International Journal of Electronics and Communications*, vol. 51, no. 1, pp. 1–8, 1997.

[19] R. Mukhtar, M. Zukerman and F. Cameron, Packet latency for type-II hybrid ARQ transmissions over a correlated error channel, in *Proceedings of European Wireless 2002 Conference*, Florence, Italy, February, pp. 107–113, 2002.

[20] P. F. Turney, An improved stop-and-wait ARQ logic for data transmission in mobile radio systems, *IEEE Transactions on Communications*, vol. 29, no. 1, pp. 68–71, 1981.

[21] B. H. Saeki and I. Rubin, An analysis of a TDMA channel using stop-and-wait, block, and select-and-repeat ARQ error control, *IEEE Transactions on Communications*, vol. 30, no. 5, pp. 1162–1173, 1982.

[22] G. Fayolle G., E. Gelenbe and G. Pujolle, An analytic evaluation of the performance of the "send and wait" protocol, *IEEE Transactions on Communications*, vol. 26, no. 3, pp. 313–319, 1978.

[23] R. Fantacci, Queueing analysis of the selective repeat ARQ protocol in wireless packet

networks, *IEEE Transactions on Vehicular Technology*, vol. 45, no. 2, pp. 258–264, May 1996.

[24] J. G. Kim and M. Krunz, Delay analysis of selective repeat ARQ for transporting Markovian sources over a wireless channel, *IEEE Transactions on Vehicular Technology*, vol. 49, no. 5, pp. 1968–1981, September 2000.

[25] D. Towsley and J. K. Wolf, On the statistical analysis of queue lengths and waiting times for statistical multiplexers with ARQ retransmission schemes, *IEEE Transactions on Communications*, vol. 27, no. 4, pp. 693–702, April 1979.

[26] D. Towsley, A statistical analysis of ARQ protocols operating in a nonindependent error environment, *IEEE Transactions on Communications*, vol. 27, no. 7, pp. 971–981, July 1981.

[27] M. Rossi, L. Badia and M. Zorzi, Exact statistics of ARQ packet delivery delay over Markov channels with finite round-trip delay, in *Proceedings of IEEE Globecom 2003*, San Francisco, CA, December 1–5, pp. 3356–3360, 2003.

[28] H. Bruneel and B. G. Kim, *Discrete-Time Models for Communication Systems Including ATM*, Kluwer Academic Publishers, Boston, 1993.

[29] K. Tworus, S. De Vuyst, S. Wittevrongel and H. Bruneel, Transmitter buffer behavior of the stop-and-wait ARQ scheme under correlated errors, in *Proceedings of the Conference on Design, Analysis and Simulation of Distributed Systems, DASD 2004*, Washington, DC, April 18–22, pp. 10–18, 2004.

[30] S. De Vuyst, K. Tworus, S. Wittevrongel and H. Bruneel, Analysis of stop-and-wait ARQ for a wireless channel, *4OR, A Quarterly Journal of Operations Research*, accepted for publication, 2008.

[31] H. Bruneel, Performance of discrete-time queueing systems, *Computers & Operations Research*, vol. 20, no. 3, pp. 303–320, 1993.

[32] H. Bruneel, B. Steyaert, E. Desmet and G. Petit, Analytic derivation of tail probabilities for queue lengths and waiting times in ATM multiserver queues, *European Journal of Operational Research*, vol. 76, pp. 563–572, 1994.

[33] D. Fiems and H. Bruneel, A note on the discretization of Little's result, *Operations Research Letters*, vol. 30, no. 1, pp. 17–18, 2002.

[34] F. R. Gantmacher, *The Theory of Matrices, Volume One*, AMS Chelsea Publishing, Providence, RI, 1959.

[35] C. D. Meyer, *Matrix Analysis and Applied Linear Algebra*, SIAM, Philadelphia, PA, 2000.

[36] E. C. Titchmarsh, *The Theory of Functions*, 2nd edition, Oxford University Press, 1939.

[37] J. Abate and W. Whitt, Numerical inversion of probability generating functions, *Operations Research Letters*, vol. 12, no. 4, pp. 245–251, 1992.

10

A Markovian Analytical Model for a Hybrid Traffic Scheduling Scheme

Lan Wang, Geyong Min, Demetres D. Kouvatsos and Xiaolong Jin

Department of Computing, School of Informatics, University of Bradford, Bradford BD7 1DP, UK; e-mail: {l.wang9, g.min, d.kouvatsos, x.jin}@bradford.ac.uk

Abstract

The provisioning of differentiated Quality-of-Service (QoS) is an important objective in the design and implementation of next generation communication networks. This chapter proposes a new analytic model for a hybrid traffic scheduling scheme, which integrates Priority Queueing (PQ) and Weighted Fair Queueing (WFQ) for QoS differentiation. Analytic expressions are derived for the performance metrics of the system including the mean number of packets in the queue, throughput, mean queueing delay, packet loss probability and fairness of individual traffic flows. The model is adopted to investigate the effect of the weight of individual traffic flows in the WFQ policy and the capacities of low priority queues on the performance of the hybrid scheduling scheme.

Keywords: Scheduling scheme, Priority Queueing (PQ), Weighted Fair Queueing (WFQ), performance modeling, Markov chain.

10.1 Introduction

With the rapid development of modern communication and networking technologies, more and more diversified multimedia applications, such as IP

D. D. Kouvatsos (ed.), Mobility Management and Quality-of-Service for Heterogeneous Networks, 229–243.

telephony, IPTV, video conferencing, interactive game, and distant learning, have emerged. These applications propose more stringent and varying Quality-of-Service (QoS) requirements on throughput, packet loss, delay and jitter, or a combination of these items. However, the traditional best-effort service model is unable to provide QoS guarantees, because it treats all traffic in an equivalent manner. Therefore, the provisioning of differentiated QoS has emerged as one of the most pressing tasks in the design and implementation of next generation communication networks.

It is well known that scheduling schemes play a significant role in supporting QoS differentiation. Numerous research efforts have been dedicated to the design and performance evaluation of various scheduling schemes, such as, Priority Queueing (PQ), Generalized Process Sharing (GPS), and Weighted Fair Queueing (WFQ). Basically, there are two types of PQ schemes depending on whether or not on-going service can be interrupted, namely, pre-emptive and non-pre-emptive priority. Pre-emptive priority can be further classified as pre-emptive resume and pre-emptive repeat priority according to whether or not the pre-empted packet can resume its processing when the buffer of the high priority traffic flow becomes empty. In a GPS system, each traffic flow is assigned a weight in order to guarantee a minimum service rate and differentiated QoS, even though other flows may be greedy in demanding service. This property offers forwarding assurance to individual traffic flows and prevents them from experiencing service starvation. WFQ is a packet-based approximation to the well-known GPS service discipline [1, 5–7] and can be readily implemented in practice. The basic idea behind WFQ is to select the packet that would complete its service earliest if the GPS scheduling mechanism was applied as the next one to be transmitted. It has been empirically verified that both GPS and WFQ can provide bounded delay, guaranteed throughput, and relative fairness to individual traffic flows.

Recently, several hybrid scheduling schemes have been proposed and have already attracted tremendous research interests from both academia and industry [1–4]. Among them an extremely promising scheme is PQ-WFQ, which combines the ordinary PQ and WFQ scheduling mechanisms in a hierarchical manner and thus can provide more diversified QoS differentiation [7]. Generally speaking, in a PQ-WFQ scheduling system one traffic flow is served with the strict high priority over the other flows. As a result, this flow experiences low loss, low delay and jitter, which are the desirable QoS requirements of real-time multimedia applications. As the first attempt to model PQ-WFQ scheduling systems, this study takes into account PQ with pre-emptive resume scheme in order to facilitate the modelling issue. For other traffic flows in the PQ-GPS system, the WFQ scheduling policy is adopted to serve them with low priority, which, as aforementioned, can provide forwarding assurance and is thus suitable for non-real-time applications.

In this chapter, a three-dimensional Markov analytic model is devised for characterizing PQ-WFQ scheduling systems with multi-class Poisson traffic. More specifically, new expressions for several performance metrics are derived, including the mean number of packets in the queue, throughput, mean queueing delay and packet loss probability as well as fairness of individual traffic flows. Finally, numerical experiments are presented for the performance evaluation of the PQ-GPS system and an investigation is undertaken on the effects of the weight ratio of the WFQ flows on the performance of the hybrid scheduling scheme.

The remainder of this paper is organized as follows. Section 10.2 describes the PQ-WFQ scheduling scheme, develops an analytic Markov model and derives analytic expressions for the aforementioned performance metrics. Section 10.3 investigates the impact of the weight of the WFQ policy and the capacities of low priority queues on the system performance. Conclusions follow in Section 10.4.

10.2 Analytical Model of PQ-WFQ

10.2.1 System Description

We consider a stable single server queueing system with PQ-WFQ scheduling scheme as shown in Figure 10.1 where three classes of traffic wait for service at three separate queues (Q_1, Q_2 and Q_3) having finite capacities L_1, L_2, and, L_3, respectively. PQ with the pre-emptive resume priority scheme provides a guaranteed priority to Class-1 traffic over the others. Packets in Q_1 are always served before those in Q_2 and Q_3 and can interrupt an on-going service of any packet from the other two traffic flows. Packets in Q_2 and Q_3 are scheduled according to the WFQ principle. The weight for each queue is fixed in this scheme. Generally speaking, real-time traffic requiring low delay is assigned a heavier weight w_2. On the other hand, the delay-insensitive data traffic is assigned with a lighter weight w_3. The arrivals of traffic flow c follow a Poisson process with average arrival rate λ_c. The inter-service time is exponentially distributed with mean μ^{-1}. It is assumed that packet sizes of each traffic flow are same. Therefore, traffic flow c ($c = 2, 3$) will achieve an average service rate $\mu_c = \mu w_c$. A space in a buffer can be released only after the packet has been served completely.

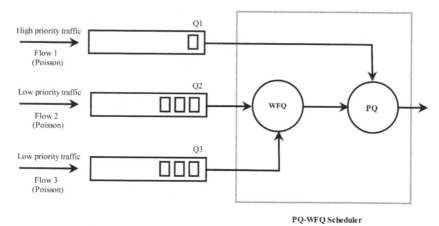

Figure 10.1   Single server queueing system with PQ–WFQ scheduling scheme.

10.2.2 Proposed Markov Model

This PQ-WFQ scheduling system subject to three classes of traffic is represented using the state transition diagram in Figure 10.2. Each State (i, j, k) $(0 \leq i \leq L_1, 0 \leq j \leq L_2, 0 \leq k \leq k \leq L_3)$ corresponds to the situation where there are i, j, and k packets of Class-1, Class-2 and Class-3, respectively, in the queue. The transitions from State (i, j, k) to $(i + 1, j, k)$, $(0 \leq i < L_1, 0 \leq j \leq L_2, 0 \leq k \leq L_3)$, from State (i, j, k) to $(i, j + 1, k)$, $(0 \leq i \leq L_1, 0 \leq j < L_2, 0 \leq k \leq L_3)$, and from State (i, j, k) to $(i, j, k + 1)$, $(0 \leq i \leq L_1, 0 \leq j \leq L_2, 0 \leq k < L_3)$, imply that a packet from Class-1, Class-2 and Class-3 traffic enters into the queue, respectively. Consequently, three corresponding transition rates are *lambda*$_1$, λ_2, and λ_3. Moreover, the rate out of State (i, j, k) to $(i - 1, j, k)$, $(1 \leq i \leq L_1 + 1, 0 \leq j \leq L_2, 0 \leq k \leq L_3)$, is equal to the service rate of Class-1 traffic, μ. Class-2 and Class-3 packets can be transmitted if and only if the queue of Class-1 traffic becomes empty. If both Class-2 and Class-3 packets are in the system, bandwidth is shared between them. Therefore, the rate out of State $(0, j, k)$ to $(0, j - 1, k)$ $(1 \leq j \leq l_2, 1 \leq j \leq l_3)$, and that out of State $(0, j, k)$ to $(0, j, k - 1)$, $(1 \leq j \leq L_2, 1 \leq j \leq L_3)$ are equal to the service rate of Class-2 traffic, μ_2 and the service rate of Class-3 traffic μ_3, respectively. Otherwise, bandwidth is used by only one traffic class. So the transition rate out of State $(0, 0, k)$, $(1 \leq k \leq L_3)$ to $(0, 0, K - 1)$ and that out of State $(0, J, 0)$, $(1 \leq j \leq L_2)$ to $(0, j - 1, 0)$ are μ.

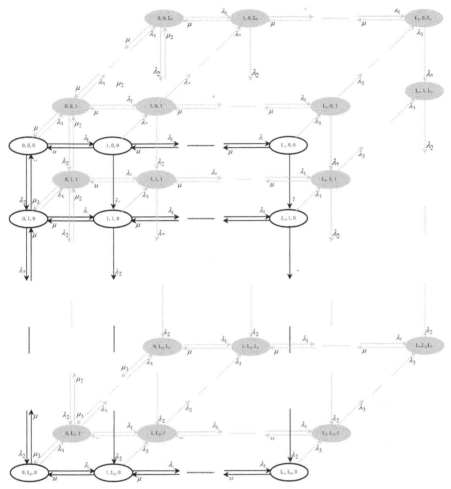

Figure 10.2 State transition diagram of the three-dimensional Markov chain of the HQ-GPS scheduling system.

The joint state probability, p_{ijk}, in the three-dimensional Markov chain can be solved using the method reported in [8]. Let $\mathbf{P}$ be the steady-state probability vector of this Markov chain,

$$\mathbf{P} = (p_{0,0,0}, \ldots, p_{0,L_2,0}, \ldots, p_{0,0,L_3}, \ldots, p_{0,L_2,L_3}).$$

The infinitesimal generator matrix $\mathbf{Q}$ of this Markov chain is of size

$$((L_1 + 1) \times (L_2 + 1) \times (L_3 + 1)) \times ((L_1 + 1) \times (L_2 + 1)).$$

The steady-state probability vector **P** satisfies the following equations:

$$\begin{cases} \mathbf{PQ} = \mathbf{0} \\ \mathbf{Pe} = 1 \end{cases},$$

(10.1)

where $\mathbf{e} = (1, 1, \ldots, 1)^T$ is a unit column vector of length $((L_1 + 1) \times (L_2+1) \times (L_3+1)) \times 1$. Solving Equation (10.1) using the approach presented in [8] yields the steady-state probability vector **P** as

$$\mathbf{P} = \boldsymbol{\alpha}(\mathbf{I} - \mathbf{X} + \mathbf{e}\boldsymbol{\alpha})^{-1},$$

(10.2)

where matrix $\mathbf{X} = \mathbf{I} + \mathbf{Q}/\beta$, $\beta \leq \min\{\mathbf{Q}_{ii}\}$ and $\boldsymbol{\alpha}$ is an arbitrary row vector of **X**.

The marginal state probability p_i^c, that i packets of Class-c, ($c = 1, 2, 3$), are in the queue can also be calculated using Equation (10.3) based on the joint state probability p_{ijk}. The probabilities, p_i^c, are useful for the following derivations of the marginal performance metrics.

$$p_m^c = \begin{cases} \sum_{k=0}^{L_3} \sum_{j=0}^{L_2} p_{mjk} & c = 1, \\ \sum_{k=0}^{L_3} \sum_{i=0}^{L_1} p_{imk} & c = 2, \\ \sum_{i=0}^{L_1} \sum_{j=0}^{L_2} p_{ijm} & c = 3. \end{cases}$$

(10.3)

10.2.3 Derivation of Performance Metrics

This section derives the analytical expressions that estimate the marginal performance including the number of packets in the queue, mean queueing delay, throughput, packet loss probability and fairness.

The mean marginal number of packets in the queue can be calculated as follow.

$$L^c = \sum_{m=0}^{L_c} (i \times p_m^c).$$

(10.4)

Throughput is commonly defined as the average rate at which packets go through the system in the steady state. Due to the equilibrium of the rates of incoming and outgoing flows in the steady state, the mean marginal throughputs, T_c, is equal to the corresponding arrival rate multiplied by the marginal probability p_m^c.

$$T^c = \lambda_c \times \sum_{m=0}^{L_c-1} p_m^c.$$

(10.5)

Table 10.1 System parameters.

L_1	L_2	L_3	λ_1	λ_2	λ_3	μ
5	5	5	14	10	3	28

The expressions for marginal mean queueing delay can be derived using Little's Law [9].

$$D^c = \frac{L^c}{T^c}. \tag{10.6}$$

As an arriving packet from Class-c can be dropped only when its queue becomes full, the packet loss probability for Class-c equals to the marginal state probability that there is no space available in the queue.

$$PLP^c = p^c_{L_c}. \tag{10.7}$$

Finally, we adopt Jain's fairness index [10] to calculate the fairness of the two classes of traffic in the system as follows

$$F = \frac{\left(\sum_{i=1}^{3} T^i\right)^2}{3 \times \sum_{i=1}^{3} (T^i)^2} \tag{10.8}$$

10.3 Performance Analysis of the Proposed Model

10.3.1 Effects of the Traffic Weight

This section focuses on investigating the impacts of the weight, on the marginal performance metrics. Figures 10.3–10.7 illustrate, respectively, the mean marginal number of packets in the queue, throughput, mean queueing delay, packet loss probability and fairness against the weight w_2. Performance results depicted in these figures are presented for the following cases: The capacity of each queue is set to be $L_1 = L_2 = L_3 = 5$. The arrival rates of traffic flow c ($c = 1, 2, 3$) are 14, 10 and 3. The mean service rate μ, for Class-1 traffic is 28. All these parameters are listed in Table 10.1.

Furthermore, Table 10.2 presents the parameters of the weights and the mean service rates for Class-2 and Class-3 traffic flows. The weight w_2 rises from 10 to 90% and thus reduces from 90 to 10%.

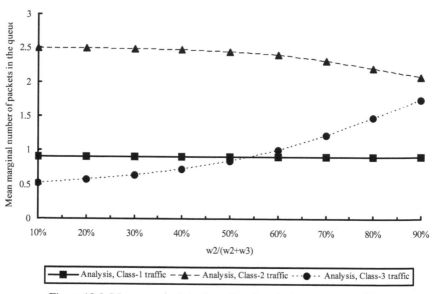

Figure 10.3 Mean marginal number of packets in the queue vs weight.

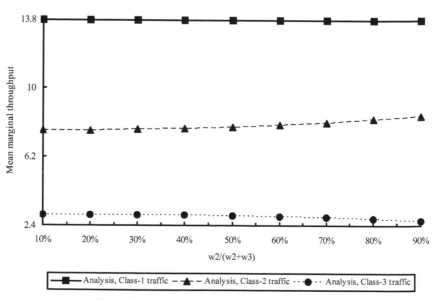

Figure 10.4 Mean marginal throughput vs weight.

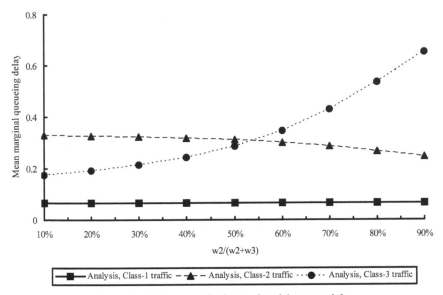

Figure 10.5 Mean marginal queueing delay vs weight.

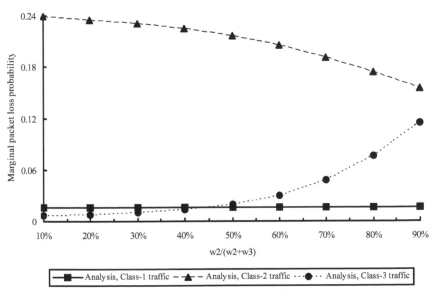

Figure 10.6 Marginal packet loss probability vs weight.

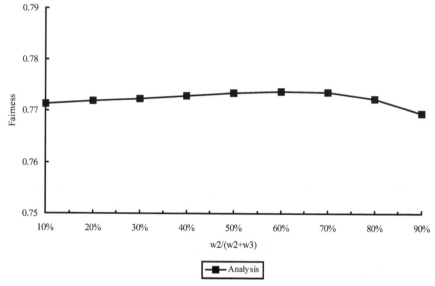

Figure 10.7 Fairness vs weight.

Table 10.2 Parameters of the weights and the mean service rates for Class-2 and Class-3 traffic flows as well as the corresponding weight ratio.

w_2	w_3	μ_2	μ_3
0.1	0.9	2.8	25.2
0.2	0.8	5.6	22.4
0.3	0.7	8.4	19.6
0.4	0.6	11.2	16.8
0.5	0.5	14	14
0.6	0.4	16.8	11.2
0.7	0.3	19.6	8.4
0.8	0.2	22.4	5.6
0.9	0.1	25.2	2.8

Figure 10.3 shows that as the weight w_3 increases the mean number of Class-2 packets tends to decrease, while that of Class-3 packets increases. This is because the mean service rate of Class-2 traffic increases. On the other hand, the mean service rate of Class-3 traffic reduces. Figure 10.4 illustrates that the mean marginal throughput of Class-2 traffic increases but that of Class-3 traffic decreases with the growth of the weight ratio. The remarkable increases in the queueing delay and packet loss probability of Class-3 traffic is clearly shown in Figures 10.5 and 10.6, respectively. Figure 10.7 reveals that the fairness increases to a maximum value (i.e., when) and then decreases with the growth of the weight ratio. Moreover, Figures 10.3–10.7 also show the minor changes in the QoS of Class-1 traffic as this traffic class is served with the highest priority.

10.3.2 Effects of the Queue Capacity

This section evaluates the effects of the capacities of low priority queues (i.e., Queues-2 and -3) on the performance metrics. We set the sum of capacities of low priority Queues-2 and -3 to be 20. The x-axis in Figures 10.8–10.12 represent that the capacity of Queue-2, L_2, rises from 2 to 18 by 2 and meanwhile that of Queue-3, L_3, decreases from 18 to 2. Generally speaking, the highest priority queue (i.e., Queue-1) has a small capacity enough to provide the best possible performance and save resources (i.e., space-saving). So the capacity of Queue-1 is set to be $L_1 = 5$. The arrival rate (λ_c) of each traffic flow c and the mean service rate (μ) can be found in Tables 10.1 and 10.2. According to the analysis above in Section 10.3.1, the PQ-WFQ scheduling provides the best fairness when the weights w_2 and w_3 are 60 and 40%, respectively. As and have not changed compared to the case in Section 10.3.1, this result still holds in this case. Effects of the capacity of Queue-2 are investigated based on the aforementioned assumptions.

Figures 10.8–10.12 depict, respectively, the mean marginal number of packets in the queue, throughput, mean queueing delay, packet loss probability and fairness against the capacities of low priority Queues-2 and 3. Figures 10.8 and 10.10, respectively, reveal that the mean number of Class-2 packets and its delay increase almost linearly when L_2 rises from 2 to 16 (i.e., L_3 decreases from 18 to 4), whilst it decreases slightly afterwards. It is obvious that the growth of L_2 can raise the mean number of Class-2 packets when L_2 is unchanged. In addition, the reduction of L_3 causes the decrease in the mean number of Class-2 packets when L_2 is unchanged. Consequently, the reason of the aforementioned increases in the mean number of Class-2

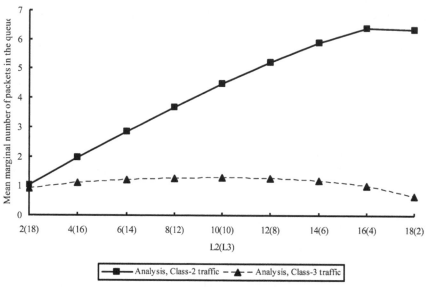

Figure 10.8 Mean marginal number of packets in the queue vs capacity of Queue-2.

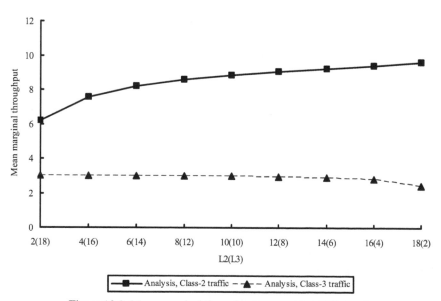

Figure 10.9 Mean marginal throughput vs capacity of Queue-2.

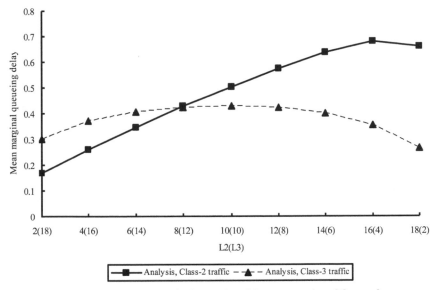

Figure 10.10 Mean marginal queueing delay vs capacity of Queue-2.

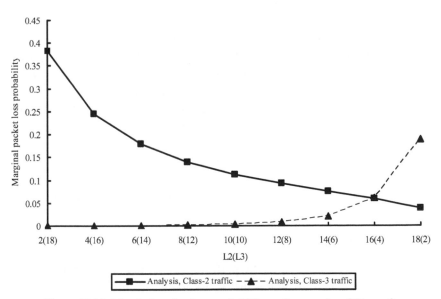

Figure 10.11 Marginal packet loss probability vs the capacity of Queue-2.

Analysis

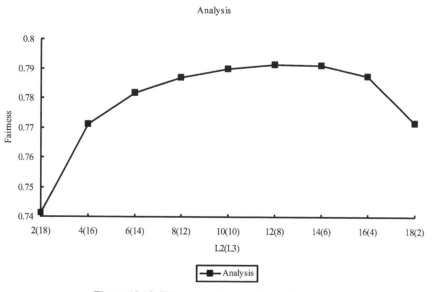

Figure 10.12 Fairness vs the capacity of Queue-2.

packets and its delay are because the effect of L_2 plays a major role. And *vice versa*, the mean number of Class-2 packets and its delay decrease in that the effect of L_3 takes the dominant position. On the other hand, the mean number of Class-3 packets and its delay increase to a maximum value (i.e., when $L_2 = L_3 = 10$) and then decreases. Similarly, this is because that the growth of L_2 increases the mean number of Class-3 packets and its delay more remarkably than the reduction of L_3 decreases them until $L_2 = L_3 = 10$. Figure 10.9 illustrates that the mean marginal throughput of Class-2 traffic increases remarkably but that of Class-3 traffic decreases slightly with the growth of the capacity of Queue-2. The remarkable decreases in the packet loss probability of Class-2 traffic and increases in that of Class-3 are clearly shown in Figure 10.11. Figure 10.12 demonstrates that the best fairness is provided when the capacities of Queue-2 and 3 are 12 and 8, respectively.

10.4 Conclusions

This chapter presents a new analytic performance model of an efficient hybrid scheduling scheme that integrates the well-known PQ and WFQ. To this end, a three-dimensional Markov model was developed to assist the derivation of the performance metrics of individual traffic flows in the scheduling system.

Moreover, the analytic model has been used to evaluate the impact of the weight ratio on important marginal performance metrics such as the mean number of packets in the queue, throughput, mean queueing delay and packet loss probability as well as fairness. Performance results have shown that the weight ratio and the capacities of low priority queues affect significantly the performance of traffic flow with the low priority.

Acknowledgements

This work is supported in part by the EC NoE Euro-FGI (NoE 028022), the EC IST project VITAL (IST-034284 STREP) and the UK EPSRC Grant EP/C525027/1.

References

[1] S. I. Maniatis, E. G. Nikolouzou and I. S. Venieris, Qos issues in the converged 3G wireless and wired networks, *IEEE Communications Magazine*, vol. 40, no. 8, pp. 44–53, 2002.

[2] J. Chen, W. Jiao and Q. Guo, An integrated QoS control architecture for IEEE 802.16 broadband wireless access systems, in *Proc. IEEE Global Telecommunications Conf. (GLOBECOM'05)*, IEEE Press, pp. 3330–3335, 2005.

[3] N. Liu, X, Li, C. Pei and B. Yang, Delay character of a novel architecture for IEEE 802.16 systems, in *Proc. 6th Int. Conf. Parallel and Distributed Computing, Applications and Technologies*, IEEE Press, pp. 293–296, 2005.

[4] K. Wongthavarawat and A. Ganz, Packet scheduling for QoS support in IEEE 802.16 broadband wireless access systems, *Int. Journal of Communication Systems*, vol. 16, no. 1, pp. 81–96, 2003.

[5] T. Engel, H. Granzer, B. F. Koch, M. Winter, P. Sampatakos, I. S. Venieris, H. Hussmann, F. Ricciato and S. Salsano, AQUILA: Adaptive resource control for QoS using an IP-based layered architecture, *IEEE Communications Magazine*, vol. 41, no. 1, pp. 46–53, 2003.

[6] X. Jin and G. Min, Analytical modelling of hybrid PQ-GPS scheduling systems under long-range dependent traffic, in *Proc. 21th Int. Conf. Advanced Information Networking and Applications (AINA'07)*, pp. 1006–1013, 2007.

[7] A. K. Parekh and R. G. Gallager. A generalized processor sharing approach to flow control in integrated services networks: The single-node case. *IEEE/ACM Transactions on Networking*, vol. 1, no. 3, pp. 344–357, 1993.

[8] W. Fischer and K. Meier-Hellstern, The Markov-modulated Poisson Process (MMPP) cookbook, *Performance Evaluation*, vol. 18, no. 2, pp. 149–171, 1993.

[9] L. Kleinrock, *Queueing Systems: Compute Applications*, vol. 1, John Wiley & Sons, New York, 1975.

[10] R. Jain, *The Art of Computer Systems Performance Analysis*, John Wiley and Sons, New York, 1991.

11

Delay Analysis of Go-Back-N ARQ for Correlated Error Channels

Koen De Turck and Sabine Wittevrongel

Stochastic Modeling and Analysis of Communication Systems (SMACS) Research Group, Department of Telecommunications and Information Processing, Ghent University, Sint-Pietersnieuwstraat 41, 9000 Gent, Belgium; e-mail: {kdeturck, sw}@telin.ugent.be

Abstract

We investigate the performance of the Go-Back-N ARQ (Automatic Repeat reQuest) protocol over a wireless channel. Data packets are sent from transmitter to receiver over the wireless transmission channel. When a packet is received, the receiver checks whether it has been received correctly or not, and sends a feedback message to notify the transmitter of the condition of that packet. When the transmitter is notified of a transmission error, the incorrectly received packet is sent again, as well as every following packet.

Our modeling assumptions are based on two convictions. On the one hand, a good view of the performance of an ARQ protocol not only requires an analysis of the throughput, but also of the buffer behavior. Therefore, we offer a complete queueing analysis of the transmitter buffer, in addition to a throughput analysis. Secondly, due to the highly variable nature of the error process in wireless networks, we have to take error correlation explicitly into account.

Hence, we model the channel by means of a general Markov chain with M states and a fixed error probability in every state. The transmitter buffer is modeled as a discrete-time queue with infinite storage capacity and independent and identically distributed packet arrivals from slot to slot.

D. D. Kouvatsos (ed.), Mobility Management and Quality-of-Service for Heterogeneous Networks, 245–267.

We find concise expressions for the probability generating functions of the unfinished work and the packet delay of the transmitter buffer. Furthermore, we show explicit expressions for the mean and the variance of both system characteristics and we derive some heavy-load approximations. Finally, we provide some numerical examples.

Keywords: Wireless telecommunication, queueing theory, performance evaluation, generating functions, automatic repeat request.

11.1 Introduction

A popular way of protecting against transmission errors is the so-called Automatic Repeat reQuest (ARQ). For this mechanism to work, the transmitter adds a simple error checking code to each packet so the receiver can detect the most common transmission errors. When a packet is received, the receiver checks whether it has been received correctly or not, and sends a feedback message to notify the transmitter of the condition of that packet. That is, a positive acknowledgement (ACK) is sent in case of a correct transmission; a negative acknowledgement (NAK) is sent in case of a transmission error. This organization requires a bi-directional channel between the sender and the receiver. Since not all arriving packets can be sent immediately and moreover the transmitter has to keep a copy of each packet until it is correctly transmitted, a buffer to store packets is required at the transmitter side.

Various types of ARQ protocols have been proposed in the literature [1]. They differ in the way the transmissions and retransmissions of packets are organized. In this paper, we focus on the Go-Back-N ARQ protocol (GBN-ARQ). Its operation is illustrated in Figure 11.1. In case of GBN-ARQ, the transmitter keeps on sending packets to the receiver without interruptions until a NAK is received. Upon reception of a NAK, the incorrectly received packet is sent again, as well as every following packet. We introduce the term feedback delay, which consists of (1) the time during which a packet is travelling through the channel, (2) the processing time of that packet in the receiver, and (3) the time during which the feedback message is travelling back.

The performance of the GBN-ARQ protocol has been investigated before. However, in many existing studies it is assumed that transmission errors occur independently, which leads to a considerably simpler analysis (see e.g. [2–4]). Such an assumption is not realistic in case of non-stationary wireless transmission channels due to e.g. fading effects, see [5] for details. In particular,

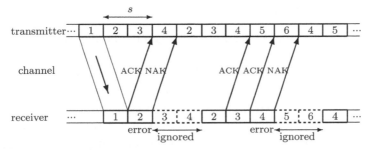

Figure 11.1 Operation of the Go-Back-N protocol (feedback delay $s = 2$).

the time-varying nature of the channel will result in correlation in the occurrence of transmission errors. In this paper, we therefore model the channel by means of a Markov chain with M states and with a fixed error probability in every state. A special case of our model, where $M = 2$, is known as the Gilbert–Elliott model [6, 7]. Previous work on the behavior of GBN-ARQ for correlated errors is reported in [8], where the throughput and the transmitter buffer content are studied. In our paper, on the other hand, we present a full analysis of another important performance metric, namely the packet delay. Other related work is given in [5] where the performance of Stop-and-Wait ARQ over a dynamic two-state channel is investigated. The selective-repeat protocol with correlated errors is analyzed in [9] under the assumption of so-called ideal ARQ, where the feedback delay is neglected and acknowledgement messages are received instantly.

The paper is organized as follows. The modeling assumptions are given in Section 11.2. In Section 11.3, we introduce the *probability generating matrix* (pgm) of the service time, which plays a crucial role in the further analysis. Next, we derive the distribution of the unfinished work at an arbitrary slot boundary in Section 11.4, as a preparatory step for the analysis of the *probability generating function* (pgf) of the delay in Section 11.5. The actual computation of the moments of the packet delay is elaborated on in Section 11.6. We also derive some heavy-load approximations there. We introduce a generalization of the popular Gilbert–Elliott channel model, which we dub the 'cyclic' channel model in Section 11.7. In Section 11.8, we provide numerical examples to show the soundness of our approximations and the impact of the error correlation. Finally, conclusions are drawn in Section 11.9.

11.2 Modeling Assumptions

We set out to analyze the behavior of the transmitter buffer for the GBN-ARQ protocol. Throughout the analysis, we assume that the data to be transmitted are divided into fixed-length packets. The time axis is likewise assumed to be divided into fixed-length slots, where a slot corresponds to the time needed to transmit one packet. Synchronous transmission is used, i.e., transmission always starts at the beginning of a slot. Packets are transmitted according to a first-come-first-served discipline. After a constant period of s slots, an acknowledgement message from the receiver indicating whether or not the packet was correctly received, arrives at the transmitter (see Figure 11.1). This interval of s slots is referred to as the feedback delay of the channel. It is assumed that no errors occur in the acknowledgement messages. This is a reasonable assumption as the information content of an acknowledge-ment message is low and can thus be heavily protected. Moreover, if we do not make this assumption, then we cannot ensure completely reliable communication, but only minimize the risk, see the so-called Two Generals' Problem [10].

Let the random variable a_k denote the number of packets arriving at the transmitter during slot k. The a_ks are assumed to be independent and iden-tically distributed (iid) variables with mean $E[A]$, variance $Var[A]$ and pgf $A(z)$. The transmitter buffer is assumed to have an infinite storage capacity and packets are stored in this buffer until they (and their predecessors) are successfully transmitted over the channel. Note that due to the synchronous transmission mode, a packet arriving in an empty buffer is not transmitted until the next slot.

The error process in the channel is modeled in this paper by means of a general Markov chain with M states. Let the random variable c_k denote the channel state at slot boundary k (i.e. at the beginning of slot k), then the entries of the transition probability matrix $\mathbf{q}$ associated with the Markov chain are given by

$$q_{ij} \doteq \Pr[c_{k+1} = j \mid c_k = i]. \tag{11.1}$$

State i has an error probability of e_i. When there is an error, the packet sent during that slot will be incorrectly received, and the transmitter will receive a NAK message s slots later. We define the matrix $\mathbf{e}$ as the diagonal matrix with elements $e_1, \ldots, e_M$. We also introduce the notation $\bar{x} \doteq 1 - x$. Likewise, $\bar{\mathbf{e}}$ is a diagonal matrix with elements $\bar{e}_1, \ldots, \bar{e}_M$.

In the rest of this paper, we will often rely on a trick that was introduced by Towsley and Wolf [2]. The trick is to consider a slightly modified sys-service time of packet $\mathcal{P}$

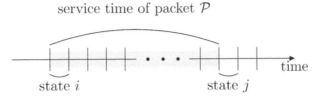

Figure 11.2 Service time starting in state i and followed by state j.

tem in which a packet immediately leaves the transmitter buffer after being successfully transmitted, instead of staying another s slots until the positive acknowledgement is received. Note that the delay in the modified system of every packet is exactly s slots less than in the original system, which makes it easy to convert results for the delay obtained for the modified system into results for the original system, as will be elaborated on later.

11.3 Distribution of the Service Time

By putting the evolution of the service time into a convenient form, we set the stage for the queueing analysis proper. As we will see, the results in this section will greatly simplify the derivations in the subsequent sections.

The service time of a packet is defined as the time interval (expressed in slots) that starts at the slot boundary where the packet is transmitted for the first time and ends with the slot boundary where the packet leaves the system. We consider the modified system, so packets leave the buffer at the end of the slot where they are correctly transmitted. Remark that this also means that in the modified system, service times do not overlap. Service times in the real system are s slots longer, and may overlap.

In view of the above modeling assumptions, the exact distribution of the service time of a packet will depend on the channel state in the slot during which the service of the packet starts. Therefore, consecutive service times in the considered model are not iid, unlike in the case of uncorrelated transmission errors. In order to study the service time, let us introduce a pgm $\mathbf{S}(z)$ of dimension $M \times M$. The element $[\mathbf{S}(z)]_{ij}$ is the partial conditional pgf of the service time that is followed by a slot with channel state j, given that the service time starts in channel state i (see Figure 11.2), that is

$$[\mathbf{S}(z)]_{ij} = \mathrm{E}[z^S \, 1(c_{k'} = j) \mid c_k = i], \qquad (11.2)$$

where the random variable S denotes a service time, k and k' denote the slot boundary where the service starts and ends respectively, and $1(.)$ is the indicator function of the event between brackets. Note that a service time has some kind of generalized geometric property: when there is no error, the service finishes after exactly one slot; when there is an error, a new transmission starts after $s + 1$ slots, and this can be seen as the start of a completely new service time, but this time starting in state c_{k+s+1}. Indeed, when there is an error, the total service lasts $s + 1$ slots plus a remaining service time. Given the state c_{k+s+1} at the start of this remaining service time, the mechanism that determines the length of the remaining service time is completely equivalent to the one that determines the length of a full service time and therefore this remaining service time also has pgm $\mathbf{S}(z)$, i.e. it has the pgm of a completely new service time. From these observations, we can see that

$$[\mathbf{S}(z)]_{ij} = z \Pr[\text{no error at } k, c_{k'} = j \mid c_k = i] \tag{11.3}$$

$$+ z^{s+1} \sum_{i'=1}^{M} \Pr[\text{error at } k, c_{k+s+1} = i' \mid c_k = i]$$

$$\times \mathbb{E}[z^S 1(c_{k'} = j) \mid c_{k+s+1} = i'] \tag{11.4}$$

$$= z(1 - e_i)[\mathbf{q}]_{ij} + z^{s+1} \sum_{i'=1}^{M} e_i [\mathbf{q}^{s+1}]_{ii'} [\mathbf{S}(z)]_{i'j}, \tag{11.5}$$

where we have used the fact that the $(s + 1)$-step transition probabilities for the Markov chain of the channel are given by the transition matrix $\mathbf{q}^{s+1}$. Exploiting the matrix notation, we can write pgm $\mathbf{S}(z)$ succinctly as

$$\mathbf{S}(z) = \bar{\mathbf{e}}\mathbf{q}z + \mathbf{e}\mathbf{q}^{s+1}z^{s+1}\mathbf{S}(z). \tag{11.6}$$

From the above relation, we can derive $\mathbf{S}(z)$ as

$$\mathbf{S}(z) = (\mathbf{I} - \mathbf{e}\mathbf{q}^{s+1}z^{s+1})^{-1}\bar{\mathbf{e}}\mathbf{q}z, \tag{11.7}$$

where $\mathbf{I}$ denotes the $M \times M$ identity matrix.

The advantage of capturing the distribution of the service time in the pgm $\mathbf{S}(z)$ is that we can express the pgm of the length of n subsequent service times simply as $\mathbf{S}(z)^n$. Equation (11.7) also allows us to compute the throughput of the protocol. The throughput η is defined as the inverse of the mean service time under assumption that there are always packets available (heavy-traffic assumption). Notice that the matrix $\mathbf{S}(1)$ is the transition probability matrix that records the channel state transition between the start and the end

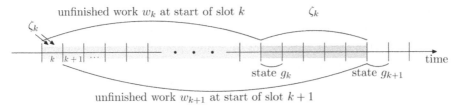

Figure 11.3 Definitions of w_k, ζ_k and g_k.

of a service. The steady-state probability vector π is given by

$$\pi = \pi S(1) \quad \text{and} \quad \pi 1 = 1, \tag{11.8}$$

where 1 is an $N \times 1$ column vector of ones. The vector π records the probabilities of finding the channel in a particular state at the beginning of a service, if new services start uninterruptedly. Now, $[S'(1)]_{ij}$ equals $E[S \, 1(c_{k'} = j) \mid c_k = i]$, i.e. $[S'(1)]_{ij}$ denotes the mean length of a service time that is followed by state j, given that it starts in state i. Therefore, we have that

$$\pi S'(1)1 = \sum_{i=1}^{M} \sum_{j=1}^{M} \Pr[c_k = i] \, E[S \, 1(c_{k'} = j) \mid c_k = i]$$

$$= E[S]. \tag{11.9}$$

That is, $\pi S'(1)1$ is the expected length of a service time, given that there are no gaps between services. This is equal to the reciproque of the throughput η:

$$\eta = \frac{1}{\pi S'(1)1}. \tag{11.10}$$

11.4 Distribution of the Unfinished Work

In this section we derive an expression for the pgf of the unfinished work w_k at the beginning of an arbitrary slot k. This is the time in slots needed to serve the packets that are present in the buffer at the beginning of slot k. We again consider the modified system here.

Let ζ_k denote the amount of work (in slots) that enters the buffer during slot k, i.e. the sum of the service times of all packets arriving in slot k. Then the unfinished work evolves according to the following system equation:

$$w_{k+1} = (w_k - 1)^+ + \zeta_k, \tag{11.11}$$

where $(\ldots)^+ = \max(0, \ldots)$. The random variables w_k and ζ_k on the right-hand side of (11.11) are not independent, because both depend on the state of the channel. This means that the set $\{w_k\}$ does not form a Markov chain. We can make it a Markov chain by adding an extra variable. The most convenient choice is the variable g_k, which indicates the state of the channel in the earliest slot where work arriving during slot k could start. So, when $w_k = 0$, g_k corresponds to the channel state during slot $k + 1$. The case where $w_k > 0$ is illustrated in Figure 11.3. We have that

$$
g_{k+1} = \begin{cases} g_k^{(1)} & \text{when } w_k = \zeta_k = 0, \\ g_k^{(\zeta_k)} & \text{otherwise,} \end{cases} \tag{11.12}
$$

where the notation $g_k^{(n)}$ stands for the state of the channel n slots after the slot with state g_k. From (11.11) and (11.12), it is easily seen that the set of random variables $\{w_k, g_k\}$ indeed forms a Markov chain.

Let us now define the partial pgfs

$$
W_{j,k}(z) = \sum_{n=0}^{\infty} \Pr[w_k = n, g_k = j]z^n,
$$

and introduce the vector notation $\mathbf{W}_k(z) = (W_{1,k}(z), \ldots, W_{M,k}(z))$. Then it is possible to rewrite the above system equations in the z-domain as:

$$
\mathbf{W}_{k+1}(z) = \frac{1}{z}(\mathbf{W}_k(z) - \mathbf{W}_k(0))A(\mathbf{S}(z)) + \mathbf{W}_k(0)\left[A(\mathbf{S}(z)) - A(0)\mathbf{I} + A(0)\mathbf{q}\right].
$$

Here we have introduced the convenient shorthand notation $A(\mathbf{S}(z))$ to denote a matrix that is a power series expansion in the pgm $\mathbf{S}(z)$ with the same coefficients as the power series expansion in z of $A(z)$. Similar notations will also be used later in this paper. Note that

$$
\boldsymbol{\pi} A(\mathbf{S}(1)) = \boldsymbol{\pi} \sum_{k=0}^{\infty} \Pr[a = k]\mathbf{S}(1)^k = \boldsymbol{\pi}, \tag{11.13}
$$

which shows that transition matrices $\mathbf{S}(1)$ and $A(\mathbf{S}(1))$ have the same steady-state probability vector. The system will reach an equilibrium if the average amount of work per slot (commonly called the load ρ) is strictly smaller than 1. The load ρ satisfies

$$
\rho = \mathrm{E}[S]\,\mathrm{E}[A] = \boldsymbol{\pi} A'(1)\mathbf{S}'(1)\mathbf{1}. \tag{11.14}
$$

In the steady state both $\mathbf{W}_k(z)$ and $\mathbf{W}_{k+1}(z)$ will converge to a common limiting vector $\mathbf{W}(z) = (W_1(z), \ldots, W_M(z))$. Taking limits for $k \to \infty$ and solving the resulting equation for $\mathbf{W}(z)$, we obtain

$$\mathbf{W}(z)[z\mathbf{I} - A(\mathbf{S}(z))] = \mathbf{W}(0)\big[(z-1)A(\mathbf{S}(z)) + zA(0)(\mathbf{q} - \mathbf{I})\big]. \quad (11.15)$$

Note that this formula still contains the unknown vector $\mathbf{W}(0)$, whose computation is the subject of a huge number of studies [11, 12]. The proposed algorithms fall apart into two broad categories. The first class are the so-called spectral methods, which are based on the property that there are a number of values z_i inside the unit circle, such that $z_i\mathbf{I} - A(\mathbf{S}(z_i))$ is a singular matrix, and such that the corresponding right null spaces $\mathbf{R}_i$ have together a total rank of $M - 1$ [11]. Then we can form a set of $M - 1$ equations

$$\mathbf{0} = \mathbf{W}(0)\big[(z_i - 1)A(\mathbf{S}(z_i)) + z_iA(0)(\mathbf{q} - \mathbf{I})\big]\mathbf{R}_i. \quad (11.16)$$

The last equation for $\mathbf{W}(0)$ is supplied by the well-known property of single-server queues that utilization equals the load. The utilization equals $1 - \mathbf{W}(0)\mathbf{1}$, while the load is given by equation (11.14), so that we complete the set of equations by stating that

$$\mathbf{W}(0)\mathbf{1} = 1 - \rho. \quad (11.17)$$

Some authors question the numerical stability of this method, especially if there are z_i which have a corresponding null space of a high dimension, and they give preference to the second method.

This other method is based on the weak canonical Wiener–Hopf factorization of the Laurent series $\mathbf{I} - \frac{A(\mathbf{S}(z))}{z}$, which says that this expression can be factorized under mild conditions into

$$\mathbf{I} - \frac{A(\mathbf{S}(z))}{z} = \mathbf{F}(z)\left(\mathbf{I} - \frac{1}{z}\mathbf{G}\right). \quad (11.18)$$

Here power series $\mathbf{F}(z)$ has no roots inside the unit circle, whereas the roots of expression $\mathbf{I} - \frac{1}{z}\mathbf{G}$ lie inside or on the unit circle. It can be shown [12] that matrix $\mathbf{G}$ is the well-known fundamental matrix as it occurs in M/G/1-type Markov chains, for which many algorithms have been developed. If we substitute the Wiener–Hopf factorization into equation (11.15), and identify the terms in z^{-1}, we find after some manipulations that

$$\mathbf{W}(0) = \mathbf{W}(0)(\mathbf{G} + A(0)(\mathbf{I} - \mathbf{q})). \quad (11.19)$$

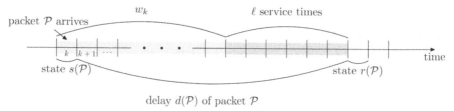

Figure 11.4 Illustration of the various components of the delay.

Hence the computation of the boundary probabilities is reduced to a linear system of equations, which must again be supplemented by equation (11.17).

11.5 Distribution of the Packet Delay

Now that we have found an expression for the pgf of the unfinished work, it is rather straightforward to analyze the delay of an arbitrary packet $\mathcal{P}$ arriving in slot k. Let us first focus again on the modified system. We denote the channel state during the arrival slot of $\mathcal{P}$ as $s(\mathcal{P})$. The channel state in the first slot after the service of $\mathcal{P}$ is denoted by $r(\mathcal{P})$.

The delay $d(\mathcal{P})$ of packet $\mathcal{P}$ consists of two components, as illustrated in Figure 11.4: (a) the remaining number of slots from slot $k + 1$ onwards needed to execute the unfinished work present in the system at the beginning of slot k, i.e., $(w_k - 1)^+$, and (b) the service times of the ℓ packets arriving in the buffer in slot k that will be served no later than (but including) $\mathcal{P}$. The distribution of the unfinished work has been derived in the previous section. Owing to the fact that $\mathcal{P}$ is an arbitrary packet, the pgf $L(z)$ of ℓ can be derived as (see e.g. [13]):

$$L(z) = \frac{z(1 - A(z))}{E[A](1 - z)}. \tag{11.20}$$

Now let $D_j(z)$ denote the partial pgf of the packet delay, provided that $r(\mathcal{P}) = j$. Then we can write, again using vector notation,

$$\mathbf{D}(z) = \frac{1}{z}(\mathbf{W}(z) + (z - 1)\mathbf{W}(0))L(\mathbf{S}(z)). \tag{11.21}$$

The unconditional pgf $\tilde{D}(z)$ of the packet delay in the modified system is given by $\tilde{D}(z) = \mathbf{D}(z)\mathbf{1}$. From (11.21), it then follows that

$$z\tilde{D}(z) = (\mathbf{W}(z) + (z - 1)\mathbf{W}(0))L(\mathbf{S}(z))\mathbf{1}. \tag{11.22}$$

From this, we can finally obtain the pgf of the delay $D(z)$ in the real system:

$$D(z) = z^s \tilde{D}(z). \qquad (11.23)$$

11.6 Computing Moments of the Packet Delay

In this section, we explain in detail how to compute the mean and higher-order central moments of the packet delay. Once we have the boundary probabilities in the form of vector $\mathbf{W}(0)$, only a moderate amount of numerical computation is needed to get the first few moments of the delay. We illustrate this by computing the first two central moments, as higher-order moments are obtained by similar means.

Notice that the mean and the variance of the delay in the real system can be trivially obtained in function of the same quantities of the modified system:

$$E[D] = E[\tilde{D}] + s \quad \text{and} \quad \text{Var}[D] = \text{Var}[\tilde{D}]. \qquad (11.24)$$

Let us now compute the first two derivatives of $\tilde{D}(z)$ as given in equation (11.22), and evaluate them for $z = 1$:

$$\tilde{D}'(1) + 1 = (\mathbf{W}'(1) + \mathbf{W}(0))\mathbf{1} + \mathbf{W}(1)\ell'(1); \qquad (11.25)$$

$$\tilde{D}''(1) + 2\tilde{D}'(1) = \mathbf{W}''(1)\mathbf{1} + 2(\mathbf{W}'(1) + \mathbf{W}(0))\ell'(1) + \mathbf{W}(1)\ell''(1), \qquad (11.26)$$

where we introduced the column vector $\ell(z) = L(\mathbf{S}(z))\mathbf{1}$. Note that $\ell(1) = \mathbf{1}$. In view of the fact that $\text{Var}[D] = \tilde{D}''(1) + \tilde{D}'(1) - \tilde{D}'(1)^2$, we find after some further manipulations that

$$E[D] = E[W] - \rho + \mathbf{W}(1)\ell'(1) + s; \qquad (11.27)$$

$$\begin{aligned} \text{Var}[D] = \text{Var}[W] &+ 2(\mathbf{W}'(1) + \mathbf{W}(0))\ell'(1) + \mathbf{W}(1)\ell''(1) \\ &- (1 - \rho + \mathbf{W}(1)\ell'(1))(2\,E[W] - \rho + \mathbf{W}(1)\ell'(1)), \end{aligned} \qquad (11.28)$$

where we introduced $E[W] = \mathbf{W}'(1)\mathbf{1}$ and $\text{Var}[W] = \mathbf{W}''(1)\mathbf{1} + \mathbf{W}'(1)\mathbf{1} - (\mathbf{W}'(1)\mathbf{1})^2$. So far, we have reduced the problem of finding the moments of the delay into the problem of finding the row vectors $\mathbf{W}(1)$, $\mathbf{W}'(1)$ and $\mathbf{W}''(1)$ associated with the unfinished work, and the column vectors $\ell'(1)$ and $\ell''(1)$, which are connected to the amount of work that arrives during an arrival slot.

Let us first focus our attention on the unknowns $\mathbf{W}(1)$, $\mathbf{W}'(1)$ and $\mathbf{W}''(1)$. When we evaluate equation (11.15) for $z = 1$ and do the same for its first two derivatives in z, we obtain

$$\mathbf{W}(1)(\mathbf{I} - \mathbf{A}(1)) = \mathbf{W}(0)A(0)(\mathbf{q} - \mathbf{I}); \tag{11.29}$$

$$\mathbf{W}'(1)(\mathbf{I} - \mathbf{A}(1)) + \mathbf{W}(1)(\mathbf{I} - \mathbf{A}'(1)) = \mathbf{W}(0)(A(1) + A(0)(\mathbf{q} - \mathbf{I})); \tag{11.30}$$

$$\mathbf{W}''(1)(\mathbf{I} - \mathbf{A}(1)) + 2\mathbf{W}'(1)(\mathbf{I} - \mathbf{A}'(1)) - \mathbf{W}(1)\mathbf{A}''(1) = 2\mathbf{W}(0)\mathbf{A}'(1), \tag{11.31}$$

where we have introduced $\mathbf{A}(z) = A(\mathbf{S}(z))$ for reasons of convenience. We see that $\mathbf{W}(1)$, $\mathbf{W}'(1)$ and generally $\mathbf{W}^{(n)}(1)$ satisfy a linear equation of the form $\mathbf{x}(\mathbf{I} - \mathbf{A}(1)) = \mathbf{b}$. However, as $\mathbf{A}(1)$ is a stochastic matrix, the system of equations is singular, and hence we need an extra equation for each $\mathbf{W}^{(n)}(1)$. Note that the solution of an inhomogeneous system of equations can generally be written as the sum of one particular solution of the inhomogeneous system, plus a solution of the homogeneous system, which in this case has the form $c\boldsymbol{\pi}$, for an arbitrary constant c. We define the particular solution of the problem with the help of the group inverse [14], which is the subject of the Appendix. The useful property of the group inverse for our purposes is that the group inverse $(\mathbf{I} - \mathbf{A}(1))^{\#}$ gives us the solution for $\mathbf{x}$ of the equation $\mathbf{x}(\mathbf{I} - \mathbf{A}(1)) = \mathbf{b}$, such that $\mathbf{x}\mathbf{1} = 0$. Let $\tilde{\mathbf{W}}(1)$, $\tilde{\mathbf{W}}'(1)$ and $\tilde{\mathbf{W}}''(1)$ denote the particular solutions of (11.29)–(11.31) found by means of the group inverse, i.e.

$$\tilde{\mathbf{W}}(1) = \mathbf{W}(0)A(0)(\mathbf{q} - \mathbf{I})(\mathbf{I} - \mathbf{A}(1))^{\#}; \tag{11.32}$$

$$\tilde{\mathbf{W}}'(1) = (\mathbf{W}(0)(A(1) + A(0)(\mathbf{q} - \mathbf{I})) - \mathbf{W}(1)(\mathbf{I} - \mathbf{A}'(1)))(\mathbf{I} - \mathbf{A}(1))^{\#}; \tag{11.33}$$

$$\tilde{\mathbf{W}}''(1) = (2\mathbf{W}(0)\mathbf{A}'(1) - 2\mathbf{W}'(1)(\mathbf{I} - \mathbf{A}'(1)) + \mathbf{W}(1)\mathbf{A}''(1))(\mathbf{I} - \mathbf{A}(1))^{\#}. \tag{11.34}$$

Then we can write

$$\mathbf{W}(1) = \tilde{\mathbf{W}}(1) + \mathbf{W}(1)\mathbf{1}\boldsymbol{\pi}; \tag{11.35}$$

$$\mathbf{W}'(1) = \tilde{\mathbf{W}}'(1) + \mathbf{W}'(1)\mathbf{1}\boldsymbol{\pi}; \tag{11.36}$$

$$\mathbf{W}''(1) = \tilde{\mathbf{W}}''(1) + \mathbf{W}''(1)\mathbf{1}\boldsymbol{\pi}, \tag{11.37}$$

where we still have to determine the scalars $\mathbf{W}(1)\mathbf{1}$, $\mathbf{W}'(1)\mathbf{1}$ and $\mathbf{W}''(1)\mathbf{1}$. Note that the normalization condition learns us that $\mathbf{W}(1)\mathbf{1} = 1$. Furthermore, when we substitute equation (11.36) into equation (11.31), and postmultiply by the column vector $\mathbf{1}$, we find

$$2(\tilde{\mathbf{W}}'(1) + \mathbf{W}'(1)\mathbf{1}\pi)(\mathbf{I} - \mathbf{A}'(1))\mathbf{1} - \mathbf{W}(1)\mathbf{A}''(1)\mathbf{1} = 2\mathbf{W}(0)\mathbf{A}'(1)\mathbf{1}. \quad (11.38)$$

Since $\pi(\mathbf{I} - \mathbf{A}'(1))\mathbf{1} = 1 - \rho$, this leads to

$$2(1 - \rho)\mathbf{W}'(1)\mathbf{1} = 2\mathbf{W}(0)\mathbf{A}'(1)\mathbf{1} + \mathbf{W}(1)\mathbf{A}''(1)\mathbf{1} - 2\tilde{\mathbf{W}}'(1)(\mathbf{I} - \mathbf{A}'(1))\mathbf{1}. \quad (11.39)$$

Similarly, from the evaluation of the third derivative of equation (11.15) for $z = 1$ we can infer the value for $\mathbf{W}''(1)\mathbf{1}$.

Having expounded on the computation of the derivatives of $\mathbf{W}(z)$ in $z = 1$, we now set up a similar scheme for the computation of $\boldsymbol{\ell}'(1)$ and $\boldsymbol{\ell}''(1)$. Note that

$$E[A](\mathbf{S}(z) - \mathbf{I})\boldsymbol{\ell}(z) = \mathbf{S}(z)(\mathbf{A}(z) - \mathbf{I})\mathbf{1}. \quad (11.40)$$

The first two derivations of this expression with respect to z, evaluated in $z = 1$, are equal to

$$E[A](\mathbf{S}'(1)\mathbf{1} + (\mathbf{S}(1) - \mathbf{I})\boldsymbol{\ell}'(1)) = \mathbf{S}(1)\mathbf{A}'(1)\mathbf{1}; \quad (11.41)$$

$$E[A](\mathbf{S}''(1)\mathbf{1} + 2\mathbf{S}'(1)\boldsymbol{\ell}'(1) + (\mathbf{S}(1) - \mathbf{I})\boldsymbol{\ell}''(1))$$
$$= 2\mathbf{S}'(1)\mathbf{A}'(1)\mathbf{1} + \mathbf{S}(1)\mathbf{A}''(1)\mathbf{1}. \quad (11.42)$$

Employing again the technique of the group inverse, we establish that

$$\boldsymbol{\ell}'(1) = \tilde{\boldsymbol{\ell}}'(1) + \mathbf{1}\pi\boldsymbol{\ell}'(1), \quad (11.43)$$

where

$$\tilde{\boldsymbol{\ell}}'(1) = \frac{1}{E[A]}(\mathbf{I} - \mathbf{S}(1))^{\#}(E[A]\mathbf{S}'(1)\mathbf{1} - \mathbf{S}(1)\mathbf{A}'(1)\mathbf{1}). \quad (11.44)$$

We find the scalar $\pi\boldsymbol{\ell}'(1)$ by plugging the previous equation into equation (11.42), and premultiplying by π:

$$2\rho\pi\boldsymbol{\ell}'(1) = 2\pi\mathbf{S}'(1)\mathbf{A}'(1)\mathbf{1} + \pi\mathbf{A}''(1)\mathbf{1} - E[A]\pi\mathbf{S}''(1)\mathbf{1} - 2E[A]\pi\mathbf{S}'(1)\tilde{\boldsymbol{\ell}}'(1). \quad (11.45)$$

We omit the entirely analogous derivation of $\boldsymbol{\ell}''(1)$.

The computational complexity of our method compares favorably to alternatives. After we have computed 2 group inverses of matrices of order $M \times M$, we only have to compute matrix-vector multiplications and vector additions, which have complexities $O(M^2)$ and $O(M)$ respectively. In practice, the real cost is the computation of the boundary probability vector $\mathbf{W}(0)$, which is almost always perforce an iterative procedure. We therefore elaborate on an approximative method for heavy loads (i.e. $\rho \to 1$) which permits to skip the computation of $\mathbf{W}(0)$ altogether.

The idea behind this approximation is that as the load approaches one, $\mathbf{W}(0)$ becomes almost equal to the zero vector, so that the terms in which $\mathbf{W}(0)$ occurs, vanish for loads close to one. Let us mark the system characteristics as they hold for heavy-load conditions by a subscript h, e.g. $E[D_h]$, $\mathbf{W}_h(1)$, $\mathbf{W}'_h(1)$, ... First note that the vector $\mathbf{W}_h(1)$ in that case approaches the vector π. This can also be intuited from the fact that for heavy loads, empty periods get scarce, and hence services follow each other without gaps, which is exactly the definition of the steady-state vector π. Similar simplifications in equation (11.39) lead to

$$E[W_h] = \frac{1}{2(1-\rho)}(\pi \mathbf{A}''(1)\mathbf{1} + 2\pi \mathbf{A}'(1)(\mathbf{I} - \mathbf{A}(1))^{\#}\mathbf{A}'(1)\mathbf{1}). \quad (11.46)$$

The extra terms in the expression (11.27) for the mean delay $E[D]$ can be approximated as well. This leads to the following heavy-load limit for the mean packet delay:

$$E[D_h] = E[W_h] - \rho + \pi \boldsymbol{\ell}'(1) + s, \quad (11.47)$$

where $\pi \boldsymbol{\ell}'(1)$ follows from (11.45). From the numerical results, it is observed that $E[D_h]$ is dominated by the contribution of $E[W_h]$. Note that $E[W_h]$ quantifies the delay incurred by the buffer, whereas the other terms quantify the delay incurred by other packets arriving during the same slot, and the delay incurred by the feedback channel. It is clear that the delay due to buffering will dwarf the other terms in case of high loads. Note that the computational complexity of the heavy-load mean delay is not so much greater than the complexity of the throughput, yet it offers a lot of additional insight in the performance.

11.7 Channel Models

A vast amount of papers have been written on the modeling of communication channels in general [6, 7] and wireless communication channels in particular [15–17]. Many of the proposed models fit into the discrete Markovian framework we have used in this paper.

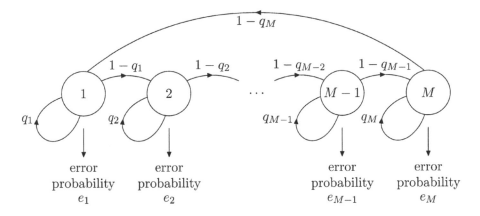

Figure 11.5 The Markov chain of a cyclic channel model.

It is intuitively clear that a larger number of background states can more faithfully model the behavior of real communication channels. However, there are also disadvantages. A model with a large number of background states needs in general a large number of parameters (order M^2). If these have to be estimated from observations, one needs a lot of data to reliably estimate all these parameters. Moreover, it gets difficult to assign an interpretation to the parameters, which is a disadvantage for numerical examples.

11.7.1 Cyclic Channel Model

We propose a channel model that is very versatile yet simple enough to be reasoned about: we allow only cyclic transitions of the channel state, and the sojourn time in each state i, $1 \leq i \leq M$ is (shifted) geometrically distributed with parameter q_i. This chain is illustrated in Figure 11.5. We have the following transition matrix:

$$\mathbf{q} = \begin{pmatrix} q_1 & \bar{q}_1 & & & \\ & q_2 & \bar{q}_2 & & \\ & & \ddots & \ddots & \\ & & & q_{M-1} & \bar{q}_{M-1} \\ \bar{q}_M & & & & q_M \end{pmatrix}. \tag{11.48}$$

State i has error probability e_i. Hence, $2M$ parameters define the channel model (M error probabilities and M transition probabilities). As the mean sojourn time in state i equals $\bar{q}_i^{-1}$, the mean length of an entire cycle (a visit to all states 1 to M, in order) equals

$$L_c = \sum_{j=1}^{M} \frac{1}{1 - q_j}. \tag{11.49}$$

This is an important characteristic of the burstiness of the channel: the higher the mean cycle length, the burstier the channel. The steady-state probabilities σ_i of the channel states are easily obtained as

$$\sigma_i = \frac{\bar{q}_i^{-1}}{L_c}. \tag{11.50}$$

Instead of using the transition probabilities to describe the channel model, we can also use the steady-state probability vector and the mean cycle length, which may give a somewhat more intuitive description. However, not all cycle lengths are possible for a given probability vector. Considering that $\bar{q}_i \leq 1$, for all i, $1 \leq i \leq M$, we see from equation (11.50) that $L_c \geq \sigma_i^{-1}$ for all i, $1 \leq i \leq M$. Hence the minimal cycle length $L_c^{\min}$ equals

$$L_c^{\min} = \frac{1}{\min_i \sigma_i}. \tag{11.51}$$

In the numerical examples, we characterize a cyclic channel model by the cycle length L_c, and two vectors σ and e_c, which record respectively the steady-state probabilities σ_i and the error probabilities e_i.

11.7.2 Gilbert–Elliott Channel Model

For $M = 2$, the cyclic channel model reduces to a Gilbert–Elliott channel model [6, 7], which is the staple model in wireless communication modeling.

The two states of the Gilbert–Elliott model are usually labelled 1 and 2, or 'GOOD' and 'BAD' (see Figure 11.6). The parameters e_1 and e_2 are the error probabilities of the channel in resp. state 1 and 2. Of course, the designations GOOD and BAD make only sense when $e_1 < e_2$, but this is not a requirement for the analysis.

The transitions in the Gilbert–Elliott model are completely defined by the parameters q_1 and q_2, where

$$q_i = \Pr[\text{state in next slot is } i \mid \text{state in current slot is } i].$$

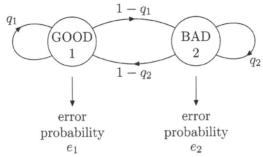

Figure 11.6 The Gilbert–Elliott channel model.

Rather than using q_1 and q_2, we define the parameters

$$\sigma = \frac{1 - q_1}{2 - q_1 - q_2} \quad \text{and} \quad K = \frac{1}{2 - q_1 - q_2} \tag{11.52}$$

to be understood as follows: σ is the fraction of the time that the system is in state 2, while the parameter K can be seen as a measure for the mean lengths of 1- and 2-periods. Specifically, the mean length of a 2-period is K/σ, of a 1-period it is $K/\bar{\sigma}$. Therefore, the factor K can be seen as a measure for the *absolute* lengths of the 1-periods and 2-periods, while σ characterizes their *relative* lengths. The parameter K thus characterizes the degree of correlation in the channel state and is therefore referred to further on as the correlation factor, the value $K = 1$ corresponds to an uncorrelated channel state from slot to slot.

Note that in this model, the sojourn times in both states are geometrically distributed, while some measurements have indicated [17] that even though good periods can be modeled more or less faithfully with geometric lengths, bad periods cannot. A cyclic channel with more than one state designated to model the bad periods, can offer a solution to this problem.

11.8 Numerical Results and Discussion

In this section, we provide some numerical examples. First, we show some probability mass functions of the delay in Figures 11.7 and 11.8, for a Gilbert–Elliott channel model and geometrically distributed batch arrivals with mean $E[A] = 0.3$. The figures show the influence of the correlation factor K, and the feedback delay s respectively. These curves have been obtained from the pgf $D(z)$ by means of an inversion method explained in [18].

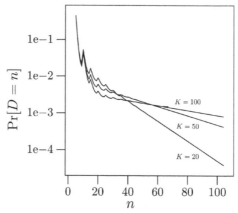

Figure 11.7 Logarithmic plot of Prob[$D = n$] for $s = 4, \sigma = 0.2, e_1 = 0.05, e_2 = 0.5$ and $K = 20, 50, 100$.

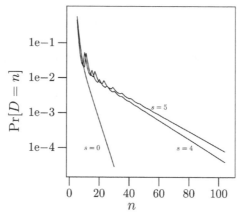

Figure 11.8 Logarithmic plot of Prob[$D = n$] for $\sigma = 0.2, e_1 = 0.05, e_2 = 0.5, K = 20$ and $s = 0, 4, 5$.

We see that the packet delay over a more correlated channel has a heavier tail, which shows that correlation of the channel has an important influence on the performance of the system. A larger feedback delay also has a negative effect on the delay performance: we see a heavier tail for larger values of s.

The huge impact of the correlation is also observed in Figure 11.9, where we plot the mean delay E[D] against the cycle length L_c. We used a Poisson arrival process, with pgf $A(z) = \exp(\lambda(z-1))$, with $\lambda = 1/4$, feedback delay $s = 6$, a four-state cyclic channel model with steady-state probability vector

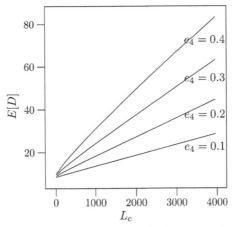

Figure 11.9 Mean delay versus cycle length L_c, with Poisson arrivals ($\lambda = 1/4$), $s = 6$, $\sigma = (0.4, 0.2, 0.2, 0.2)$ and $\mathbf{e}_c = (0.0, 0.1, 0.1, e_4)$, for different values of e_4.

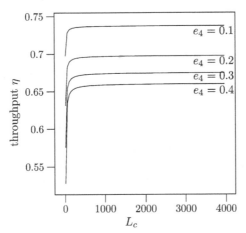

Figure 11.10 Throughput versus cycle length L_c, with Poisson arrivals ($\lambda = 1/4$), $s = 6$, $\sigma = (0.4, 0.2, 0.2, 0.2)$ and $\mathbf{e}_c = (0.0, 0.1, 0.1, e_4)$, for different values of e_4.

$\sigma = (0.4, 0.2, 0.2, 0.2)$, and fixed error probabilities for the first three states (respectively $e_1 = 0$, $e_2 = 0.1$, $e_3 = 0.1$). We let the error probability e_4 take on different values, as indicated in the plot. We see an almost linear increase of the mean delay in function of the cycle length. Hence bursty channels deteriorate the performance of the channel tremendously, although the throughput is actually an increasing function for the very same parameters

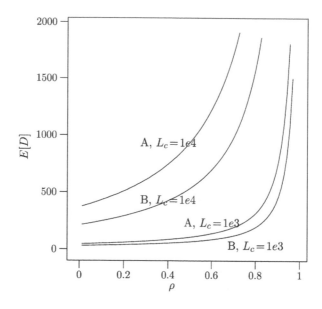

Figure 11.11 A plot of the mean delay versus the load, with feedback delay $s = 6$ for a two-state channel model A ($\mathbf{e}_c = (0.0, 0.3)$, and $\boldsymbol{\sigma} = (0.2, 0.8)$) and a nine-state channel model B ($\mathbf{e}_c = (0.0, 0.3, \ldots, 0.3)$ and $\boldsymbol{\sigma} = (0.2, 0.1, \ldots, 0.1)$). The load is varied by varying parameter λ of the Poisson arrival process.

(see Figure 11.10). This makes a strong case for the fact that the throughput is not sufficient to evaluate the performance of a retransmission protocol.

Next, we show that bad periods with non-geometric lengths have indeed a different performance than bad periods with geometric lengths in Figure 11.11. We compare channel model A with parameters $\mathbf{e}_c = (0.0, 0.3)$, and $\boldsymbol{\sigma} = (0.2, 0.8)$ with channel model B with parameters $\mathbf{e}_c = (0.0, 0.3, \ldots, 0.3)$ and $\boldsymbol{\sigma} = (0.2, 0.1, \ldots, 0.1)$. That is, we compare a Gilbert–Elliott model where the bad period length is geometrically distributed, with a model where the bad period consists of 8 consecutive geometrically distributed subperiods. Note that the error probability in a bad slot in both cases equals 0.3, and moreover, the average length of a bad period is equal to $0.8L_c$ in both cases. In the second model however, the length of a bad period has a significantly lower variance, and (hence) shows a markedly better performance than the Gilbert–Elliott model. The difference is particularly large for more correlated channels (longer L_c). This figure shows

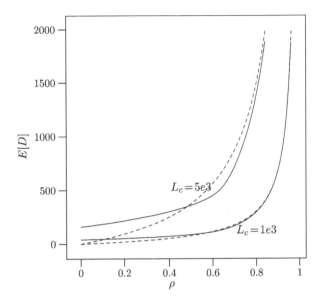

Figure 11.12 A plot of the mean delay versus the load, with the heavy-load approximation indicated with dashed lines, for a four-state model, with $\sigma = (0.4, 0.3, 0.2, 0.1)$ and
$\mathbf{e}_c = (0.0, 0.3, 0.3, 0.3)$ for $s = 6$ and $L_c = 1000, 5000$. The load is varied by varying
parameter λ of the Poisson arrival process.

that more complicated models may indeed be necessary in order to faithfully
capture the performance of a protocol over a wireless link.

Finally, we demonstrate the soundness of our heavy-load approximation
in Figure 11.12. The computationally inexpensive approximations prove to
be accurate for loads higher than 0.7 and sometimes even well before that
point.

11.9 Conclusions

We have studied the transmitter buffer behavior for a Go-Back-N ARQ protocol over a time-varying channel. In particular, we have analyzed and found
expressions for the steady-state distributions of the unfinished work and of
the packet delay, as well as some very efficient heavy-load approximations.
Finally, by means of some examples we have discussed the influence of the
model parameters on the delay performance.

Appendix

A Group Inverse of a Matrix

In this appendix, we elaborate on the technique of the group inverse, which constitutes an alternative for spectral decomposition, although some results [19] hint at a close relationship between the two techniques.

The group inverse of a matrix $\mathbf{M}$ is the unique matrix $\mathbf{M}^{\#}$ which satisfies $\mathbf{M}\mathbf{M}^{\#}\mathbf{M} = \mathbf{M}$, $\mathbf{M}^{\#}\mathbf{M}\mathbf{M}^{\#} = \mathbf{M}^{\#}$, and $\mathbf{M}\mathbf{M}^{\#} = \mathbf{M}^{\#}\mathbf{M}$. We refer to [14] for more details and recall here only that for a transition matrix $\mathbf{A}(1)$ of an irreducible Markov chain the group inverse of $\mathbf{I} - \mathbf{A}(1)$ equals

$$(\mathbf{I} - \mathbf{A}(1))^{\#} = (\mathbf{I} - \mathbf{A}(1) + \mathbf{1}\boldsymbol{\pi})^{-1} - \mathbf{1}\boldsymbol{\pi}, \qquad (11.53)$$

where $\boldsymbol{\pi}$ is the stationary probability vector of $\mathbf{A}(1)$. It can readily be seen that $(\mathbf{I} - \mathbf{A}(1))^{\#}\mathbf{1} = \mathbf{0}$ and $\boldsymbol{\pi}(\mathbf{I} - \mathbf{A}(1))^{\#} = \mathbf{0}$. In general, the solution of $\mathbf{x}(\mathbf{I} - \mathbf{A}(1)) = \mathbf{b}$ is given by $\mathbf{x} = \mathbf{b}(\mathbf{I} - \mathbf{A}(1))^{\#} + \mathbf{x}\mathbf{1}\boldsymbol{\pi}$. Note that the group inverse does not necessarily have to be computed via equation (11.53). Indeed, an alternative algorithm can be found in [20], where one does not have to compute $\boldsymbol{\pi}$ first, but rather $\boldsymbol{\pi}$ is obtained as a simple by-product of the group-inverse computation.

It is beyond the scope of this article to show the similarities of and differences between group-inverse and spectral-decomposition techniques. An advantage of the group inverse is that it needs fewer assumptions. It is enough to assume that the matrix $\mathbf{A}(1)$ has a simple eigenvalue in 1 (ergodicity of the corresponding Markov chain is a sufficient condition for that) to ensure that the group inverse is given by equation (11.53), whereas spectral decomposition techniques often need additional assumptions, such as the assumption that $\mathbf{A}(1)$ be diagonalizable. Another tricky aspect that the use of the group inverse manages to circumvent is the normalization of the eigenvectors.

References

[1] C. T. Bhunia, ARQ – Review and modifications, *IETE Technical Review*, vol. 18, pp. 381–401, 2001.

[2] D. Towsley and J. K. Wolf, On the statistical analysis of queue lengths and waiting times for statistical multiplexers with ARQ retransmission schemes, *IEEE Transactions on Communications*, vol. 25, pp. 693–703, 1979.

[3] A. G. Konheim, A queueing analysis of two ARQ protocols, *IEEE Transactions on Communications*, vol. 28, pp. 1004–1014, 1980.

[4] M. De Munnynck, S. Wittevrongel, A. Lootens and H. Bruneel, Queueing analysis of some continuous ARQ strategies with repeated transmissions, *Electronics Letters*, vol. 38, pp. 1295–1297, 2002.

[5] S. De Vuyst, S. Wittevrongel and H. Bruneel, Queueing delay of Stop-and-Wait ARQ over a wireless Markovian channel, in *Mobility Management and Quality-of-Service for Heterogeneous Networks*, D. Kouvatsos (Ed.), River Publishers, 2009 (this volume).

[6] E. N. Gilbert, Capacity of a burst-noise channel, *The Bell System Technical Journal*, vol. 39, pp. 1253–1265, 1960.

[7] E. O. Elliott, Estimates of error rate for codes on burst-noise channels, *Bell System Technical Journal*, vol. 42, pp. 1977–1997, 1963.

[8] D. Towsley, A statistical analysis of ARQ protocols operating in a non-independent error environment, *IEEE Transactions on Communications*, vol. 27, pp. 971–981, 1981.

[9] J. G. Kim and M. Krunz, Delay analysis of selective repeat ARQ for transporting Markovian sources over a wireless channel, *IEEE Transactions on Vehicular Technology*, vol. 49, pp. 1968–1981, 2000.

[10] E. A. Akkoyunlu, K. Ekanadham and R. V. Huber, Some constraints and tradeoffs in the design of network communications, in *Proceedings of the Fifth ACM Symposium on Operating Systems Principles*, Austin, TX, 19–21 November, pp. 67–74, 1975.

[11] H. R. Gail, S. L., Hantler and B. A. Taylor, Spectral analysis of $M/G/1$ and $G/M/1$ type Markov chains, *Advances in Applied Probability*, vol. 28, pp. 114–165, 1996.

[12] D. A. Bini, G. Latouche and B. Meini, *Numerical Methods for Structured Markov Chains*, Oxford University Press, 2005.

[13] H. Bruneel, Buffers with stochastic output interruptions, *Electronics Letters*, vol. 19, pp. 735–737, 1983.

[14] S. L. Campbell and C. D. Meyer, *Generalized Inverses of Linear Transformations*, Dover Publications, 1991.

[15] Y. Y. Kim and S. Q. Li, Capturing important statistics of a fading/shadowing channel for network performance analysis, *IEEE Journal on Selected Areas in Communications*, vol. 17, pp. 888–901, 1999.

[16] W. Kumwilaisak, C.-C. J. Kuo and W. Dapeng, Fading channel modeling via variable-length Markov chain technique, *IEEE Transactions on Vehicular Technology*, vol. 57, pp. 1338–1358, 2008.

[17] J. McDougall and S. Miller, Sensitivity of wireless network simulations to a two-state Markov model channel approximation, in *Proceedings of GLOBECOM 2003*, San Francisco, CA, 1–5 December, pp. 697–701, 2003.

[18] J. Abate and W. Whitt, Numerical inversion of probability generating functions, *Operations Research Letters*, vol. 12, pp. 245–251, 1992.

[19] C. D. Meyer and G. W. Stewart, Derivatives and perturbations of eigenvectors, *SIAM Journal on Numerical Analysis*, vol. 25, pp. 679–691, 1988.

[20] C. D. Meyer, The role of the group generalized inverse in the theory of finite Markov chains, *SIAM Review*, vol. 17, pp. 443–464, 1975.

12

Queueing Network Topology for Modelling Cellular/Wireless LAN Interworking Systems

Guozhi Song, Laurie Cuthbert and John Schormans

Networks Research Group, School of Electronic Engineering and Computer Science, Queen Mary University of London, Mile End, London E1 4NS, U.K.; e-mail: guozhi.song@elec.qmul.ac.uk

Abstract

A network of queues (or queueing network) is a powerful mathematical tool in the performance evaluation of many complex systems. It has been used in the modelling of hierarchical structure cellular wireless networks with much success, and some researchers have begun to introduce queueing network modelling in the study of cellular/WLAN interworking systems. In the process of queueing network modelling, obtaining the network topology of a system is usually the first step in the construction of a good queueing network model. In this paper, we first review the current research situation and then model an integrated system of cellular and wireless local area networks (WLANs) as an open network of Erlang loss systems. A new network topological representation of the system is also proposed to facilitate the analysis of handover traffic rates. The performance measures of blocking probabilities and dropping probabilities are calculated using the Erlang fixed-point method. An example scenario of three cellular cells and three WLANs interworking is set up, and the numerical results show a clear agreement with the realistic situation.

Keywords: Cellular/WLAN interworking, Erlang loss system, queueing network, topology.

D. D. Kouvatsos (ed.), Mobility Management and Quality-of-Service for Heterogeneous Networks, 269–285.

12.1 Introduction

Queueing theory has a long history for its application in the research of tele-communication systems starting from Erlang, who gave birth to queueing theory and founded teletraffic engineering in his two papers [1, 2] on the calculation of performance probabilities in telephony published in the early 20th century. Jackson first developed the theory of queueing networks in his study of a special type of network of queues [3], named after him as Jackson networks. A queueing network is a network of queues where the departures from one queue enter the next queue. They can be classified into two categories: open queueing networks and closed queueing networks. Open queueing networks have an external input and an external final destination. Closed queueing networks are completely contained and the customers circulate continually, never leaving the network. As an important branch of queueing theory, queueing networks have been used extensively in the modelling and analysis of both wired and wireless networks.

Next-generation wireless communications systems are envisaged to be heterogeneous in nature, with the ability to seamlessly integrate a wide variety of wireless access network technologies that have been developed during the last decade to offer a comprehensive and flexible service to the wireless customers. Among all these techniques cellular networks (including various 2G, 3G technologies such as GSM/GPRS, UMTS/WCDMA, etc.) and wireless local area networks (WLANs) provide very good complementary features in terms of coverage, capacity and mobility support. Thus cellular/WLAN interworking has drawn a lot of attention from both industry and academia as a good candidate for a future integrated wireless solution.

Cellular networks originally aim at providing a high-quality circuit-switched voice service to mobile users within wide areas with coverage of several kilometres and high mobility support, and they have been well deployed around the world and have evolved into the Third Generation (3G) [4]. Three major standards for 3G mobile/wireless cellular networks are the Universal Mobile Telecommunication System (UMTS), CDMA20000 and TD-SCDMA. While WLANs usually operate at unlicensed frequency bands, they were originally aimed at providing relatively high rate data services with lower cost. The majority of current research on cellular/WLAN interworking focuses on relatively high-level issues such as the integration architecture [5–7]. The system is studied from the perspectives of access control, mobility management, security and billing etc. Another research focus of cellular/WLAN integration is the issue of RAN (Radio Access Network) selection and handover decision [8, 9]. Here the objective is to determine the conditions under which the handover should be performed, and which radio access network should be selected to obtain a better performance. Handover is the process of switching connections among networks; it is either horizontal (between networks with the same access technology), or vertical (between networks with different access technology). Handover makes mobility possible, but also makes the mobile-level topology of cellular/WLAN integrated systems unstable and unappreciated in the study of such systems.

In the interworking system between cellular networks and WLANs, there is a hierarchical two-tier overlaying structure [10], similar to a two-tier hierarchical cellular structure, in which small-size microcells overlay with large macrocells to achieve and maintain good network performance over a given service area taking advantage of the high capacity of the microcells and the high coverage of the macrocells. This two-tier structure has been proved to be able to significantly improve the overall system performance. Quite a lot of research has been done on hierarchical cellular networks using queueing theory in modelling, as summarized by Rappaport in [11]. He discussed the inherent load-balancing capability of the hierarchical overlay structure of an integrated system of both microcellular and macrocellular networks, and developed a good analytical model for teletraffic performance (including handover) using queueing theory. Xiao in [12] analyzed of the performance of a hierarchical cellular network that allows the queueing of both overflow slow-mobility calls (from the lower layer microcells) and macrocell handover fast-mobility calls that are blocked due to lack of free resources at the macrocell with the consideration of call repeat phenomenon using both analytical and simulation techniques. By replacing the microcells with WLANs, many methods used in hierarchical cellular network can actually be applied in a cellular/WLAN interworking system.

Due to the difficulties in mathematically modelling complicated traffic features of a cellular/WLAN interworking system, the majority of research in the literature has been performed using simulation. For example, the work we have done in [13] to compare the performance of vertical handover between WCDMA/WLAN and TD-SCDMA/WLAN is based on a simulation model developed using Matlab. However, a highly detailed simulation model implies a large number of parameters and the meticulous implementation of every detail of a specific network increases the complexity of the model and reduces the efficiency, and hence such a model can usually be applied to only very limited situations. Also a simulation model built with specific simulation tools such as OPNET, NS2 and Matlab will inevitably inherit any error that might be contained in these products.

Thus an analytic model of a network would be attractive as it would be able to evaluate network performance under a wide range of conditions, and be computed comparatively easily. It is also essential to the understanding of the underlying principles, and it has the advantage that it can incorporate numerical optimization techniques for network design [14]. It also will avoid the inherited errors introduced from any simulation development products.

So research on an analytic model of cellular/WLAN interworking is still a hot topic. Wei first introduced the method for hierarchical cellular networks into a cellular/WLAN interworking system and analyzed in [10, 16], via a rigid cell cluster structure with only one WLAN in each overlaid cellular cell. She developed her analytical model based on two-dimensional Markov processes and models the cellular/WLAN integrated network as a multi-class loss system. She then used an approximation method to obtain the steady-state probabilities and thus evaluate QoS metrics in terms of call blocking/dropping probabilities and mean data response time. Enrique [15] further developed an analytic model using birth-death process analysis. The model he proposed loosens the constraint on the number of WLANs in each cellular cell. But the complexity of deriving each handover traffic flow among different networks has not been addressed.

In this paper, we extend the modelling method of Enrique and model a cellular/WLAN interworking system as an open network of loss systems with the help of a new network topology scheme to facilitate the analysis of the complex handover traffic flows. The performance measures of the system then can be easily computed using an Erlang fixed-point method. An example interworking system with 3 cellular cells and 3 WLANs is evaluated based on this analytic model. Numerical results are given and discussed.

The remainder of this paper is organized as follows. In Section 12.2 the network deployment of the cellular/WLAN interworking system model is discussed with the introduction of the new network topology scheme and traffic flow analysis in each node and link components, while the study of a cellular/WLAN interworking scenario is given and the results are shown and commented in Section 12.3. Section 12.4 concludes the paper.

12.2 System Model

12.2.1 Network Deployment and Topology

We consider a cellular/WLAN interworking system in which arbitrary numbers of overlaying WLANs are deployed in hot-spot areas inside each cell of the cellular network, as shown in Figure 12.1. Cellular cells are adjacent to each other providing the overall service coverage, while WLANs can be either adjacent or separate from each other with limited coverage to hot-spot areas like shopping mall, airport or a busy office. The mobile terminals are all dual-mode, i.e. they have wireless access interface to both cellular network and WLAN, and can thus switch between these two different networks without interrupting the service.

Such a system can be mapped into the network topology graph as shown in Figure 12.2, with nodes representing each cellular cell and WLAN, and links representing the handover traffic among them. Handover can only happen when there is a link between two nodes, indicating that they are either adjacent or overlaid to each other. There are two types of links in the graph. One shown as a continuous line represents horizontal handover between homogeneous networks, and the other shown as a dashed line represents vertical handover between heterogeneous networks. We can see that the two-tier overlay structure and the handover relation among different networks are well illustrated in this way.

Different from the dynamic physical network topology used in the study of Mobile Ad-hoc NETworks (MANETs), in which each mobile terminal is modelled as a node and the links are the route the packets travel between any two mobile terminals in the signal range, the network topology here is a static logical traffic topology depicting the handover traffic relationship among different networks. Such wireless topologies are defined as *mobile-level topologies* in this paper. Network topology was rarely used in the study of cellular wireless networks or infrastructure WLANs because the physical mobile-level network topology of each cell in cellular network or Basic Service Set (BSS) in WLAN would be just a simple star structure. But when it is introduced into BS (Base Station)/AP (Access Point)-level we obtain a very different topology structure, which proved to be very useful in the analysis of complex handover traffic flows as we will find out later. This model focuses on the traffic of each base station in the cellular networks and each access point in the WLANs. We define this wireless topology as *AP/BS-level topology* as opposed to mobile-level topology.

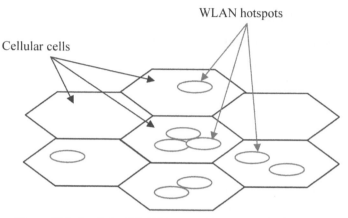

Figure 12.1 A cellular/WLAN interworking system deployment.

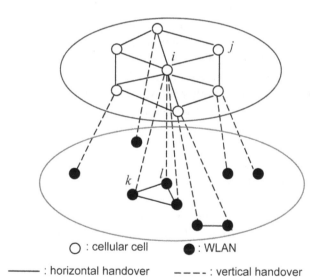

Figure 12.2 Cellular/WLAN interworking system network topology.

We notice that the mobility of mobile terminals is reflected by the handover rates between adjacent nodes, so the movement of each individual mobile terminal is not our concern. Therefore the network topological structure of the system is fixed and it does not evolve with time.

Let us define C_i^a as the set of cellular cell nodes connected to cell node i, i.e., all the cells adjacent to cell i; W_i^o as the set of WLAN nodes connected to cell node i, i.e., all the WLANs inside the coverage of cell i (overlaid

WLAN); W_k^a as the set of WLAN nodes connected to WLAN node k, i.e., all the WLANs adjacent to WLAN k, and C_k^o as the set of cellular cell nodes connected to WLAN node k, i.e., all the overlaying cells of WLAN k.

The capacity of each cell is C_i^c, and the capacity of each WLAN k is C_k^w. Let $R_i^c (R_k^w)$ denote the number of guard channels reserved in cell i (WLAN k) for the handover call. A new call can only be admitted in cell i (WLAN k) if the number of the connections n_i in cell i (WLAN k) is less than $C_i^c - R_i^c$.

The use of reservation channels or guard channels is analogous to the use of trunk reservation in fixed-wire circuit switched networks to give priority to fresh traffic over overflow traffic [14]. So each node can be modelled as an Erlang loss system. The service time at a node would be the channel holding time in the corresponding network the node represented.

12.2.2 Traffic Flow Analysis

The traffic flows in each cellular node and WLAN node are illustrated in Figure 12.3. We can observe that the total arrival rate at cell i is the sum of the arrival rate of new calls, handover calls (including horizontal handover calls from adjacent cells and vertical handover calls from the WLANs inside the cell) and overflow calls from WLANs within the cell. Similarly, the total arrival rate at WLAN k is the sum of the arrival rate of new calls, handover calls (including horizontal handover calls from adjacent WLANs and vertical handover calls from the overlaid cellular cell) and overflow calls from the overlaid cell.

Let us analyze the traffic in the each node in the system. Based on the conservation of traffic flow, traffic flows into a node and out of the same node should be equal. The overall incoming traffic to each node consists of new traffic, horizontal handover traffic, vertical handover traffic, and overflow traffic as shown in Figure 12.3. As an open network, the new traffic $\lambda_i^{(cn)}$ $(\lambda_k^{(wn)})$ comes from outside, while handover traffic, and overflow traffic moving among the nodes along the links, are always in the system. Traffic flows leave the system when calls complete.

$$\lambda_i^c(n_i^c) = \begin{cases} \lambda_i^{(cn)} + \sum_{i \in C_i^a} \lambda_{ij}^{(cc)} + \sum_{k \in W_i^o} \lambda_{ki}^{(wc)} + \sum_{l \in W_i^o} \lambda_{li}^{(wo)}, & n_i^c \leq C_i^c - R_i^c \\ \sum_{j \in C_i^a} \lambda_{ji}^{(cc)} + \sum_{k \in W_i^o} \lambda_{ki}^{(wc)} + \sum_{l \in W_i^o} \lambda_{li}^{(wo)}, & n_i^c > C_i^c - R_i^c \end{cases}$$

(12.1)

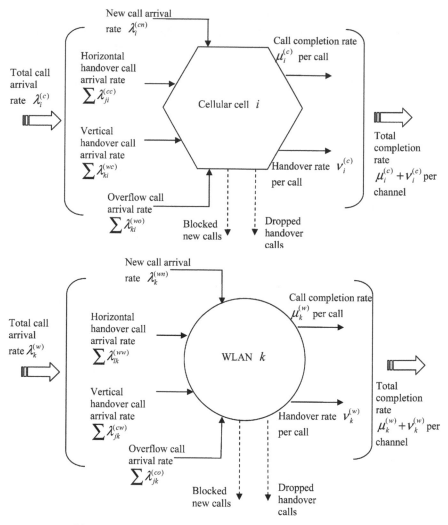

Figure 12.3 Traffic flows in cellular cell node and WLAN node.

Similarly, the overall traffic in WLAN k is given as

$$
\lambda_k^w(m_i^w) = \begin{cases} \lambda_k^{(wn)} + \displaystyle\sum_{l \in E_k^a} \lambda_{lk}^{(ww)} + \sum_{j \in C_k^o} \lambda_{jk}^{(cw)} + \sum_{g \in C_k^o} \lambda_{gk}^{(co)}, & m_k \le C_k^w - R_k^w \\[2ex] \displaystyle\sum_{l \in W_k^a} \lambda_{lk}^{(ww)} + \sum_{j \in C_k^o} \lambda_{jk}^{(cw)} + \sum_{g \in C_k^o} \lambda_{gk}^{(co)}, & m_k > C_k^w - R_k^w \end{cases}
$$

$$(12.2)$$

The total rate of channel completions at each node due to either a channel becoming free because the call has been completed normally or because it is handed over to another node is therefore $\mu_i^{(c)} + v_i^{(c)}$ in cell i ($\mu_k^{(w)} + v_k^{(w)}$ in WLAN k), i.e., $1/(\mu_i^{(c)} + v_i^{(c)})$ is the mean channel holding time at cell i ($1/(\mu_k^{(w)} + v_k^{(w)})$ in WLAN k).

Let us assume that a call in cellular cell i may either finish and leave the system at the end of its holding time with a probability of $T_i^{(c)}$, or still move within the system and continue in an adjacent cell or WLAN node with probability $1 - T_i^{(c)}$ which is the sum of probabilities that the call goes to any adjacent nodes. So we have

$$1 - T_i^{(c)} = \sum_{j \in C_i^a} H_{ij}^{(cc)} + \sum_{k \in W_i^o} H_{ik}^{(cw)}, \qquad (12.3)$$

where $H_{ij}^{(cc)}$ is the probability of attempting a horizontal handover to adjacent cell j, and $H_{ik}^{(cw)}$ is the probability of attempting a vertical handover to WLAN k inside cell i.

Similarly, in WLAN we can obtain

$$1 - T_k^{(w)} = \sum_{l \in W_k^a} H_{kl}^{(ww)} + \sum_{i \in C_k^o} H_{ki}^{(wc)}, \qquad (12.4)$$

where $H_{kl}^{(ww)}$ is the probability of attempting a horizontal handover to adjacent WLAN l, $H_{ki}^{(cw)}$ is the probability of attempting a vertical handover to overlaying cell i.

As each node of the system is modelled as an Erlang loss system, the number of calls in a node evolves according to an irreducible birth-death process with birth rate λ_i^c in cellular cell node, λ_k^w in WLAN node, and death rate $(\mu_i^{(c)} + v_i^{(c)}) \cdot n_i^c$ in cellular cell node, $(\mu_k^{(w)} + v_k^{(w)}) \cdot n_k^w$ in WLAN node.

12.2.3 Handover Rates

We now focus on the traffic on the links, i.e., handover traffic and overflow traffic, to derive the handover rates in (13.1) and (13.2).

The horizontal handover rate $\lambda_{ji}^{(cc)}$ on link L_{ij} is the handover traffic of cell j offered to cell i, for the adjacent cells i and j. It is given as all the non-blocked (dropped) traffic in cell j multiplied by a handover probability $H_{ji}^{(cc)}$.

$$\lambda_{ji}^{(cc)} = H_{ji}^{(cc)} \cdot \left[\lambda_j^{(cn)} \cdot (1 - B_j^{(c)}) \right.$$

$$\left. + \left(\sum_{x \in C_j^a} \lambda_{xj}^{(cc)} + \sum_{y \in W_j^o} \lambda_{yj}^{(wc)} + \sum_{z \in W_j^0} \lambda_{zj}^{(wo)} \right) \cdot (1 - D_j^{(c)}) \right], \tag{12.5}$$

where $B_j^{(c)}$ and $D_j^{(c)}$ are the new call blocking and handover dropping probabilities in cell j respectively.

In (13.1) and (13.2), $\lambda_{zj}^{(wo)}$ is the overflow traffic, which is the handover traffic not accepted in the WLAN due to the lack of resource and thus overflowed to cell j.

$$\lambda_{zj}^{(wo)} = D_z^{(w)} \cdot \left(\lambda_{jz}^{(cw)} + \sum_{l \in W_z^a} \lambda_{lz}^{(ww)} \right). \tag{12.6}$$

The vertical handover rate on link L_{ki} is the handover traffic of WLAN k offered to an overlay cell

$$\lambda_{ki}^{(wc)} = H_{ki}^{(wc)} \cdot \left[\lambda_k^{(wn)} \cdot (1 - B_k^{(w)}) \right.$$

$$\left. + \left(\sum_{x \in W_k^a} \lambda_{xk}^{(ww)} + \sum_{y \in C_k^o} \lambda_{yk}^{(cw)} + \sum_{z \in C_k^o} \lambda_{zk}^{(co)} \right) \cdot (1 - D_k^{(w)}) \right], \tag{12.7}$$

where $B_k^{(w)}$ and $D_k^{(w)}$ are respectively the new call blocking and handover dropping probabilities in WLAN k.

In equation (13.5), $\lambda_{zk}^{(co)}$ is the handover traffic that is not accepted in cell z due to the lack of resource and thus overflowed to WLAN k

$$\lambda_{zk}^{(co)} = D_z^{(c)} \cdot \left(\lambda_{kz}^{(wc)} + \sum_{l \in C_z^a} \lambda_{lz}^{(cc)} \right) \cdot \gamma_{zk}, \tag{12.8}$$

where γ_{zk} is the coverage factor between WLAN k and overlay cell z, i.e. the ratio between the radio coverage area of WLAN k and the radio coverage area of cell z with $0 < \gamma_{zk} \le 1$.

The horizontal handover rate $\lambda_{lk}^{(ww)}$ is the handover traffic of WLAN l offered to WLAN k, for adjacent WLAN l and k. It is given as

$$\lambda_{lk}^{(ww)} = H_{lk}^{(ww)} \cdot \left[\lambda_l^{(wn)}(1 - B_l^w) \right. \tag{12.9}$$

$$\left. + \left(\sum_{x \in W_l^a} \lambda_{xl}^{(ww)} + \sum_{y \in C_l^o} \lambda_{yl}^{(cw)} + \sum_{z \in C_l^o} \lambda_{zl}^{(co)} \right) \cdot (1 - D_l^w) \right],$$

where $B_k^{(w)}$ and $D_k^{(w)}$ are the new call blocking and handover dropping probabilities in WLAN k respectively.

Vertical handover rate of cell j offered to WLAN k is

$$\lambda_{jk}^{(cw)} = H_{jk}^{(cw)} \cdot \left[\lambda_j^{(cn)}(1 - B_j^c)\gamma_{jk} \right. \tag{12.10}$$

$$\left. + \left(\sum_{x \in C_j^a} \lambda_{xj}^{(cc)} \gamma_{jk} + \sum_{y \in W_j^o} v_{yj}^{(wc)} \gamma_{jk} + \sum_{z \in W_j^o} \lambda_{zj}^{(wo)} \right) \cdot (1 - D_j^c) \right].$$

We cannot observe that overflow traffic is actually the handover traffic that can be admitted into the destination node thus bounced back.

With the help of the network topology given in Section 13.2.1, we can easily identify the horizontal handover and vertical handover traffic existed in the interworking system and write the handover traffic equations.

12.2.4 Performance Metrics

The performance metrics of concern in our cellular/WLAN interworking system are the new call blocking probabilities and the handover call dropping probabilities in each cellular cell and WLAN. As the admission policy adopted is a cut-off policy, with reservation channels allocated for the handover traffic, the handover traffic has higher priority than the new call traffic. So we expect lower handover dropping probabilities than new call blocking probabilities.

Because both cellular cell and WLAN nodes are modelled as Erlang loss systems, the steady-state probability of calls in cellular cell i is given by

$$\pi_i^c(n_i) = \pi_i^c(0) \prod_{l=0}^{n_i} \frac{\lambda_i^c}{n_i \cdot (\mu_i^{(c)} + v_i^{(c)})}, \quad n_i = 1, 2, \ldots, C_i^c, \tag{12.11}$$

where $\pi_i^c(0)$ is the normalization constant, obtained as

$$\pi_i^c(0) = \left[\prod_{l=0}^{C_i^c} \frac{\lambda_I^c}{n_i \cdot (\mu_i^{(c)} + v_i^{(c)})} \right]^{-1}. \qquad (12.12)$$

Thus the blocking and dropping probabilities in cell i are

$$B_i^{(c)} = \sum_{n_i}^{C_i^c - 1} \pi_i^c(n_i) + \pi_i^c(C_i^c), \qquad (12.13)$$

$$D_i^{(c)} = \pi_i^c(C_i^c). \qquad (12.14)$$

Similarly, the steady-state probability of calls in WLAN k can be given by

$$\pi_k^w(m_k) = \pi_k^w(0) \prod_{l=0}^{m_k} \frac{\lambda_k^w}{m_k \cdot (\mu_k^{(w)} + v_k^{(w)})}, \qquad m_k = 1, 2, \ldots, C_k^w, \quad (12.15)$$

where $\pi_k^w(0)$ is the normalization constant, obtained as

$$\pi_k^w(0) = \left[\prod_{l=0}^{C_k^w} \frac{\lambda_k^w}{(\mu_k^{(w)} + v_k^{(w)})} \right]^{-1}. \qquad (12.16)$$

We can calculate the blocking and dropping probabilities in WLAN k as

$$B_k^{(w)} = \sum_{K_k^w = C_k^w - R_k^w}^{C_c^w - 1} \pi_k^w(m_k) + \pi_k^w(C_k^w), \qquad (12.17)$$

$$D_k^{(w)} = \pi_k^w(C_k^w). \qquad (12.18)$$

Blocking and dropping probabilities and overall call arrival rate at each node are interdependent with each other, through handover rates, as we can observe from (13.1) to (13.18). This model gives a set of non-linear equations which can be solved numerically by successive substitution to obtain the blocking and dropping probabilities in the system.

There is a fairly broad class of approximation techniques to calculate the blocking and dropping probabilities often referred as reduced load approximations as described in [17]. The general approach is to diminish the arrival rates of offered traffic to a subnetwork (which may be a single link) by a factor

Table 12.1 Erlang fixed-point algorithm for calculating blocking and dropping probabilities.

1: set initial values to $B_i^{(c)}$, $D_i^{(c)}$, $B_k^{(w)}$, $D_k^{(w)}$

2: set $0 < \varepsilon << 1$

3: **while** $\sum_i B_{i_{new}}^{(c)} - B_i^{(c)} + \sum_i B_{i_{new}}^{(c)} - B_i^{(c)} > \varepsilon$ **and** $\sum_i B_{i_{new}}^{(c)} - B_i^{(c)} + \sum_i B_{i_{new}}^{(c)} - B_i^{(c)} > \varepsilon$

4:      determine the handover rates $\lambda_{ji}^{(cc)}$, $\lambda_{lk}^{(ww)}$, $\lambda_{ki}^{(wc)}$, $\lambda_{jk}^{(cw)}$

        by solving system of equations (3)-(8)

5:      determine the total traffic to cell i (λ_i^c) and WLAN (λ_k^w)

6:      calculate the steady state distribution $\pi_i^c(n_i)$ and $\pi_k^w(m_k)$

7:      compute the blocking and dropping probabilities $B_{i_{new}}^{(c)}$, $D_{i_{new}}^{(c)}$, $B_{k_{new}}^{(w)}$, $D_{k_{new}}^{(w)}$

8: **end while**

9: Output ($B_i^{(c)}$, $D_i^{(c)}$, $B_k^{(w)}$, $D_k^{(w)}$)

equal to the probability that a new call on that route would not be blocked on the other link of its path. Typically this method leads to a set of fixed point equations for which there exists a (not necessarily unique) solution. The underlying assumption is that of independent blocking between the individual subnetworks and, although invalid in most non-trivial situations, does yield particularly good results when traffic correlations are small.

In Table 12.1, we give the Erlang fixed-point algorithm for calculating blocking and dropping probabilities, which is a member of the reduced load class. It each time use the calculated new probabilities $B_{i_{new}}^{(c)}$, $D_{i_{new}}^{(c)}$, $B_{k_{new}}^{(w)}$, $D_{k_{new}}^{(w)}$ to estimate the target probabilities $B_i^{(c)}$, $D_i^{(c)}$, $B_k^{(w)}$, $D_k^{(w)}$, and these two set of values are compared. When the difference between them converges to a satisfactory degree, defined by ε, we have our final results.

12.3 Numerical Results and Discussion

We now evaluate a scenario of cellular/WLAN interworking system consisting of 3 cellular cells and 3 overlay WLANs with the deployment and network topology as shown in Figure 12.4. We assume that all cells in the cellular network have capacity $C_i^c = 30$ units of bandwidth, and all WLANs have capacity $C_k^w = 60$ units of bandwidth. The mean channel holding time in the cellular network is 1 min, and the mean channel holding time in the WLANs is 4 min. The coverage factor R_{jk} is 0.75. The mobility of customers given by (13.3) in the cellular network is $T_i^{(c)} = 0.4$, and the mobility in the WLAN, given by (13.4) is $T_k = 0.4$. These values define a mobility level of 60% (i.e.,

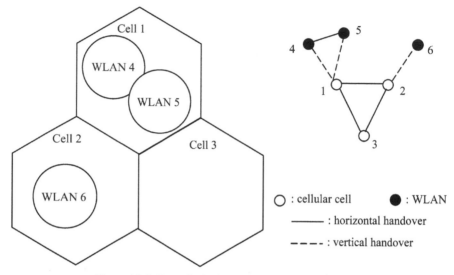

Figure 12.4 Scenario deployment and network topology.

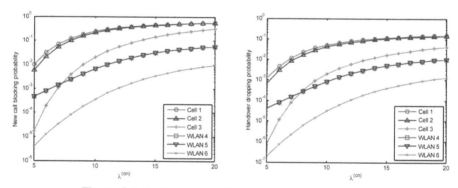

Figure 12.5 Call blocking and handover dropping probabilities.

an average 60% of customers perform handovers). All these parameters were used in [15].

Figure 12.5 shows the call blocking probabilities and handover dropping probabilities for each cellular cell and WLAN node in the interworking system when the new call arrival rate of cellular cell $\lambda_i^{(cn)}$ is increased from 5 to 20 calls per minute, and the $\lambda_k^{(wn)}$ fixed to 5 calls per minute. We can observe that due to the distribution of the nodes being asymmetric the performance of each node is different. Except for the new traffic, both cell 1 and cell 2 have overlay WLANs inside, which brings extra handover traffic, so the call

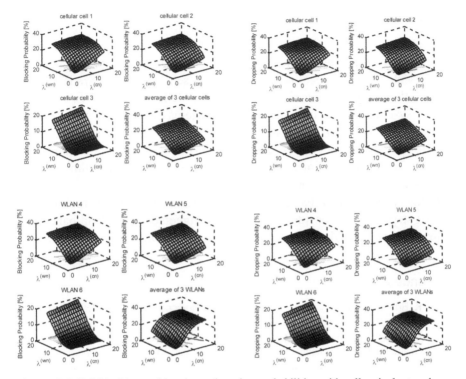

Figure 12.6 Call blocking and handover dropping probabilities with call arrival rates change in both cellular networks and WLANs.

blocking and dropping probabilities are much higher than in cell 3 where there is no overlay WLAN.

Also there are two WLANs inside cell 1 and one WLAN inside cell 2, so traffic in cell 1 is higher than in cell 2, thus the blocking and dropping probabilities are higher. Because of the locally symmetric position of WLAN node 4 and 5 they have the same blocking and dropping probability pattern, which is to be expected.

In Figure 12.6 we changed the call arrival rates of both cellular cells and WLANs and plotted the new call blocking probabilities and handover dropping probabilities for each cellular cell, WLAN and their average. So we can have a good picture of the effect of traffic increase of both cellular and WLAN upon the performance of the integrated system.

12.4 Conclusion

In this paper, we reviewed the current research on the application of queueing network theory in the modelling of cellular/WLAN interworking systems. We then presented a new logical network topology scheme in the BS/AP level to facilitate the analysis of handover rates and modelled cellular/WLAN interworking system as a network of Erlang loss systems to derive the performance metrics of the interworking system. Our numerical results for an example scenario show the model can be used in a realistic situation. The method can be used in the analysis of any arbitrary number of cellular cells with overlaid WLANs with arbitrary topological structure.

In the current model, handover is only triggered by the user mobility. In the next step, we can consider the handover triggered by the load balancing requirement as well. Multi-class traffic scenarios can also be considered.

References

[1] A. K. Erlang, The theory of probabilities and telephone conversations, *Nyt Tidsskrift for Matematik B*, vol. 20, pp. 33–39, 1909.

[2] A. K. Erlang, Solution of some problems in the theory of probabilities of significance in automatic telephone exchanges, *Elektrotkeknikeren*, vol. 13, pp. 5–13, 1917.

[3] J. R. Jackson, Jobshop-like queueing systems, *Management Science*, vol. 10, pp. 131–142, 1963.

[4] W. Song, W. Zhuang and A. Saleh, Interworking of 3G cellular networks and wireless LANs, *Int. J. Wireless and Mobile Computing*, vol. 2, no. 4, pp. 237–247, 2007.

[5] M. M. Buddhikot, G. Chandranmenon, S. Han, Y.-W. Lee, S. Miller and L. Salgarelli, Design and implementation of a WLAN/cdma2000 interworking architecture, *IEEE Commun. Mag.*, vol. 41, no. 11, pp. 90–100, November 2003.

[6] A. K. Salkintzis, Interworking techniques and architectures for WLAN/3G integration toward 4G mobile data networks, *IEEE Wireless Commun. Mag.*, vol. 11, no. 3, pp. 50–61, June 2004.

[7] M. Bernaschi, F. Cacace, G. Iannello, S. Za and A. Pescape, Seamless internetworking of WLANs and cellular networks: architecture and performance issues in a Mobile IPv6 scenario, *IEEE Wireless Commun. Mag.*, vol. 12, no. 3, pp. 73–80, June 2005.

[8] E. Stevens-Navarro, Y. Lin and V. W. S. Wong, An MDP-based vertical handoff decision algorithm for heterogeneous wireless networks, *IEEE Trans. Vehicular Technol.*, vol. 57, no. 2, pp. 1243–1254, March 2008.

[9] Alkhwlani, M.M., Ayesh, A., Access network selection using combined fuzzy control and MCDM in heterogeneous networks, in *Proceedings of the International Conference on Computer Engineering & Systems, ICCES'07*, 27–29 November, pp. 108–113, 2007.

[10] W. Song, H. Jiang and W. Zhuang, Performance analysis of the WLAN-first scheme in cellular/WLAN interworking, *IEEE Trans. Wireless Commun.*, vol. 6, no. 5, pp. 1932–1943, May 2007.

[11] S. S. Rappaport and L. Hu, Microcellular communication systems with hierarchical macrocell overlays: Traffic performance models and analysis, *Proc. IEEE*, vol. 82, no. 9, September, 1994.

[12] X. Liu and A. O. Fapojuwo, Performance analysis of hierarchical cellular networks with queueing and user retrials, *Int. J. Commun. System*, vol. 19, pp. 699–721, 2006.

[13] S. Jin, G. Song and L. Cuthbert, Performance comparison of vertical handover between WCDMA/WLAN and TD-SCDMA/WLAN interworking systems, in *Proceedings of AISPC'08 – 2nd Annual IEEE Student Paper Conference*, Aalborg, Denmark, February, pp. 1–5, 2008.

[14] D. E. Everitt, Traffic engineering of the radio interface for cellular mobile networks, *Proc. IEEE*, vol. 82, no. 9, pp. 1371–1382, September 1994.

[15] E. Stevens-Navarro, and V. W. S. Wong, Resource sharing in an integrated wireless cellular/WLAN system, in *Proceedings of CCEC'07*, Washington, DC, November, pp. 631–634, 2007.

[16] W. Song, H. Jiang, W. Zhuang, and A. Saleh, Call Admission Control for Integrated Voice/Data Services Cellular/WLAN Interworking, in *Proceedings of IEEE ICC'06*, Istanbul, Turkey, June, pp. 5480–5485, 2006.

[17] M. R. Thompson, and P. K. Pollett, A reduced load approximation accounting for link interactions in a loss network, *J. Appl. Math. Decision Sci.*, vol. 7, no. 4, pp. 229–248, 2003.

PART FOUR
ACCESS NETWORKS COVERAGE

13

Radio-over-Fiber to Increase Effective Coverage of Motorway Access Networks

Weixi Xing[1], Demetres D. Kouvatsos[2] and Yue Li[1]

[1]*Institute of Advanced Telecommunications, University of Wales, Swansea SA2 8PP, U.K.; e-mail: {w.xing, y.li}@swansea.ac.uk*
[2]*PERFORM – Networks and Performance Engineering Research Unit, Department of Computing, School of Informatics, University of Bradford, Bradford BD7 1DP, West Yorkshire, U.K.; e-mail: d.d.kouvatsos@scm.brad.ac.uk*

Abstract

WiMAX is a standards-based wireless technology that provides high-throughput broadband connections over long distance. However, deploying conventional size of WiMAX base station (BS) on the motorway may cause low effective coverage and wireless resource waste as a large proportion of wireless coverage areas may not have any traffic flow at all. This paper develops an optimized cell size based on the width of motorway and the overlapping length between adjacent motor-picocells. Consequently, it carries out system design and implementation through a combination of Time Division Multiple Access (TDMA) and radio-over-fiber (RoF) technologies. It is observed that, by converting conventional cells into motor-picocells of remote antenna units (RAUs), the effective coverage can be greatly improved. Furthermore, four handover indicators are introduced to deal with frequent handover requests evolved among RAUs governed by single BS and also handovers between different BSs. In this context, the novel concept of time slot distributor (TSD) is introduced to avoid co-channel interferences and make the system cost-effective.

D. D. Kouvatsos (ed.), Mobility Management and Quality-of-Service for Heterogeneous Networks, 289–306.

Keywords: Worldwide Interoperability for Microwave Access (WiMAX), Internet protocol (IP), Medium Access Control (MAC), Time Division Multiple Access (TDMA), radio-over-fiber (RoF), quality-of-service (QoS), mobile host (MH), road vehicle communication (RVC), motor-picocell, remote antenna unit (RAU), efficient coverage scheme, time slot distributor (TSD), fast handover.

13.1 Introduction

Worldwide Interoperability for Microwave Access (WiMAX) is leading the new trend of wireless communications, and has proven to be a cost-effective fixed wireless alternative to cable and digital subscriber loop services [1]. WiMAX is particular value in rural areas and developing countries, which frequently lack optical fiber or copper-wire infrastructures for broadband services. As providers are unwilling to install the necessary equipment for regions with little profit potential, wireless communications can address this problem [2].

Mobile WiMAX based on the IEEE 802.16e enables wireless communication systems to address mobile roadband services at vehicular speeds greater than 120 km/h with quality-of-service (QoS) guaranteed comparable to broadband wireline access alternatives. The WiMAX forum has defined the design requirements of the network architecture necessary for implementing an end-to-end Mobile WiMAX network [3, 4]. Jang et al. [5] has described two possible deployment architectures of 802.16 networks and the mobile host's (MH) handover over it. That deployment has been shown in Figure 13.1, which incorporates two Internet protocol (IP) subnets, an access router (AR) and several base stations (BSs) form a single subnet. Note that in this case, the movement between BSs does not always require IP mobility. The handover from BS1 to BS2, or within same subnet, can be carried out using link layer mobility without IP mobility. However, the handover from BS5 to BS6 may require IP mobility since they belong to the two different respective subnets.

Figure 13.2 represents an alternative 802.16e deployment where a subnet consists of only single AR and single BS. In this case, a BS may be integrated with an AR, composing one box in view of implementation. Every handover in this architecture means a change of subnet, resulting in IP handovers.

In 802.16e, the handover procedure is conceptually divided into two steps: 'handover preparation' and 'handover execution' [6]. The handover preparation begins with a decision at the MH or BS. During the handover

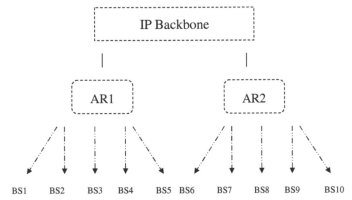

Figure 13.1 The 802.16e deployment architecture in a centralized manner.

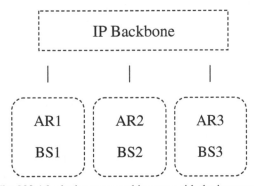

Figure 13.2 The 802.16e deployment architecture with the integrated BS and AR.

preparation, neighbouring BSs are compared by standard metrics such as signal strength or QoS parameters and the target BS is selected among them. If necessary, the MH may try to associate (i.e., initial ranging) with candidate BSs to expedite a potential future handover. Once the MH decides handover, it may notify its intent by sending a request message to the serving BS, which then sends a reply containing the recommended BSs to the MH after negotiating with candidates. When the target BS is decided, the serving BS may confirm the handover to target BS over backbone. After handover preparation, handover execution occurs. However, when the MH moves to different subnet, it should re-configure new IP address and re-establish IP connection. To resume the active session of the previous link, the MH should also perform IP handover.

Kim et al. [7] proposed the road vehicle communication (RVC) system as an effective tool to utilize the infrastructure deployed along the motorway and provide on road vehicle access networks with intelligent transportation systems. This will make telecommunication based on real-time moving images between the roadside and vehicles possible in the near future, enabling the occupants to access service through Internet [8]. For this purpose, RVC systems indicate that data rate of about 2–10 Mbps per MH will be required, in which not only voice, data but also multimedia services such as real time video under high mobility conditions are necessary [9]. To support higher data rate, some system designs suggest millimeter-wave (mm-wave) bands such as 36 or 60 GHz bands [10, 11]. These designs are characterized by very small cell size (less than hundred meters) due to millimeter-wave bands having higher free space propagation loss as compared to conventional microwave bands, whilst numerous BSs are required to cover long roads and high user mobility. For the system to be cost-effective, the adoption of centralized network architecture is recommended so that resource management could be made more efficient. In particular, a large number of BSs, which need to be deployed along the road and serve as remote antenna units (RAU) for MH, are interconnected with a control station (CS) that perform all processing such as modulation/demodulation, routing, medium access control (MAC) and so on through radio-over-fiber (RoF) technologies.

As a consequence, mobility management clearly becomes very significant. The system must support a fast and simple handover procedure, as for example, in a motorway scenario where a vehicle with an ongoing communication session is running at a speed of 120 km/h and the cell size is 100 m, then it will request handover every three second. In addition, if the overlapping area between two adjacent cells is 10 m, handover must be done within 0.3 sec. In Abe et al. [8], a virtual cellular zone concept was introduced, namely all small BSs connecting to the same CS form a large cellular zone in which handover is done between the large cellular zones, not between the small BSs zones. This can dramatically reduce the times of handover, but a drawback of the system is that data cannot be properly received in the overlapping region between small BSs because of co-channel interference. A medium access control (MAC) protocol for RoF RVC system has been proposed in [12]. It is based on reservation slotted ALOHA and dynamic slot assignment. This architecture also assumes simulcasting from all the BSs connected to a CS, having co-channel interference problem in overlapping areas.

However, few researches have addressed the problem on how to deploy WiMAX on a motorway scenario, which leaves unanswered a number of crucial questions such as: What kind of system architecture is more suitable for motorway characteristics? How to utilize motorway infrastructure to create a cost effective system? Note that conventional WiMAX cell size ranges from several km to tens of km in radius where overlapping region is so large that there is enough time to treat handover. But when applying conventional size BS to motorway scenario at one dimension, it will inevitably lead to large areas of unused coverage that have no traffic flow at all. How to overcome the contradiction between conventional cell size and low effective wireless single coverage? Is there an optimized cell size available?

This paper employs a millimeter-wave bands system structure into microwave bands of WiMAX, so as to overcome the low effective coverage problem when deploying conventional size of WiMAX BS in a motorway scenario. In this context, while millimeter-wave bands architecture employs one CS controlling a number of BSs, a further division is suggested of one BS into quite a number of picocells of RAU.

Utilizing the infrastructure of motorway to deploy WiMAX needs balance among system processing capacity, cell size as well as cost [8]. Suppose the system processing capacity of one BS can be reflected by its covering area, increasing the effective coverage can get its potential processing capacity fully utilized. Through a systematic calculation, an optimized cell size with the best effective coverage can be figured out based on the width of motorway and the length of overlapping, which are predetermined. However, new challenging system requirements can be raised, namely the system should be cost-effective and support a fast and simple handover procedure as well as be able to overcome co-channel interference. To address these issues,, a new system structure is proposed by combining RoF with time division multiple access (TDMA) and time division duplex (TDD) technologies. To this end, a time slot distributor (TSD) is introduced to avoid co-channel interference and many remote access units (RAUs) can share one headend to reduce cost further. To complement this system design, a simplified handover procedure is presented with four handover indicators, while maintaining the burden of resource management at the BS remains unchanged.

This paper is organized as follows. In Section 13.2 a theoretical calculation is performed to get an optimized cell size based on certain motorway width and overlapping distance. The new system design and implementation are presented in Section 13.3. The related frame structure and handover scheme are presented in Section 13.4. Numerical results and conclusions follow in Sections 13.5 and 13.6, respectively.

13.2 Analytic Solution of Motor-Picocell System

As it was mentioned earlier, a conventional WiMAX cell size ranges from several kilometers to several tens of kilometers in radius. applying a fixed radius at one dimension to the motorway scenario will inevitably lead to large areas of unused coverage. Even if the cell size is reduced to 500 m in radius, taking into account a six lane of motorway normally not more than 32 m in width, there are still hundreds of thousand square meters of unused covering areas on each side of the motorway, where no vehicular traffic flow at all. In this section, the calculation of the effective coverage rate is determined analytically and it is based on the following notation.

Notation

R	conventional cell radius
r	motor-picocell radius
θ	central angle of the conventional cell
α	central angle of the motor-picocell
Cov_R	effective measure parameter based on R
Cov_r	effective measure parameter based on r
d	width of the mortoway
D	length of the overlap
l	width of the overlap
n	number of the motor-picocells of the ECMAN system (cluster size)

The effective parameter, namely Cov_R, is defined as effective coverage of motorway access networks (ECMAN) and is expressed by

$$\mathrm{Cov}_R = \frac{\mathrm{Area}_{\mathrm{motorway}}}{\mathrm{Area}_{\mathrm{BS}}} = \frac{2\pi R^2 \frac{\alpha}{360} + R^2 \sin\theta}{\pi R^2} \tag{13.1}$$

where $\mathrm{Area}_{\mathrm{motorway}}$ is the wireless signal coverage on the motorway (cf. area 1 in Figure 13.4). R is radius of the BS, θ is central angle. Please note that θ is denoted by the width of the motorway and radius of the conventional cell. As it can be observed in Figure 13.3, as the the radius of the conventional cell size is reduced, the effective coverage Cov_R is improved. In particular, for a radius less than 100 m, the Cov_R increases sharply.

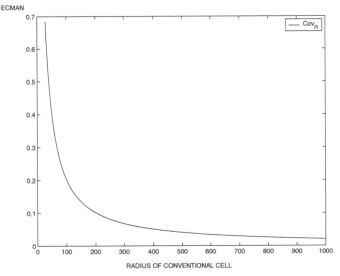

Figure 13.3 General coverage of the motor-picocells system.

To increase the efficiency of the radio signal on a motorway scenario is very important task towards improving system performance and making it cost effective. A flat approach to conventional WiMAX cell is adopted herewith by narrowing the cover of motorway on both sides whilst stretching the cover along the motorway. This could be achieved by dividing a conventional BS into a number of picocells and deploy them along the motorway (see Figure 13.4).

It is assumed that the area does not change after the convertion, i.e., the processing capacity of BS is the same as before but it can be fully utilized through this convertion. However, note that a certain degree of overlapping areas among motor-picocells is needed so as to guarantee vehicles with an ongoing communication session not to be dropped. To this end, the following equation clearly holds:

$$\pi R^2 = n\pi r^2 - 2(n-1)\left(\pi r^2 \frac{\alpha}{360} - \frac{1}{2}r^2 \sin\alpha\right) \qquad (13.2)$$

where the area of conventional cell equals to the sum areas of motor-picocells. R is the radius of the conventional cell and r is the radius of the motor-picocell. In this context, the enhanced effective coverage, Cov_r, is defined

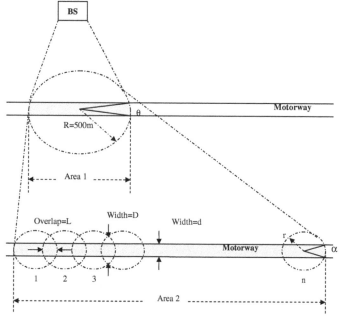

Figure 13.4 Motor-picocells scheme.

as

$$\text{Cov}_r = \frac{d\left(2nr - 2(n-1)\left(r - 1/2\sqrt{4r^2 - d^2}\right)\right)}{\pi R^2} \qquad (13.3)$$

where n is the number of the motor-picocells (size of cluster), r is the radius of the motor-picocells, and d is the width of the motorway.

However, in a real situation this converting process is limited by a number of factors. The first factor is the selection of the length of overlapping area (L in Figure 13.5) because handover is related to the speed of vehicles, e.g., L must be long enough for the vehicles to handle handover. For this purpose, the width of the overlapping area (D in Figure 13.5) must be larger than that of the motorway (d in Figure 13.5), e.g.,

$$D = d + \Delta d \qquad (13.4)$$

where Δd is the guarantee distance. So we can easily get

$$r^2 = \left(\frac{D}{2}\right)^2 + \left(r - \frac{l}{2}\right)^2 \qquad (13.5)$$

and also

$$\sin\frac{\alpha}{2} = \frac{D/2}{r} \tag{13.6}$$

Then

$$r = \frac{D^2 + l^2}{4l} \tag{13.7}$$

From the conditions above, we can abstract the following general analytical solution:

$$\text{Cov} = \frac{D\left(2nr - 2(n-1)\left(r - \frac{1}{2}\sqrt{4r^2 - D^2}\right)\right)}{\pi R^2} \tag{13.8}$$

in which n can be expressed as

$$n = \frac{\pi R^2 - 2\left(\pi r^2 \frac{\alpha}{360} - \frac{1}{2}r^2 \sin\alpha\right)}{\pi r^2 - 2\left(\pi r^2 \frac{\alpha}{360} - \frac{1}{2}r^2 \sin\alpha\right)} \tag{13.9}$$

Theoretical analysis shows that r is determined by D and L. As long as r is known, we can obtain Cov, in which n is a function of R and r.

13.3 ECMAN System Design and Implementation

The proposed ECMAN system design is based on the combination of RoF and TDMA technologies, which are used to convert a conventional size BS into a certain number of motor-picocells, each of which is allocated a particular sub-frame working on a different time slot. The system structure of converting classic cells into motor-picocells can be seen in Figure 13.5.

In this architecture, the CS is responsible for exchanging information among different BSs, whilst each BS connects with a RoF subsystem. In this RoF subsystem, the BS, which is located at the central station, connects with a headend unit including modulation and demodulation modules. The headend unit in turn connects with a certain number of RAUs of motor-picocells that are deployed along the motorway at one dimension in an overlapping manner. At the headend, radio frequency (RF) signals are converted into an optical signal by directly modulating a laser diode (LD). The obtained optical signal is carried over the downlink single mode fibre (SMF) to an RAU. As the RF amplifier would help to provide the required radio coverage, RAU converts on remote site optical signals to RF signal by using a photodiode (PD). In this context, RF circulator is used to separate transmitted signals from received signals. For uplink transmission, the wireless signal received at the RAU is

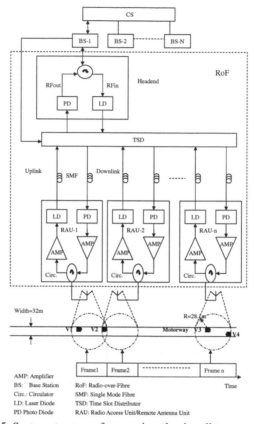

Figure 13.5 System structure of converting classic cell to motor-picocells.

first amplified and then converted into an optical signal by an LD, which, in turn, transported over uplink SMF to the headend. Another issue is the presence of co-channel interference due to the motor-picocells that belong to the same BS use the same frequency. By adopting the TDMA scheme, the frame structure of BS is divided into the same number of sub-frames to match with the motor-picocells on a different time slot.

In this converting processing, most of the parameters of the conventional cell are unchanged. The RF in the same group of motor-picocells is the same and, to avoid co-channel interference, adjacent group of motorpicocells that belong to another cell must not use the same RF channel. Therefore, while a vehicle is running within the same group of cells the same RF channels are used. The vehicle must change RF channels only when it enters a new group of motor-picocells. Whilst the previous frame structure designed for conventional cell is subdivided into frames for the motor-picocells in the group and a frame is composed of downlink and uplink portion, the size of a frame for a motor-picocell can be made proportional to the traffic demand of the cell.

It should be emphasized at this point that during a time period for frame i only the corresponding RAU_i is activated by the BS at a disjoint time period (i.e. frames). Therefore, although one RF channel is employed there is no co-channel interference between cells within a cluster of motor-picocells. If a vehicle is in a non-overlapping area, it will listen to one frame that corresponds to the cells where it is located. If it is in an overlapping area, it will listen to two frames from adjacent motor-picocells. In Figure 13.4, for example, vehicle 1 (V1) receives only frame 1, while vehicle 2 receives frames 1 and 2 since it is in the overlapping area between cells 1 and 2. Moreover, the figure also indicates the fact that a frame can support multiple vehicles as described in cell n.

The proposed system introduces TSD as a novel concept between the headend unit and RAUs, which assigns each active RAU with a unique alloc-ated time slot. TSD is controlled by its related BS and operates based on the frame structure at the bottom of Figure 13.6. Frame structure is predetermined by BS according to traffic flow and communication demand within each RAU (a detailed description of frame structure is given in Section 13.4). In this way, many RAUs can share a single BS and its electrical/optical devices to greatly reduce overall costs. TSD can be placed quite near to remote site, where all RAUs can share a pair of SMFs for downlink and uplink transmission of the signals in order to reduce the cost further. Although this system design will increase the complexity of each BS nevertheless the number of BSs needed for the whole system is decreased significantly. Normally, the cost of one RAU corresponds to no more than 3% of the BS, so, in general, the cost of whole system can be greatly reduced. This conversion is a purely transparent transmission process, it has hardly any impact on each terminal and it is associated with QoS.

Based on the theoretical analysis of Section 13.2, when $R = 500$ m, $d = 32$ m and $L = 10$ m, parameters r and n are calculated as 28.1 m and 338, respectively i.e., a conventional cell radius of 500 m can be divided into 338 motor-picocells with radius 28.1 m deployed along the motorway.

13.4 Intelligent Handover Indicators of the Effective Coverage System

Apparently, after we convert one conventional size BS into hundreds of small motor-picocells, handover frequency will increase dramatically to hundreds of times as well. Also, the motor-picocells are so small that they could not support conventional handover scheme. So we have to simplify the handover process and reduce latency time. In reality, it is to divide a large frame previously serve a limited amount of customers within the BS at a different time domain into a certain number of subframe shared by more customers.

In this context, as a vehicle is using the same RF channels of the motor-picocells of the CS, each sub-channel belongs to a different time slot associated with the conventional cell. Only if it enters a new group of motor-picocells of different RF channels, the new sub-channels could be defined at that point. Specifically, the frame of the conventional cell could be subdivided into several sub-frames, which are prepared to serve the motor-picocells. Only if it enters a new group of motor-picocells of the different RF channels, the new sub-channels could be defined since then. Moreover, the frame of the conventional cell could be subdivided into several sub-frames, which is preparing to serve those motor-picocells. In the sub-frame, it is also composed of downlink and uplink portion.

The frame structure determines the 'air time' given to RAU within the BS, i.e., the time period each RAU uniquely uses to communicate with vehicles located in its coverage area (see Figure 13.6).

Each frame belonging to a certain RAU begins with a 'beacon' field generated at the BS that consists of RAU identification (ID) number and slot assignment map specifying the start slot position and length for each vehicle. The following field is 'reservation slots' which is accessed by vehicles that have not yet reserved any slots but have the data to transmit in contention-based way. The results of a reservation trial in the previous frame is broadcasted in the 'broadcast' field, which is followed by downlink and uplink slots assigned to each vehicle as specified in the slot assignment map. In the uplink slot, there is a handover indication area consisting of four bits for fast handover within the same BS and between different BSs: the first bit is intra-handover flag for handover request within the same BS; the second bit is direction parameter of the vehicle; the third bit is for reminding handover preparation between BSs; the fourth bit is inter-handover flag for handover between BSs. This intra-handover is explained below [6, 10].

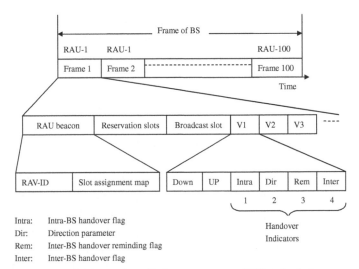

Figure 13.6 Motor-picocells frame structure of BS and sub-frame.

13.4.1 Intra-Handover in the Same BS

As all RAUs of a BS utilize the same RF channel, a vehicle entering the overlapping region between RAUs begins to receive two beacons, each containing a different RAU-ID during an allocated frame time. The vehicle sends in turn to the BS a handover request by setting the 'intra-handover flag' and, subsequently, the BS reserves bandwidth for the vehicle in the next picocell and, at the end, releases bandwidth used by the vehicle at the old cell. It should be noted that resources to handover a connection from one RAU to its successor RAU are always available. This is because the centralized MAC may adjust (i.e., shorten) the frame length of the RAU left by a vehicle and, hence, it can increase the frame duration of the successor RAU in order to provide the next vehicle with the required resources. As a result, in intra-BS handover, zero handover latency and zero handover dropping are possible; moreover, time slot is aligned to the vehicle with its fast-moving morbidity.

13.4.2 Inter-Handover between Different BSs

On a highway scenario, vehicle can only travel towards one dimension with two opposite directions. A direction parameter is introduced herewith to notify BS that vehicles in each RAU are running in different directions. After comparing the new RAU-ID with the old one, the vehicle will set the direction parameter to '1' if the ID works in an increasing base, otherwise, it will set it to '0'. Note that the BS knows the location of its first and last RAUs, which are also indicated in RAU-ID. When a vehicle's direction parameter is '1' and receives the beacon of the last RAU, it sets a third bit to '1', which means it enters the last RAU of the BS, telling CS to inform the adjacent BS to prepare the handover resources. At the same time, the vehicle will begin to scan RF channels in the next BS. If it receives the new RF channel, it sends a handover request by setting the fourth bit to '1' for inter-BS handover. If the request to the new BS is successful and there is enough bandwidth in the new BS, the vehicle can continue its communication session; otherwise, the request is dropped. Thus, unlike intra-BS handover, successful inter-BS handover involves not only changing RF channels but also bandwidth management. Through setting the pre-notify bit in advance, the CS may give the vehicle requesting handover higher priority in preference to the new connection requests. Similar process is applied when direction parameter is '0' and RAU-ID is the smallest, which means the vehicle starts to request handover between BSs in opposite direction.

The handover between two CSs, i.e. two BSs controlled by two different centralized MAC entities, is the most critical in terms of guaranteeing QoS parameters to any ongoing connection. The handover procedure is similar to inter-BS handover except that the two BSs associated with it are controlled by two different CSs. Therefore, the same handover procedure for inter-BSs handover can be applied, except that it requires the CSs to exchange control traffic for handover.

13.4.3 Initialization

When a vehicle wants to initiate communication with the system, it must first scan RF channels. After having identified the RF channel used within the cell, it must send a request for bandwidth to the CS using one of the reservation mini-slots. If the request is successful and the system has enough bandwidth to accommodate the requested bandwidth, the vehicle will be assigned the bandwidth in the next frame. For all cases, an overlapping area between two adjacent RAU is assumed to be large enough to complete the handover procedure. For example, if a vehicle is running at 120 km/h, the time it takes to run 1 m is 30 ms. Thus, when the frame time is small (e.g., 5 ms, as suggested by mobile WiMax) ten meters of overlapping area would be enough for most practical cases.

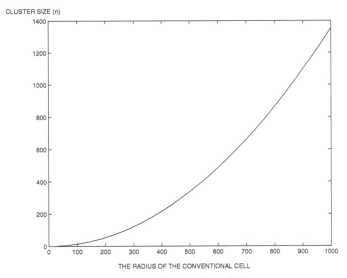

CLUSTER SIZE (n)

THE RADIUS OF THE CONVENTIONAL CELL

Figure 13.7 The varying number of the system.

13.5 Numerical Result

This section presents typical numerical experiments in order to illustrate the credibility of the proposed ECMAN system. Figures 13.7 and 13.8 illustrate the cluster size of the motor-picocells and its effective coverage, whilst Figure 13.9 presents the different utilization of conventional cell scheme versus the suggested new scheme. Typical numerical experiments illustrate the validity of the devised analytic solutions and also demonstrate their utility as simple and cost-effective performance evaluation tools for assessing the suitability of ECMAN system.

The numerical study focuses on optimum size of the motor-picocells. From equation (14.8), we can know that the radius of motor-picocell is determined by the overlap length D and the overlap width l. When $D = d$, e.g., r obtains its optimized number. That means despite the variation of radius R, the optimum motor-picocells could be defined by the width of motorway and length of overlapping distance. When $d = 32$ m, $l = 10$ m, we can get $r = 28.1$ m.

According to equation (14.9), Figure 13.7 shows a positive relationship between the radius of the conventional cell R and the size of cluster picocells, e.g. when the radius of picocell r is fixed, as R increases, n increases in a square manner.

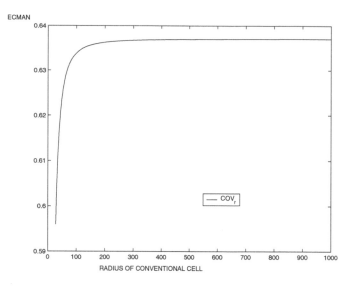

Figure 13.8 Effective coverage of the ECMAN system.

On the other hand, Figure 13.8 illustrates another picture as the demonstration of equation (14.3). Alongside the increase of the radius of conventional cell, the effective coverage increase sharply when $R < 100$ m. Between 100 and 200 m, it has an apparent curve in increasing. But after $R > 300$ m, it can just increase slightly. Even the motor-picocell system is more cost-effective than the traditional conversion cellular system. But the efficient measurement parameter is telling us not to implement millions of motor-picocell BSs. In this case, we believe the optimum number is '200' for a cost-effective system.

Figure 13.9 represents both convention scheme and our effective scheme. As far as we could see, in our scheme, the throughput is about 60% of increase with the variation of the arrival rate and service rate.

13.6 Conclusion and Future Work

This work has shown that through the combination of RoF and TDMA technologies, the conventional cell size of WiMAX BS can be converted into motor-picocells and deployed along the motorway. This system design brings two main benefits: (i) greatly improved effective coverage; and (ii) cost-effectiveness. For example, even if he conventional BS radius is decreased to

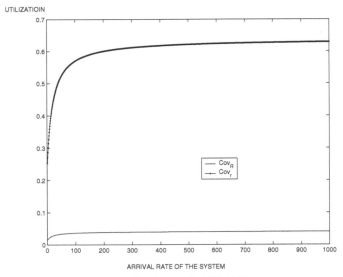

Figure 13.9 Utilization of the ECMAN system.

500 m, when motorway width is 32 m, the effective coverage is only 0.0387, lower than 4% of the total area of BS (see Figure 13.3). After the conversion, one BS is divided into 338 picocells with radius 28.1 m and effective coverage is increased to 0.6369 (see Figure 13.8), resulting into a 60% gain in total. Moreover, under the conventional system, one BS could cover 1 km of motorway with 500 m radius. However, under the new scheme, this distance has been stretched to 15 km. Clearly, at present 15 conventional BSs of 500 m radius are needed to cover 15 km. Thus, the new scheme increases significantly the effective coverage whilst it greatly reduces the overall system cost. This conversion also means the potential of system processing capacity could be fully utilized, e.g., the same BS can deal with more customers through the stretch of its effective coverage.

The questions raised by this approach are co-channel interference and dramatic increase of handover frequency. To deal with co-channel interference, the system proposes the novel concept of TSD between the headend unit and RAUs, which assigns each active RAU with a unique allocated time slot based on sub frame structure within BS. At the same time, four parameters are involved for intra-BS and inter-BS handovers in order to simplify the handover procedure on motorway scenario and make fast handover to be feasible. Performance modelling and evaluation of suitable resource management schemes are now under way and preliminary results show that the proposed architecture is efficient.

Acknowledgement

This work has been supported by the Wales Development Agency and has been partly funded by the European Union.

References

[1] D. Pareek, *WiMax: Taking Wireless to the MAX*, Auerbach Publishers, June 2006.
[2] S. J. Vaughan-Nichols, Achieving wireless broadband with WiMax, *Computer*, vol. 37, no. 6, pp. 10–13, June 2004.
[3] WiMax Forum, Mobile WiMAX – Part I: A technical overview and performance evaluation, March 2006.
[4] WiMax Forum, Mobile WiMAX – Part II: A comparative analysis, May 2006.
[5] H. Jang, J. Jee, Y. Han, S. D. Park and J. Cha, Mobile IPv6 fast handover over IEEE 802.16e networks, MIPSHOP Working Group, October 2006.
[6] H. Harada, K. Sato and M. Fujise, A radio-on-fiber based millimeter-wave road-vehicle communication system for future intelligent transport system, in *Proceedings of IEEE Vehicular Technology Conference 2001 Fall*, vol. 4, pp. 2630–2634, October 2001.
[7] H. B. Kim, M. Emmelmann, B. Rathke and A. Wolisz, A radio over fiber network architecture for road vehicle communication systems, in *Proceedings of IEEE Vehicular Technology Conference, VTC 2005 Spring*, vol. 5, pp. 2920–2924, 2005.
[8] J. M. H. Elmirghani, B. Badic, Y. Li, R. Liu, R. Mehmoodi, C. Wang, W. Xing, I. J. Garcia Zuazola and S. Jones, IRIS: An intellegent radio-fiber telematics system, in *Proceedings of ITS World Congress*, London, pp. 2105–2112, 2006.
[9] K. Abe, T. Tobana, T, Sasamori and H. Kolzumi, A study on a road-vehicle-communication system,in *Proceedings of the International Conference on Parallel and Distributed Systems*, pp. 343–348, July 2000.
[10] H. Harada, K. Sato and M. Fujise, A radio-on-fiber based millimeter-wave road-vehicle communication system by a code division multiplexing radio transmission scheme, *IEEE Trans. Intelligent Transport. Sys.*, vol. 2, no. 4, pp. 165–179, December 2001.
[11] H. Al-Raweshidy and S. Komaki (Eds.), *Radio over Fiber Technologies for Mobile Communications Networks*, Artech House, Norwood, 2002.
[12] M. Okita, H. Harada and M. Fujise, A new access protocol in radio-on-fiber based millimeter-wave road-vehicle communication system, in *Proceedings of the IEEE VTC 2001 Fall*, vol. 4, pp. 2178–2182, October 2001.

PART FIVE
QUALITY-OF-SERVICE (QoS)

PART FIVE
(SELF-SERVICE JOBS)

14

A Novel Stochastic Min-Plus System Theory Approach to Performance Evaluation and Design of Wireless Networks *

Farshid Agharebparast and Victor C. M. Leung

Department of Electrical and Computer Engineering, The University of British Columbia, Vancouver, BC, Canada V6T 1Z4; e-mail: {farshid, vleung}@ece.ubc.ca

Abstract

In this paper, we develop a novel framework for efficient stochastic quality of service (QoS) evaluations and cross-layer design in wireless networks. The methodology is based on a proposed stochastic min-plus system theory. This novel theory characterizes the second-order statistics of the system output and can be used to model or analyze any stochastic system, in particular a wireless channel, with manifold advantages. Due to its clearly similar representation to the familiar traditional stochastic system theory, it is an easy yet elaborate modeling and performance evaluation tool even with a basic familiarity of Network Calculus (NC) theory. Above all, as it only deals with the second-order statistics, it is a practical technique for establishing a packet-level cross-layer methodology for a wireless link by the knowledge of its second-order statistics at the physical layer. Finally, it conforms and can be easily coupled with NC, a sophisticated min-plus system theory for QoS evaluation and modeling of data communication networks, which is an important requirement for end-to-end QoS. We demonstrate the effectiveness of the method by using the proposed methodology to calculate the delay

---

* This manuscript is partly based on a paper published in the proceedings of the Performance Modeling and Evaluation of Heterogeneous Networks (HetNets) Conference, Ilkely, U.K., September 2006, pp. P30/(1–10).

D. D. Kouvatsos (ed.), Mobility Management and Quality-of-Service for Heterogeneous Networks, 309–339.

bounds and/or how to set the transmission rates to meet a certain delay bound, and present comparisons between analytic and simulation results. The paper concludes with a thorough discussion on the assumptions, related topics and applications with different settings.

Keywords: Wireless networking, performance modeling, stochastic min-plus systems, Network Calculus, second-order statistics, cross-layer design.

14.1 Introduction

Deterministic Network Calculus (NC) is a well-established, sophisticated min-plus system theory for deterministic quality of service evaluations and modeling of data communication networks [1]. However, in wireless networks in which a wireless link behaves like a stochastic server, deterministic NC might estimate loose QoS bounds. Currently, there are two different frameworks for stochastic or probabilistic NC extensions. $(\sigma(\theta), \rho(\theta))$-calculus and effective bandwidth method [2] is based on the constrained moment generating function of the traffic, and the other method is based on stochastically bounded burstiness [3].

In this paper, we propose and develop a novel stochastic min-plus theory as an efficient alternative method for performance evaluations and modeling, specially suited for wireless networks. The proposed methodology in particular serves as the foundation of a cross-layer design and modeling framework for wireless networks as well. In contrast with other stochastic NC methods, our proposal has the important advantage that it can easily be coupled or used with deterministic NC, an important requisite for end-to-end QoS. In addition, due to its clearly similar representation to the traditional stochastic system theory, it is much easier to use and is well-understood even with a basic familiarity of NC theory. Simply, the traffic needs to be represented by the NC's cumulative traffic characterization, and min-plus (or max-plus) convolution be used for input-output representations. Nevertheless, the method conforms with NC theory and can be respected as an integral for deterministic or stochastic NC methods.

The approach that is presented in this paper was inspired from the traditional system theory for modeling linear systems with stochastic inputs [4]. Min-plus (or its dual, Max-plus) systems, as the basis of the Network Calculus theory, are min-plus linear. We demonstrate that similar approach can be used to stochastically characterize the output of such systems, when the characteristics of the system and/or of the input are only stochastically provided. For the ease of notation and to follow the system theory model, we refer to a network element (such as a link, node, or a whole network) as a system. Similarly, the incoming traffic to the system is called input, and the outgoing traffic from the system is called output, both with bytes as their units of measure. In our wireless link, the input is the traffic transmitted to enter the link, and the output is the traffic passed through and received at the other end of the link, observed at the link layer.

The outline of the paper is as follows. In the next section, a brief overview on NC theory and the traditional system theory with a stochastic input, as a point of reference, are presented. In Section 14.3, we present second-order input-output relationships of a min-plus system with a stochastic input. Utilizing the commutative property of the convolution operator, we then modify the previous method to analyze a min-plus system with a stochastic service curve. This provides us with an efficient method for modeling communication networks with stochastically defined arrival or service processes, such as wireless channels. Based on this new system theory, in Section 14.5, we propose a packet-level performance evaluation and cross-layer framework for wireless networks when the second-order statistics of the physical layer is known. The link layer uses these statistics to calculate the second-order statistics of the traffic that will be received on the other end of the channel, and consequently the projected delays. From these information, suitable parameters of operation, such as the transmission rate, are decided upon. Two well-known channel models of AWGN and Rayleigh fading are examined. By analysis and simulation via an example, we demonstrate how effectively delay as the QoS metric of interest is calculated. Finally, after some discussions, Section 14.8 concludes the paper.

14.2 Background

A restricting and challenging characteristic of a wireless channel is that it is random and time-varying, and so it is regarded as a stochastic system. Therefore, practical modeling of wireless networks naturally requires stochastic models. In a previous work, we presented a method based on the definition of the channel's service curve using filter banks of deterministic NC elements, where the service is guaranteed with some probability [5]. Although the method is efficient and serves as a natural extension of the deterministic NC, its successful application relies on the derivation of the piece-wise service curve with a meaningful probability. Other efforts are rooted in the stochastic traffic characterization, assuming that some function of the moment generating function of the underlying process in constrained [2]. These models, such as the effective capacity model [6], are not only complicated and require the knowledge of the asymptotic log-moment generating functions, but also are mainly suited for the analysis of a standalone link.

In this paper, we present a systematic and effective methodology for the derivation of second-order statistics of the output of a stochastic min-plus system. Since our method takes a course different from those referred to by stochastic NC in the literature (such as [2] or [3]), we differentiate it by entitling it stochastic min-plus systems theory. However, the methodology conforms with NC theory, and can actually be categorized as a novel stochastic NC approach and be integrated to a NC-based analysis of a network, specially with deterministic NC. In contrast with the stochastic NC models where the moment generating of the stochastic process is considered, here we deal with the second-order statistics of mean and autocorrelation. This is a practical approach as these are usually available from the underlying physical layer models. This information may be used by the link layer to set its parameters accordingly.

The approach has many advantages. It is inspired by and is similar to the traditional stochastic system theory [4, ch. 10]; it is easy to use, easily comprehendible, systematic and scalable. And since it is familiar to engineers, it serves as a standalone approach even with a basic familiarity with min-plus theory (or NC). In terms of its applicability, the requirements are that the cumulative traffic be characterized by a non-decreasing envelope and the min-plus convolution be applied. Finally, it is based on the second-order statistics, a practical approach which can be used for efficient QoS evaluation and design of multi-service network.

In the following, we summarize the input-output relationship of a traditional linear system with a stochastic input, as a reference and a point of comparison. Then we present a concise overview for the Network Calculus theory, to describe the notations used in this paper.

14.2.1 Traditional Systems with Stochastic Inputs

In the traditional system theory, the relationship between the input $x(t)$ and the output $y(t)$ of a linear system with response $h(t)$ is characterized by the convolution operator defined by the notation:

$$y(t) = L^s[x(t)], \text{ where } L^s[x(t)] = \int x(t-s)h(s)ds. \qquad (14.1)$$

We use a superscript s with L to differentiate the system operator of the traditional systems from the corresponding min-plus notation.

For a stochastic input $x(t)$ where its mean $E\{x(t)\}$ and its autocorrelation function $R_{xx}(t_1, t_2)$ are given, similar statistics of the output are characterized based on the following two fundamental theorems.

Theorem S-I. $E\{L^s[x(t)]\} = L^s[E\{x(t)\}]$, which as well results in $\eta_y = \eta_x \otimes h$. This is an extension of the linearity of the expected values to arbitrary linear operators [4, p. 309].

Theorem S-II. The autocorrelation of the output is characterized by [4, p. 311]:

$$R_{xy}(t_1, t_2) = L_2^s[R_{xx}(t_1, t_2)],$$
$$R_{yy}(t_1, t_2) = L_1^s[R_{xy}(t_1, t_2)]. \tag{14.2}$$

L_2 means that the system operates on the variable t_2 and treats t_1 as a parameter, and L_1 is vice versa. The autocorrelation function is defined by: $R_{xx}(t_1, t_2) = E\{x(t_1)x^*(t_2)\}, t_2 \geq t_1$.

A process is wide-sense stationary (WSS), if and only if its mean is constant: $E\{x(t)\} = \eta, \forall t$ and its autocorrelation depends only on $\tau = t_2 - t_1$: $E\{x(t+\tau)x(t)\} = R_{xx}(\tau), \forall t$. Then

$$R_{xy}(\tau) = R_{xx}(\tau) \otimes h^*(-\tau),$$
$$R_{yy}(\tau) = R_{xy}(\tau) \otimes h(\tau). \tag{14.3}$$

14.2.2 Network Calculus: Min-Plus Filtering Theory

A successful modern alternative to the traditional queuing theory is based on min-plus (or its dual max-plus) algebra, usually called Network Calculus (NC). In NC theory, the traffic and the service rates are deterministically or stochastically constrained by some functions so-called arrival curves and service curves, respectively. Based on the min-plus algebra, a system theory is devised where the arrival and service curves are sufficient to characterize the output and its properties. This creates a practical and easy-to-use yet powerful framework to data networks similar to the traditional system theory whose benefits most engineers are familiar with. NC has proven to have numerous applications in solving data networking problems from a simple node to large networks. In particular, it is a natural method to be used with Internet IntServ and ATM. The reader is referred to [1, 2] for a thorough study of NC.

Min-plus algebra is based on the semi-ring of $(\Re, \wedge, +)$, which is a ring on the set of real numbers with operators of infimum $(\wedge)$ and algebraic addition $(+)$. Comparing to the normal algebra, we replace the role of addition with the infimum operator, and of multiplication with addition. In NC the input (traffic), the system (network), and the output are characterized by some non-decreasing functions and the input-output relationship is defined based on a convolution operator. If $x(t)$ is a non-decreasing function representing the input, and system's service curve is $h(t)$, then the output is defined by the min-plus convolution [7]:

$$L^{h(t)}[x(t)] = x(t) \otimes h(t) = \inf_{0 \le s \le t} (x(t-s) + h(s)), \qquad (14.4)$$

and

$$y(t) \ge L[x(t)], \text{ if } h(t) \text{ is a lower service curve,}$$

$$y(t) = L[x(t)], \text{ if } h(t) \text{ is an exact service curve,}$$

$$y(t) \le L[x(t)], \text{ if } h(t) \text{ is an upper service curve.} \qquad (14.5)$$

Max-plus algebra is the dual of min-plus, where infimum is replaced with supremum. Similar to (14.4), the max-plus convolution is defined as:

$$\widehat{L}^{h(t)}(x(t)) = x(t) \otimes h(t) = \sup_{0 \le s \le t} (x(t-s) + h(s)).$$

14.3 Min-Plus Systems with Stochastic Inputs

In this section, we introduce the theory of min-plus systems with stochastic inputs by presenting two fundamental theorems characterizing the mean and autocorrelation of the output. Due to the commutative property of the min-plus convolution, later we simply swap the role of input as the stochastic process with the service. Therefore, this section lays the foundation for the stochastic min-plus system theory presented in the following section.

In a min-plus system, the output $y(t)$ is characterized by the min-plus convolution of the input $x(t)$ and the service curve (system function) $h(t)$ (Figure 14.1). Linearity is defined in regards to the min-plus algebra, considering $(\Re, \wedge, +)$. A min-plus system is linear when:

$$L[\inf(a + u(t), b + v(t))] = \inf(a + L[u(t)], b + L[v(t)]). \qquad (14.6)$$

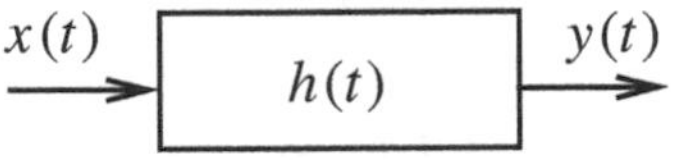

$$x(t) \quad \boxed{\quad h(t) \quad} \quad y(t)$$

$$L^h[x(t)] = x(t) \otimes h(t)$$

Figure 14.1 Min-plus system input/output representation: in a communication network, the box (system) represents a network element (link, node, network); $x(t)$ (input) represents the incoming traffic in unit of bytes; and $y(t)$ (output) represents the outgoing traffic (or received by an element at the other end) in unit of bytes.

In order to comply with the NC theory, all functions describing the input, system and output are positive and cumulative, and therefore non-decreasing. Note that for $x(t)$, we have the option of using either the actual traffic or its arrival curve. The former gives the actual instance of the output, while the latter gives an envelope for the output. Since for an arrival process $A(t)$ with an arrival curve $\alpha(t)$, we have $A(t+\tau) - A(t) \le \alpha(\tau), \forall(\tau \ge 0)$. By putting $t = 0$, we have $A(t) \le \alpha(t)$ for $\forall(t \ge 0)$. Finally using the Isotonicity of $\otimes$ [1] leads to: $y(t) \le \alpha(t) \otimes h(t)$ which means the output obeys a curve of $(\alpha \otimes h)(t)$.

For any cumulative function $x(t)$, a rate function $X(t)$ is defined. If $x(t)$ is continuous and has a derivative, then $x(t) = \int_{\tau=0}^{t} X(\tau)d\tau$ [8]. For a discrete time model, the input is mapped by sampling every time slot δ [1].

In communications systems, the statistics of $X(t)$ which is WSS, are usually known. If $X(t)$ is WSS then $E\{X(t)\} = \eta$, simply resulting $E\{x(t)\} = \eta t$, for both continuous and discrete time cases. Obviously $x(t)$ itself is not WSS, unless $\eta = 0$. Similarly for the autocorrelation of the continuous case we have:

$$R_x(t_1, t_2) = E\left\{ \int_0^{t_1} X(m)dm \int_0^{t_2} X(n)dn \right\} = \int_0^{t_1} \int_0^{t_2} R_X(m, n)dmdn.$$

For an uncorrelated process (i.e. $E\{X(m+n)X(m))\} = 0$, for $n \ne 0$), this reduces to: $R_x(t_1, t_2) = \text{VAR}(X)t_1t_2$, where $\text{VAR}(X) = R_X(0)$ is the variance of X.

Before presenting the fundamental theorems, we address some essential propositions and lemmas.

Propositions I. For a set of real numbers X, $\inf(X + c) = \inf(X) + c$, and $\inf(\beta \cdot x) = \beta \cdot \inf(x)$, $\beta > 0$.

Proof. Assume $x^* = \inf(X)$, then $x^* \le x$ for $\forall x \in X$. Therefore for $\forall x \in X$, we also have $x^* + c \le x + c$ and $\beta \cdot x^* \le \beta \cdot x$, $\forall \beta \ge 0$. So $x^* + c = \inf(X + c)$ and $\beta \cdot x^* = \inf(\beta \cdot X)$.

Proposition II. For two random variables of x and y, if $x \ge y$ (i.e., for any outcome the value of x is greater than or equal to y), then $E\{x\} \ge E\{y\}$.

Lemma I. For a set of random variables X_s, $E\{\inf_s(X_s)\} \le \inf_s(E\{X_s\})$.
   Proof. Assume that for s^o we have $X_{s^o} = \inf(X_s)$, then $X_{s^o} \le X_s$, $\forall s \in S$. Using the results of Proposition II, we have $E\{X_{s^o}\} \le E\{X_s\}$, i.e. $E\{\inf_s(X_s)\} \le \inf_s(E\{X_s\})$.

Fundamental Theorem I. In a min-plus system with service curve $h(t)$ and random process $x(t)$,

$$E\{L[x(t)]\} \le L[E\{x(t)\}]. \tag{14.7}$$

   Proof. At a given time t, define a random variable $Y^t = \inf_{0 \le s \le t}(x(t - s) + h(s))$, in which $h(t)$ is a function of time. From Lemma I, $E\{Y^t\} \le E\{(x(t - s) + h(s)\}$, $\forall(s, 0 \le s \le t)$. Thus, $E\{Y^t\} \le \inf_{0 \le s \le t}(E\{x(t - s)\} + h(s))$. From the definition stated by (15.4), we have $L[x(t)] = Y^t$ and $L[E\{x(t)\}] = \inf_{0 \le s \le t}(E\{x(t - s)\} + h(s))$, thus (14.7) follows.

Theorem I-A. In a min-plus system with an exact or upper service curve $h(t)$

$$\eta_y \le \eta_x \otimes h. \tag{14.8}$$

   Proof. By applying Theorem I in (15.5), the stated result follows.
   Similar to (15.2) in Theorem S-II, the autocorrelation functions of the output in a min-plus system are derived as follows:

Fundamental Theorem II. For a min-plus system

$$R_{xy}(t_1, t_2) \le L_2^{\eta_x(t_1)h(t_2)}[R_{xx}(t_1, t_2)], \tag{14.9}$$

and

$$R_{yy}(t_1, t_2) \le L_1^{\eta_y(t_2)h(t_1)}[R_{xy}(t_1, t_2)], \tag{14.10}$$

when $h(t)$ is an exact or upper service curve. L_2 means that the system operates on the variable t_2 and treats t_1 as a parameter, and L_1 means that the system operates on the variable t_1 and treats t_2 as a parameter.

Proof. We prove the theorem for exact service curves. For upper service curves, the proof is similar by just replacing the equalities on the first three lines with $\leq$. To prove the first equation, we start with the convolution equation and then multiply the two sides with the positive value $x(t_1)$. We apply expectation to both sides, then by using Lemma I, the first part of the theorem is proved with $t = t_2$. Similarly, we start with the convolution equation but this time we multiply the two sides with the positive value $y(t_2)$. Then we apply expectation to both sides and by using Lemma I, the second part of the theorem follows with $t = t_1$.

$$y(t) = L^{h(t)}[x(t)]$$

$$y(t)x(t_1) = \inf_s(x(t-s) + h(s))x(t_1)$$

$$E\{y(t)x(t_1)\} = E\left\{\inf_s(x(t-s)x(t_1) + h(s)x(t_1))\right\}$$

$$\leq \inf_s(E\{x(t-s)x(t_1)\} + h(s)E\{x(t_1)\})$$

$$\leq \inf_s(R_{xx}(t_1, t-s)\} + \eta_x(t_1)h(s))$$

$$R_{xx}(t_1, t_2) \leq L_2^{\eta_x(t_1)h(t_2)}[R_{xx}(t_1, t_2)]$$

and

$$y(t) = L^{h(t)}[x(t)]$$

$$y(t)y(t_2) = \inf_s(x(t-s) + h(s))y(t_2)$$

$$E\{y(t)y(t_2)\} = E\left\{\inf_s(x(t-s)y(t_2) + h(s)y(t_2))\right\}$$

$$\leq \inf_s(E\{x(t-s)y(t_2)\} + h(s)E\{y(t_2)\})$$

$$\leq \inf_s(R_{xy}(t-s, t_2)\} + \eta_y(t_2)h(s))$$

$$R_{yy}(t_1, t_2) \leq L_1^{\eta_y(t_2)h(t_1)}[R_{xy}(t_1, t_2)]$$

As demonstrated in Figure 14.2, we may use two different time axes to define the two time instances t_1 and t_2 that we need to calculate the autocorrelation. For clarity and only in this figure, we use two different notations for time, in order to differentiate between the transmission times (τ) and observation times (t). In other words, the autocorrelation can

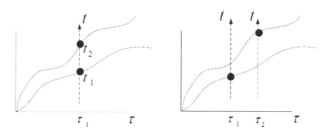

Figure 14.2 Calculation of the autocorrelation. On the left figure, the two instances of time needed for the calculation of the autocorrelation are two possible instances observed at different times but for the same transmission time; and on the right, they are at two different transmission times.

be calculated for one instance of output over actual time span during a transmission, or two different times of two different instances of output. Here we are interested in calculating the autocorrelation over a constant τ, for t_1, t_2 due to two reasons. First, we are interested in the variability of the output for a transmission. Second, this will ensure that R_{yy} is wide-sense stationary provided that the underlying physical layer process (SNR here, as explained in Section 14.5) is wide-sense stationary, which is a widely accepted and valid assumption.

Fundamental Theorems in Max-plus algebra. Since max-plus algebra is the dual of the min-plus algebra, both of the fundamental theorems proved here are applicable to max-plus systems as well. Of course, all infimum operators are replace with supremum, and the direction of inequalities are reversed for upper service curve $h(t)$. Theorem I will be $E\{\widehat{L}[x(t)]\} \geq \widehat{L}[E\{x(t)\}]$, i.e. $\eta_y \geq \eta_x * h$, which again gives a lower bound on the expected value of the output. Similarly (14.10) is rewritten as: $R_{yy}(t_1, t_2) \geq \widehat{L_1}^{\eta_y h(t)}[R_{xy}(t_1, t_2)]$.

14.4 Min-Plus Stochastic Systems

In the previous section we established a system theory for min-plus (max-plus) systems with a stochastic input. This is a logical assumption for example for a network router with a deterministic service curve but a stochastically defined characterization for the incoming traffic. Since the min-plus convolution is commutative, i.e. $(f \otimes g)(t) = (g \otimes f)(t)$, then when dealing with stochastic systems with an input constrained by an arrival curve, we can readily apply the results and theorems of the previous sections, by just

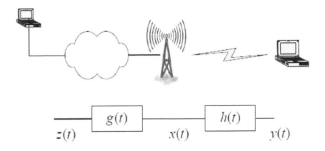

Figure 14.3 Statistical characterizations of the received traffic using the stochastic min-plus system theory.

swapping the place of $h(t)$ with $x(t)$. This will provide us with the theorems required to analyze a system with a stochastic service curve (upper or exact), and an input shaped to have a NC arrival curve. Therefore, we rewrite (14.8) as

$$\eta_y \leq \eta_h \otimes x. \tag{14.11}$$

The correlation equations of (14.9) and (14.10) are rewritten as:

$$R_{hy}(t_1, t_2) \leq L_2^{\eta_h(t_1)x(t_2)}[R_{hh}(t_1, t_2)], \tag{14.12}$$

and

$$R_{yy}(t_1, t_2) \leq L_1^{\eta_y(t_2)x(t_1)}[R_{hy}(t_1, t_2)]. \tag{14.13}$$

Figure 14.3 shows an end-to-end path from the source to the destination, in a network consisting of a wired and a wireless part. The wired network is analyzed by NC. Input $x(t)$ from the wired network is the traffic that enters the wireless network passing through a stochastically-defined wireless channel. Equipped with the theorems presented in this section, in the following we present a methodology which calculates the corresponding statistics of the output $y(t)$ (received by the destination). For the uplink path, similarly the above approach is used in the reversed order.

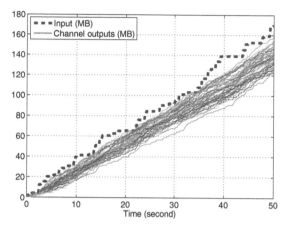

Figure 14.4 The transmitted traffic (input) and 50 different instances of the channel output due to the time-varying Rayleigh fading channel.

14.5 Wireless Channel Modeling, Performance Evaluation and Cross-Layer Design

In the previous section, we developed a general framework for modeling min-plus stochastic systems. Since a wireless channel is time-varying, the effective transmission rate available from the link layer perspective, is variable and time-varying. Therefore, for a given transmitted traffic (input), different possible channel outputs are observed. For example, Figure 14.4 plots 50 different observed outputs of the plotted traffic passing through a Rayleigh fading channel. A stop-and-wait ARQ protocol is used to compensate for packet errors, so that any point on an instance of the output corresponds to a point on the input with the same amount of traffic. Due to fading, each instance experiences a different PER versus time, thus it is different from the other possible instances of the input.

In the following, we show how the theorems presented earlier is used to calculate the statistics of such outputs. In order to do so, packet-level statistical service characteristic of the wireless channel, $h(t)$, needs to be appropriately defined. A good example of such a service characterization is presented in the following section. We then apply equations (14.11) and (14.13), to find the statistics of the output for a given arrival process passing through the wireless channel (or network, using the concatenation property). Therefore, by using the information from the current physical layer behaviour, statistical characteristics of the output is determined. By using this

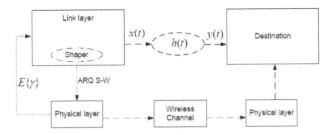

Figure 14.5 System model and building blocks.

information, the parameters of the link layer are set accordingly. For example, in Section 14.6, such information is used to set the rate parameter of the shaper at the transmitter. This is in addition to the requirements considered by the transmitter to conform to the negotiated QoS policies.

The method is a general and comprehensive. $h(t)$ can be defined as elaborate and complex as required. For a cross-layer design, this service curve should be defined in a way so that it accurately translates the statistical properties of the physical layer, which in turn is used to adjust the operation of the link layer or to allow efficient allocations of network resources. In this section, a service curve of such nature is first addressed, based on the second-order BER statistics at the physical layer. Then we introduce a framework that utilizes the theorems presented earlier for performance modeling and analysis of wireless channels. The technique is applicable for both uplink and downlink, but is in particular well-suited for the downlink where the results from the deterministic NC modeling of the previous wired parts of the path can be most easily used for an end-to-end QoS support.

14.5.1 System Model

Figure 14.5 shows the overall system model. We are interested in calculating the traffic received by the destination at the link layer. The traffic is first shaped and then sent via lower layers through a wireless Rayleigh fading channel. We assume that packets are retransmitted based on the stop-and-wait ARQ protocol in order to compensate for packet errors. Therefore, a packet stays in the queue until successfully transmitted. Similar variability of outputs is as well observed when no retransmission mechanism is used, however, in this case the delay cannot be calculated from the horizontal distance between the input and output graphs (or curves). This will be clear when calculation of delay is explained later in this section.

14.5.2 Service Curve of the Channel

The framework established earlier is general and can be used for any type of wireless network where a proper $h(t)$ is available. Depending on the physical channel behavior, the required QoS bounds and the application, there are different ways of modeling the dynamics of the wireless channel.

A reasonable way of defining $h(t)$ is by direct considerations of the available physical channel models, and to translate it to the service at the link layer. One practical approach is to characterize the instantaneous service rate of a wireless channel at the link layer by

$$r_c(t) = (1 - \text{PER}(t))\mu, \tag{14.14}$$

when the packet error rate at any given time ($\text{PER}(t)$) and the channel transmission capacity (μ) measured in packet units at Δ time interval ($\Delta \rightarrow 0$) are known [9]. μ can be set to the Shannon capacity by using $B \cdot \log(1 + \text{SNR})$, when a dedicated bandwidth of size B is available to the wireless user. This is a reasonable assumption, as the errors are accounted for by the PER. Whenever this method presents an overly optimistic bound for the channel capacity, other more elaborate methods for the calculation of the channel capacity may be used. In wireless networks where the users contend to use the wireless channel, μ can be set as the total available channel capacity, while PER is defined so that it includes packet errors due to contention as well. Equation (14.14) describes the instantaneous rate that is seen available by the link layer for successful transmissions. This equation holds whether a retransmission mechanism is implemented or not. In this paper, we assume that packets are retransmitted, based on the stop-and-wait ARQ protocol in order to compensate for packet errors.

Since the channel is time-varying, PER is a random process. The second-order statistics of PER is all we need to use (14.14). Here we consider a set of widely-used assumptions, in order to focus on describing the methodology. In Section 14.7, alternatives to these assumptions are discussed using more sophisticated methods.

The service curve of the channel is the cumulative service, which is the integral (or summation for sampled or discrete representations) of (14.14):

$$h(t) = \int_0^t r_c(t)dt = \int_0^t (1 - \text{PER}(t))\mu dt. \tag{14.15}$$

It is interesting to note that the above equation results in a stochastic rate-latency server (of the form $\beta_{r,T} = r \cdot (t - T)$), with a stochastic latency $T = \int_0^t \text{PER}(t)dt$ and rate $r = \mu$.

Calculation of Bit Error Rate (BER): Without loss of generality we assume using a QPSK modulation scheme, and no channel coding. For a $(\pi/4)$-QPSK scheme, BER is a function of SNR, denoted by [10]:

$$BER(t) = Q(\sqrt{2\text{SNR}(t)}). \qquad (14.16)$$

$Q(.)$ is the scaled complementary error function.

In the following, the distributions of SNR for two cases of Additive White Gaussian Noise (AWGN) and Rayleigh fading channels are discussed.

Case I – AWGN Channels: For an AWGN channel, the SNR is a stochastic process whose distribution depends on and is directly derived from the distribution of the additive noise which is zero-mean Gaussian.

Case II – Rayleigh Fading Channels: The instantaneous SNR per bit, γ, for an uncorrelated time-varying Rayleigh fading channel with Alamouti transmit diversity has a central chi-square distribution with two degrees of freedom. The Alamouti transmit diversity scheme is one of the proposed methods for IEEE 802.16 (WiMAX) to achieve 2-way diversity without adding an extra antenna, as described in the IEEE 802.16abc-01/53 specification. The probability density function (pdf) of such process is [11]:

$$p(\gamma) = \frac{2}{E/N_0} \exp(-\gamma 2N_0/E), \qquad (14.17)$$

where $E\{\gamma\} = E/2N_0$ is the average SNR per bit.

For the case of fully-correlated Rayleigh fading channel with Alamouti transmit diversity, the pdf is in the form of a central chi-square distribution with four degrees of freedom, denoted by:

$$p(\gamma) = \frac{\gamma}{(E/2N_0)^2} \exp(-\gamma 2N_0/E). \qquad (14.18)$$

In both cases, similar assumptions for the modulation scheme and packet error allocation are considered, so the same equations are valid and are used. Thus, given an SNR, the distribution of BER is calculated by using (14.16).

Calculation of Packet Error Rate (PER): Without loss of generality, independent bit errors are assumed which leads to uniform distributions of bit errors in each packet. For completeness, in Section 14.7, an elaborate method of PER calculation is discussed. With the above assumption, PER is defined by

$$PER(t) = 1 - (1 - BER(t))^n, \qquad (14.19)$$

where n is the number of bits in the packet.

Since we have the statistics of BER, using (14.19) the statistics of PER is calculated. The probability density function is

$$p_{PER} = \frac{d}{da} F_{PER}(a),$$

where $F_{PER}(a) = P\{PER \le a\}$. It can be easily shown that [12]:

$$F_{PER}(a) = F_B(1 - \sqrt[n]{1-a}) + (1 - F_B(1 + \sqrt[n]{1-a}))\chi(n),$$

where $\chi(n)$ is 1 if n is even, and is 0 otherwise.

In this paper, only mean and autocorrelation of the processes are needed that can be calculated either directly or from the probability density functions when available. For completeness, the distributions were presented here briefly. Direct derivations of the mean and autocorrelation functions of PER and then process $r_c(t)$ is performed by cascaded calculations of the statistics of BER, PER, r_c, from (14.16), (14.19), and (14.14), respectively.

14.5.3 Methodology

In brief, the following steps are taken:

(S1) In order to use (14.14), second-order statistics of PER need to be derived. Without loss of generality, we use the following widely-used two sub-steps (see Section 14.7 for an elaborated more realistic alternative). (a) From the physical layer model, the second-order statistics of bit error rate (BER) is given. The statistics of BER is obviously dependent on the modulation scheme and the error correcting codes (channel coding) used by the transceivers and is a function of signal to noise ratio (SNR). (b) BER is used to find the second-order statistics of the PER. This can be done by assuming uniform distributions of bit errors, or by using other methods in the presence of error dependency and channel coding (as will be explained in Section 14.7).

(S2) Then (14.14) is used to find the statistics of the channel service rate. Simply, we have

$$E\{r_c\} = (1 - E\{PER\})\mu,$$

and

$$R_{r_c}(t_1, t_2) = (1 - 2E\{\text{PER}\} + R_{\text{PER}}(t_1, t_2))\mu^2.$$

From (14.15), the NC compatible service curve of the channel (i.e. cumulative and so non-decreasing) is calculated.

The above steps provide us with statistics of a compatible service curve for the channel. The following two steps are performed, in order to calculate the delay bounds:

(S3) Equations (14.11) and (14.13) are used to obtain the second-order statistical properties of the output. These equations are applicable when $h(t)$ is an exact or upper service curve. $h(t)$ is an exact service curve, the wireless channel that is dedicated to the user is fully utilized. When the wireless channel is shared between the users in a random multiple access (such as by Aloha), $h(t)$ is an upper service curve.

(S4) Finally, NC filtering theory is used to consequently calculate the delay bound (or other QoS bounds) for a given input.

14.5.4 Calculation of Delay Bounds

The mean function gives us the expected value of the output and the autocorrelation function gives us enough statistical information about its variation. The approach we choose to follow here is based on the standard deviation function. At any given time t, the variance of the output process is $\text{Var}(y)(t) = R_{yy}(t, t) - \eta_y^2(t)$, and the standard deviation is $\text{Std}(y)(t) = \sqrt{\text{Var}(y)(t)}$. The output will vary around its mean within a bracket of $[\eta_y(t) - \kappa \cdot \text{Std}(y), \eta_y(t) + \kappa \cdot \text{Std}(y)]$ with a high probability p_1, where κ is a constant that determines p_1. For example, for a normal distribution, it is within one standard deviation around the mean with a probability of 68% and within two standard deviations around the mean with a probability of 95%. Therefore, we can use these two bounds as the worst and best possible outputs.

Having the output curve is all that is required for us to utilize NC theorems to find the second-order statistics of the delay, and therefore the statistical delay bounds. As we assumed that a stop-and-wait ARQ protocol is implemented in order to compensate for packet errors, any two points of the output with the same value represent the same amount of traffic. Then the delay that a packet transmitted at time encounters is given by

$$d(t) = \inf(\Delta \geq 0 : x(t) \leq y(t + \Delta)), \tag{14.20}$$

which graphically is the horizontal distance between the two graphs. The maximum delay is $d_{\max} = \sup(d(t))$. Therefore, from the output curves the second-order statistics of the delay are calculated. The numerical example presented in the next section will elaborate on this procedure.

14.6 Numerical Results

In the previous section, the methodology was detailed via four steps. In this section, we use a numerical example for a Rayleigh fading channel to compare the analytic results with simulation results, using MATLAB. Following the four steps of the methodology, equations (14.11) and (14.13) are used to first obtain the second-order statistical properties of the output. Having the statistics of the output, delay bounds are calculated.

The nature and values of the parameters, as well as the assumptions used in the numerical example are as follows. The application generates a traffic with exponentially distributed packet length and interarrival times. The traffic is then shaped by a leaky bucket controller to conform with an arrival curve of the affine function $at+b$, where a is the rate and b is the bucket size (so-called burst tolerance). The parameters of the shaper have to be defined according to the negotiated QoS policies. The actual rate of the traffic source, can be less, equal or larger than that of the shaper. For better presentation of the results and since the arrival curve is a good QoS benchmark, we have set the value so that the shaped traffic be close to the shaping envelope. An instance of the traffic from the source, shaped traffic, and their accumulative representations are shown in Figure 14.6, subplots (a), (b) and (c), respectively.

The channel is assumed to be uncorrelated Rayleigh fading with Alamouti transmit diversity as discussed in Section 14.5, with average SNR, $E\{\gamma\}$ and rate, μ. This is an effective diversity method used in WiMAX. A $(\pi/4)$-QPSK modulation and no channel coding are assumed. The values of the parameters used in the example are summarized in Table 14.1.

Figure 14.7 is the three dimensional plot of the autocorrelation of the service and the output. Since the underlying process is a wide-sense stationary process, these are accordingly WSS. The x axis is the transmission time and the y axis represents 50 different instances of observations with its mirror towards which collectively span through $[-49, 50]$.

Figure 14.8a shows the traffic (arrival) passing through the channel, and a set of 50 instances of possible service curves that occurred due to the time-varying fading channel. Figure 14.8b shows the mean output calculated from

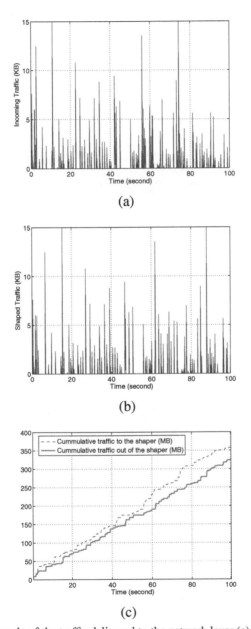

(a)

(b)

(c)

Figure 14.6 An example of the traffic delivered to the network layer (a), shaped traffic ready to be transmitted through the channel (b), and cumulative representations of traffic in a and b (c)

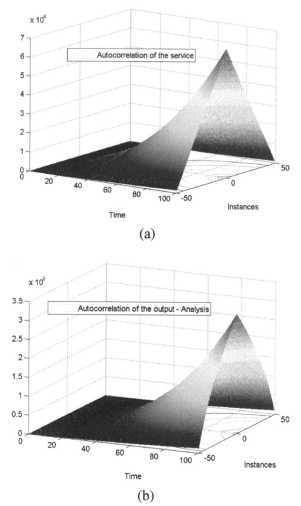

Figure 14.7 3D plots of the autocorrelation of the service (a) and the output (b). The x axis is the transmission time and the y axis represents 50 different instances of observations (with its mirror towards −50).

the analytical approach and compares it with the mean of the actual outputs from the simulation.

Knowing the input and output, the delay bound can be easily calculated. First, we use the graph of the mean output to calculate mean delay. Mean delay is the maximum horizontal distance between the input and the mean

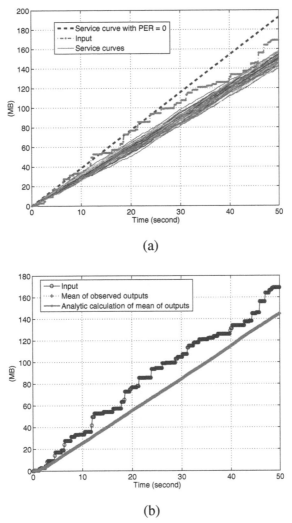

(a)

(b)

Figure 14.8 (a) Traffic and 50 instances of channel service curves. (b) Comparison of analytic and simulation results for mean outputs.

output. Similarly, the standard deviation is used to plot the best and worst output curves. Following the same procedure, the worst delay is calculated which is the maximum horizontal distance between the input and the worst output. Figure 14.9 shows the mean, worst and best outputs and illustrates the above procedure to graphically calculate the mean and worst delays.

Table 14.1 Parameters used in the numerical example.

Parameter	Notation	Value
Mean interarrival times	MIntArr	0.7 (ms)
Mean packet size	MPkSz	3 (KB)
Shaper rate	a	MPkSz/MIntArr (MB/s)
Shaper burst tolerance	b	$5 \times$ MPkSz
Average SNR	$E\{\gamma\}$	10 (dB)
Channel rate	μ	$[0.5\ 1.2] \times a$

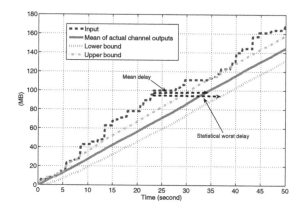

Figure 14.9 Graphical calculations of mean and worst delays from the mean and worst case outputs.

Obviously, the shaper rate (value of the parameter a) relative to μ is an important factor in the value of the observed delay. Figure 14.10 plots mean and worst delays for different values of the ratio of the channel and shaper rates (μ/a), within a wide range of [0.5, 1.2]. The figure compares the analytic and simulation values with a 95% confidence interval, considering 30 iterations per each value of the above ratio, with 50 instances of channel service curves.

Figure 14.11 demonstrates the plots of delay when the other two effective parameters, b or $E\{\gamma\}$, are changing. Figure 14.11a shows the delay versus the b parameter of the shaper, and Figure 14.11b plots the delay for a wide range of values of average SNR ($E\{\gamma\}$ in dB).

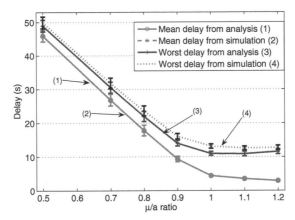

Figure 14.10 Mean and worst delays for different values of μ/α with a 95% confidence interval (30 iterations of 50 instances each).

Therefore, the link layer shapes the traffic with a rate that not only conforms to the negotiated QoS policy, but also with consideration of the wireless channel condition. Hence, the value of the parameter a is reconsidered and adjusted implementing the above cross-layer design. Depending on how often the second-order statistics of the physical layer is acquired, the value of a can be adjusted accordingly. This is so-called an impedance matching for the channel condition [14], in addition to traffic and congestion conditions, and QoS policies. Since in our scenario, the channel is under a wide-sense stationary condition, only $E\{\gamma\}$ needs to be passed up from the physical layer. As discussed earlier, this is a practical assumption for a fading channel, which obeys a chi-square distribution.

The figures show a close match of the analytical and simulation results. Our analysis also finds similar results in the case of a fully correlated time-varying Rayleigh fading channel with Alamouti transmit diversity, which was discussed in Section 14.5.

14.7 Discussions

In this section, we present some discussions to elaborate on applying the proposed method for applications with different settings. So far, we have assumed a QPSK modulation scheme, no channel coding, and uniform distribution of errors. Here we discuss how to incorporate channel coding and/or

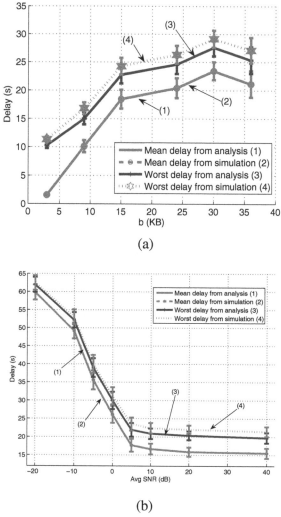

Figure 14.11 (a) Delay versus b (shapers burst tolerance), when $E\{\gamma\}$ and μ/α are 10 dB and 0.8, respectively. (b) Delay versus average SNR ($E\{\gamma\}$) in dB when μ/α is 0.8; with a 95% confidence interval (30 iterations of 50 instances each).

other modulation schemes in the model. In addition, an elaborate method for the derivation of PER, considering bit error dependency, is given.

Some topics of interest that further expand the scope of the proposed methodology are also discussed. A short note on how to consider mobility is presented; Analysis of systems with both stochastic traffic and service curve is addressed; And the statistical service guarantee in NC in regards to the proposed method is discussed. Finally, Slope domain representations of the fundamental theorems are derived.

Detailed PER Calculations: So far in this paper we assumed that the bit errors in a packet occurs independently and no channel coding is implemented. This was to be able to clearly explain the methodology without getting involved with irrelevant details. However, channel coding plays a significant role in wireless communication [15]. For example deployment of a convolutional code can lower BER by two orders of magnitude [16, p. 260]. In this section, however, we revise our assumptions and consider error dependency under the presence of a convolutional channel coding, to demonstrate that the same methodology applies.

As a replacement for (14.19), equation (14.22) that was introduced in [17] is used. Although these equations look similar, (14.22) not only considers error dependency but also takes into account convolutional coding as an error correction code.

The procedure for PER calculations stays almost similar. The statistics of SNR is first calculated. Then instead of finding BER, a new function called EER (error event rate) is used, from which PER is calculated. We take the following steps:

1. *EER(SNR)* is defined by $EER(SNR) \approx A_{d_{\text{free}}} \exp(R \cdot SNR \cdot d_{\text{free}})$, where $A_{d_{\text{free}}}$ is the number of paths of length d_{free} in the code trellis and R is the code rate.

2. From *EER(SNR)* the value of another function $\lambda(SNR)$ is defined by

$$\lambda(s) = \frac{EER(s)}{1 - [(\upsilon + 1) + \frac{1}{n_c(s/2 - \sqrt{2sr_r} + r_c)}]EER(s)}. \tag{14.21}$$

$\lambda(SNR)$ is the mean of a Geometric distribution characterizing errorless period lengths.

3. Finally, the PER is calculated from:

$$PER(t) = 1 - (1 - \lambda(SNR(t)))^n. \tag{14.22}$$

It is interesting to note that the above formula looks similar to (14.19). We only replace BER with a new function, $\lambda(SNR)$.

Under similar consideration as of our previous numerical and simulation settings, we derive the mean and worst delays using (14.22). Based on [18] for

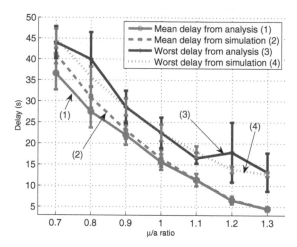

Figure 14.12 Delay versus different values of shaping rate with the consideration of convolutional channel codes and bit error dependency.

a UMTS system, a convolutional code with rate 1/2 and $\upsilon = 9$ are assumed. For this code $d_{\text{free}} = 12$. We assume that the value of n_c is equal to the mean packet length and $A_{d_{\text{free}}} = 1$.

Figure 14.12 shows analytical and simulation results for the delay. Naturally the observed delays are worse due to the bit error dependency (as compared with those in Figure 14.10), but the methodology is still applicable.

Effect of the Modulation Scheme: Up till now in this paper, we assumed a QPSK modulation scheme. Simply, for a different modulation scheme only the equation that described BER as a function of SNR needs to be changed. For example, for the DBPSK modulation scheme, (14.16) is replaced with: $\text{BER}(t) = 0.5 \exp(-\text{SNR}(t))$.

Systems with Both Stochastic Input and Service: When the service and the input are both stochastically defined, the theorems are modified as follows and can be used accordingly.

(A) The mean of the output is:

$$\eta_y \leq \eta_x \otimes \eta_h. \tag{14.23}$$

Proof. Similar to the proof of Theorem I, we have

$$E\{L[x(t)]\} = E\{x(t) \otimes h(t)\}$$

$$= E\left\{\inf_{0 \leq s \leq t} (x(t-s) + h(s))\right\}$$

$$\leq \inf_{0 \leq s \leq t} (E\{x(t-s) + h(s)\})$$

$$\leq E\{x(t)\} \otimes E\{h(t)\}$$

(B) Similarly, the autocorrelation function can be calculated from

$$R_{yy}(t_1, t_2) \leq L_1^{R_{hy}(t_1, t_2)} [R_{xy}(t_1, t_2)]. \tag{14.24}$$

Note that this equation is slightly different from (15.11), as we need to calculate R_{hy} as well as R_{xy}.

Proof.

$$y(t) = L^{h(t)}[x(t)]$$

$$y(t)x(t_1) = \inf_s (x(t-s) + h(s))x(t_1)$$

$$E\{y(t)x(t_1)\} = E\left\{\inf_s(x(t-s)x(t_1) + h(s)x(t_1))\right\}$$

$$\leq \inf_s(E\{x(t-s)x(t_1)\} + E\{h(s)x(t_1)\})$$

$$\leq \inf_s(R_{xx}(t_1, t-s) + R_{xh}(t_1, s))$$

$$R_{xy}(t_1, t) \leq R_{xx}(t_1, t) \otimes R_{xh}(t_1, t)$$

$$R_{xy}(t_1, t_2) \leq L_2^{R_{xh}(t_1, t_2)}[R_{xx}(t_1, t_2)]$$

and

$$y(t) = L^{h(t)}[x(t)]$$

$$y(t)h(t_1) = \inf_s(x(t-s) + h(s))h(t_1)$$

$$E\{y(t)h(t_1)\} = E\left\{\inf_s(x(t-s)h(t_1) + h(s)h(t_1))\right\}$$

$$\leq \inf_s(E\{x(t-s)h(t_1)\} + E\{h(s)h(t_1)\})$$

$$\leq \inf_s(R_{hx}(t_1, t-s) + R_{hh}(t_1, s))$$

$$R_{hy}(t_1, t) \leq R_{hx}(t_1, t) \otimes R_{hh}(t_1, t)$$

$$R_{hy}(t_1, t_2) \leq L_2^{R_{hh}(t_1, t_2)}[R_{hx}(t_1, t_2)]$$

And finally $R_{yy}(t_1, t_2)$ is

$$y(t) = L^{h(t)}[x(t)]$$

$$y(t)y(t_2) = \inf_{s}(x(t-s) + h(s))y(t_2)$$

$$E\{y(t)y(t_2)\} = E\left\{\inf_{s}(x(t-s)y(t_2) + h(s)y(t_2))\right\}$$

$$\leq \inf_{s}(E\{x(t-s)y(t_2)\} + E\{h(s)y(t_2)\})$$

$$\leq \inf_{s}(R_{xy}(t-s, t_2) + R_{hy}(s, t_2))$$

$$R_{yy}(t, t_2) \leq R_{xy}(t, t_2) \otimes R_{hy}(t, t_2)$$

$$R_{yy}(t_1, t_2) \leq L_1^{R_{hy}(t_1, t_2)}[R_{xy}(t_1, t_2)]$$

Mobility: The proposed method is practical, regardless whether the transmitter and/or receiver are fixed or mobile. Since all the equations in the paper are time-dependant, the location of the node is already incorporated. In this approach, both large-scale path loss and shadowing, and small-scale fading is considered. Path loss is mainly dependent on the receiver-transmitter distance from which the average SNR (expectation of SNR) is calculated. To account for mobility, this value in equation (14.17) (or (14.18)) would be time varying and needs to be reconsidered at any given time.

Statistical Guarantees: Our method fits in and can be used with the statistical NC [19, 20], in which the QoS evaluations are based on a statistical exact (or upper) service curve $h^\varepsilon(t)$, satisfying

$$\Pr\{y(t) = x(t) \otimes h^\varepsilon(t)\} \geq 1 - \varepsilon, \qquad (14.25)$$

since such service curve for the channel is derived. All QoS bound theorems for delay, backlog and output bound of NC are as well valid under the above definition, and can be used to calculate QoS metrics of interest [19].

Slope Domain Representations: In an earlier paper [21], we presented Slope domain modeling of networks which serves in regard to min-plus systems like Fourier domain to the traditional system theory. In system theory, Fourier spectrum is used for spectral representations of random signals. Power spectrum of a random process is the Fourier transform of its autocorrelation function [4, p. 416]:

$$PS(\omega) = \int_{-\infty}^{\infty} R(\tau)e^{-j\omega\tau}\,d\tau.$$

Then, the power spectrum function of the output is

$$PS_{yy}(\omega) = PS_{xy}(\omega)H(\omega) = PS_{xx}(\omega)|H(\omega)^2|.$$

Using Slope domain we can define a similar concept for min-plus algebra. The Slope transform of a min-plus function $x(t)$ is defined by

$$S_x(\alpha) = S\{x(t)\}(\alpha) = \inf_{t \in \Re}(x(t) - \alpha t). \tag{14.26}$$

This transformation maps min-plus functions into a domain with properties very similar to Fourier domain, which generally facilities modeling of min-plus systems. The reader is referred to [21] for further information.

Similar to Power Spectrum, we introduce the Slope transform of the autocorrelation functions as Load Spectrum, using the notation $S_{yy}(\alpha) = S\{R_{yy}(\tau)\}(\alpha)$. An immediate advantage of using Slope domain analysis is that all the convolution operators are translated to algebraic additions. Secondly, the Load spectrum demonstrates how traffic load is distributed over different rates (slopes). The slope transform of (14.13) leads to:

$$S_{hy}(\alpha) \leq S_{hh}(\alpha) + \eta_h X(\alpha/\eta_h),$$
$$S_{yy}(\alpha) \leq S_{hy}(\alpha) + \eta_h X(\alpha/\eta_h), \tag{14.27}$$

which finally results in

$$S_{yy}(\alpha) \leq S_{hh}(\alpha) + \eta_h X(\alpha/\eta_h) + \eta_y X(\alpha/\eta_y). \tag{14.28}$$

The autocorrelation of the output is then calculated by inverse Slope transform of $S_{yy}(\alpha)$.

14.8 Conclusions

In this paper, we introduced a stochastic min-plus methodology which not only serves as a general extension to the Network Calculus theory for efficient modeling and performance evaluations of wireless networks, but also as a foundation for cross layer design at the transmitter. We presented a simple-to-use yet elaborate framework from which the second-order statistics of the traffic that passes through a channel are calculated, using the second-order statistics of the physical layer. This information were used to calculate delay bounds or to set the transmission rate in order to meet a desired delay as the QoS metric of interest. By analysis and simulation the effectiveness of the framework was demonstrated.

Acknowledgements

This work was supported by a grant by TELUS and the Natural Sciences and Engineering Research Council of Canada under grant CRDPJ 341254-06, by an NSERC PGS-B postgraduate scholarship, and by a University of British Columbia Graduate Fellowship.

References

[1] J.-Y. Le Boudec and P. Thiran, *Network Calculus: A Theory of Deterministic Queueing Systems for the Internet*, Lecture Notes in Computer Science, vol. 2050, Springer-Verlag, January 2002.

[2] C.-S. Chang, *Performance Guarantees in Communication Networks*, Springer-Verlag, 2000.

[3] Y. Liu, C.-K. Tham and Y. Jiang, A stochastic Network Calculus, Technical Report ECE-CCN-0301, National University of Singapore, November 2004.

[4] A. Papoulis, *Probability, Random Variables and Stochastic Processes*, Third Edition, McGraw-Hill, 1991.

[5] F. Agharebparast and V. C. M. Leung, Modeling wireless link layer by Network Calculus for efficient evaluations of multimedia quality of service, in *Proc. IEEE ICC'05*, Seoul, Korea, pp. 1256–1260, May 2005.

[6] D. Wu and R. Negi, Effective capacity: A wireless link model for support of quality of service, *IEEE Trans. Wireless Communications*, vol. 2, no. 4, pp. 630–643, July 2003.

[7] A. Kumar, D. Manjunath and J. Kuri, *Communication Networking: An Analytical Approach*, Morgan Kaufmann, 2004.

[8] R. L. Cruz, A calculus for network delay. I. Network elements in isolation, *IEEE Trans. Information Theory*, vol. 37, no. 1, pp. 114–131, January 1991.

[9] Y. Y. Kim and S.-Q. Li, Modeling multipath fading channel dynamics for packet data performance analysis, *Wireless Networks*, vol. 6, no. 6, pp. 481–492, 2000.

[10] T. S. Rappaport, *Wireless Communications Principles and Practice*, Pearson Education, 2002.

[11] A. Vielmon, Y. Li and J. R. Barry, Performance of Alamouti transmit diversity over time-varying Rayleigh fading channels, *IEEE Trans. Wireless Communications*, vol. 3, no. 5, pp. 1369–1373, September 2004.

[12] S. Ross, *A First Course in Probability*, MacMillan, 1976.

[13] M. Schwartz, *Broadband Integrated Networks*, Prentice Hall, 1996.

[14] S. Shakkottai, T. S. Rappaport and P. C. Karlsson, Cross-layer design for wireless networks, *IEEE Communications Magazine*, vol. 41, no. 10, pp. 74–80, October 2003.

[15] S. B. Wicker, *Error Control Systems for Digital Communication and Storage*, Prentice Hall, 1995.

[16] H. V. Poor and G. W. Wornell, *Wireless Communications, Signal Processing Perspectives*, Prentice Hall, 1998.

[17] R. Khalili and K. Salamatian, A new analytic approach to evaluation of packet error rate in wireless networks, in *Proc. IEEE Communication Networks and Services Research Conference*, pp. 333–338, May 2005.

[18] ETSI TS 125 212 V7.0.0 (2006-03) Technical Specification, Universal Mobile Telecommunications System (UMTS); Multiplexing and channel coding (FDD), 3GPP TS 25.212 version 7.0.0 Release 7.

[19] J. Liebeherr, S. Patek and A Burchart, A calculus for end-to-end statistical service guarantees, Technical report CS-2001-19, University of Virginia, 2001.

[20] F. Agharebparast and V. Leung, Link-layer modeling of a wireless channel using stochastic network calculus, in *Proceedings IEEE CCECE'04*, Niagara Falls, Ontario, Canada, pp. 1923–1926, May 2004.

[21] F. Agharebparast and V. C. M. Leung, Slope domain modeling and analysis of data communication networks: A Network Calculus complement, in *Proc. IEEE ICC'06*, Istanbul, Turkey, pp. 591–596, June 2006.

15

Constrained-Path Discovery by Selective Diffusion

Karel De Vogeleer, Dragos Ilie and Adrian Popescu

Department of Telecommunication Systems, School of Engineering,
Blekinge Institute of Technology, Karlskrona, Sweden;
e-mail: {karel.de.vogeleer, dragos.ilie, adrian.popescu}@bth.se

Abstract

The demand for live and interactive multimedia services over the Internet raises questions on how well the Internet Protocol's (IP) best-effort effort service can be adapted to provide adequate end-to-end quality of service (QoS) for the users. Although the Internet community has developed two different IP-based QoS architectures, neither has been widely deployed. Overlay networks are seen as a step to address the demand for end-to-end QoS until a better solution can be obtained.

As part of the telecommunication research at Blekinge Institute of Technology (BTH) in Karlskrona, Sweden, we are investigating new theories and algorithms concerning QoS routing. We are in the process of developing Overlay Routing Protocol (ORP), a framework for overlay QoS routing consisting of two protocols: Route Discovery Protocol (RDP) and Route Management Protocol (RMP). In this paper we describe RDP and provide preliminary simulation results for it. The results indicate that the RDP scales almost linearly with the number of nodes in the network. The system's ability to find feasible paths is intimately related to the time-to-live (TTL) value used in RDP messages: a large TTL value increases the chance of finding a feasible path at the cost of a higher volume of RDP traffic.

D. D. Kouvatsos (ed.), Mobility Management and Quality-of-Service for Hetero-
geneous Networks, 341–358.

Keywords: Overlay, routing, quality of service (QoS), Peer-to-Peer.

15.1 Introduction

The predominant form of Internet routing is a combination of shortest-path routing for intradomain environments coupled with policy-based routing for interdomain communication. For the past ten years it has been argued that Internet must incorporate elements of QoS in order to be used as a platform for multimedia distribution. In particular, live or interactive multimedia communications place stringent constraints on the path between sender and receiver. Examples of such constraints are constraints on available bandwidth, packet delay, packet delay variation and packet loss rate.

Two major IP-based QoS architectures have been developed so far: Integrated Services (IntServ) [1] and Differentiated Services (DiffServ) [2]. Neither architecture has been widely deployed due to lack of a viable economical solution for network operators, poor backwards compatibility with existing technology and difficulties in the interaction between different network operators [3–5]. Additionally, we would like to mention the Asynchronous Transfer Mode (ATM) Private Network-to-Network Interface (PNNI) protocol, which has support for QoS routing [6, 7]. Although ATM failed to become the technology of choice for end-nodes due to the emergence of cheap Ethernet cards, the research into ATM technology has yielded valuable results for the Internet community.

There seems to be little hope for wide QoS deployment implemented at network layer, at least in the near future. To cope with this problem several researchers have investigated the possibility to deploy QoS in overlay networks on top of IP [8–12].

An overlay network utilizes the services of an existing network in an attempt to implement new or better services. An example of an overlay network is shown in Figure 15.1. The physical interconnections of three autonomous systems (ASs) are depicted at the bottom of the figure. The grey circles denote nodes that use the physical interconnections to construct virtual paths used by the overlay network at the top of the figure. Nodes participating in the overlay network perform active measurements of particular QoS metrics associated with the virtual paths. The results from the measurements can be used in rerouting of overlay traffic, in load balancing or for traffic shaping.

The work presented in this paper is part of the Routing in Overlay Networks (ROVER) project at BTH to implement a framework for overlay QoS routing called ORP [13]. ORP is part of a larger goal to research and develop

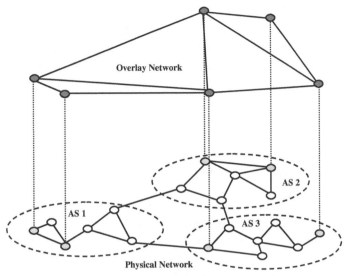

Figure 15.1 Overlay network.

a QoS layer on top of the transport layer. The main idea is to combine ORP together with additional QoS mechanisms, such as resource allocation and admission control, into a QoS layer. User applications that use the QoS layer can obtain soft QoS guarantees. These applications run on end-hosts without any specific privileges such as the ability to control the internals of TCP/IP stack, the operating system, or other applications that do not use the QoS layer. In terms of the OSI protocol stack, the QoS layer is a sub-layer of the application layer. Applications may choose to use it or to bypass it.

The QoS layer implements *per-flow* QoS resource management. In contrast to IntServ and DiffServ, we envision that it is mostly end-nodes in access networks that take part in the routing protocol. IP routers are not required to take part or be aware of the QoS routing protocol running on the end-nodes. In other words, we propose a QoS layer on top of the best-effort service provided by IP. Since a best-effort service leaves room for uncertainties regarding the resource allocation, we aim only for soft QoS guarantees.

The ORP framework consists of two protocols: Route Discovery Protocol (RDP) and Route Management Protocol (RMP).

RDP is used to find network paths subject to various QoS constraints [14, 15]. To achieve this goal, RDP uses a form of selective diffusion in which a node that receives a path request forwards the request only on outgoing links that do not violate the QoS constraints. Eventually, the request may reach the destination node if there is at least one path satisfying the constraints. At that point a reply message containing information about the complete path is sent back to the requesting node. The RDP is based on ideas presented in [16–18].

The purpose of the RMP is to alleviate changes in the path QoS metrics, due to node and traffic dynamics. This is done through a combination of route repair techniques and optimization algorithms for traffic flow allocation on bifurcated paths. The purpose of the flow allocation is to spread the demand on multiple paths towards the destination [19]. The design of RMP is influenced by ideas presented in [20, 21]. Since RMP is currently under development [22], only information pertaining to RDP is presented in this article.

The remaining of this paper is organized as follows. Section 15.2 introduces the format of the messages used by RDP. The diffusion process used for constrained-path discovery is presented in Section 15.3. The OMNeT++ simulation model for RDP is presented in Section 15.4. The simulation testbed is presented in Section 15.5. In Section 15.6 we analyze several simulation results from a performance perspective. This is followed by plans for future work in Section 15.7.

15.2 RDP Message Format

Since ORP messages can carry user data we switch freely between the terms *packet* and *message*. We assume that the transport layer below ORP allows a node to send packets to any known peer in the overlay.

All ORP messages start with the generic header shown in Figure 15.2. Field values in the packet header are arranged in network byte order. The following elements are included in the ORP packet header:

Version ORP protocol version. At the moment of writing the protocol is at version 1.

Type ORP packet type.

Field value	Packet type
0	reserved
1	control packet (CP)
2	acknowledgement packet (AP)
3	data packet (DP)
4	used by the RMP

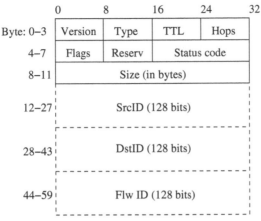

Figure 15.2 ORP generic packet header.

TTL Time-to-live, denoting how many overlay hops the packet is allowed to travel.

Hops Indicates the amount of links the packet already has passed. If the value in the Hops field equals the value in the TTL field the packet is dropped.

Flags Bitfield arranged as |0|0|E|0|D|C|B|R|, where 0 denotes unused bits.

  E indicates the node is leaving the overlay and all routes associated with *SrcID* should be rerouted or deleted.

  D indicates that the path associated with the *FlwID* should be deleted.

  C denotes a route change.

  B denotes a bidirectional route request.

  R indicates a redundant AP.

Reserv Reserved for future use.

Status Used to exchange status codes among nodes.

Size Packet size in bytes excluding the generic header.

SrcID UUID denoting the source node of the packet.[1]

DstID UUID denoting the destination node of the packet.

FlwID UUID of the flow to which this packet belongs.

---

[1] The ORP UUIDs are defined as specified in [23].

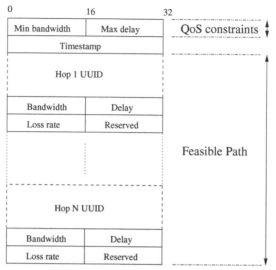

Figure 15.3 QoS map.

RDP uses two different kinds of packets: control packets (CPs) and acknowledgement packets (APs).

A CP begins with the generic header followed by a data structure called *QoS map*, as shown in Figure 15.3. The QoS map starts with the *QoS constraints* for the requested path. ORP currently supports two type of QoS constraints: *minimum bandwidth* specified in kilobytes per second and *maximum path delay*, specified in milliseconds. We plan to integrate additional constraint types in future ORP versions. The *timestamp*, in Coordinated Universial Time (UTC) format, indicates the time when the QoS map was sent to the next hop.

Following the path QoS constraints comes the *feasible path* explored so far by the CP in question. Each node that forwards the CP appends an entry to the feasible path. The entry consists of the UUID of the downstream node and a set of QoS metrics associated with the link on which the packet is forwarded. Statistics currently supported by ORP are *bandwidth* (expressed in kilobytes per second), *delay* (expressed in milliseconds) and *packet loss rate*. The packet loss rate is a fraction with the accuracy $1/(2^{16} - 1)$. A loss rate of 0 indicates that no packets are lost whereas a loss rate of $(2^{16} - 1)$ denotes that all packets are lost. The use of the last field is not defined yet. The manner in which the the QoS metrics are computed is not within the scope of this paper.

When the destination node receives a CP it assembles an AP by copying from the CP the fields SrcID, DstID, FlwID and the feasible path. Then, the AP is sent back to the source node over the reverse feasible path.[2] The purpose of APs is to inform nodes on the feasible path that a complete route to the destination has been found.

When a route has been established between two nodes, the source node can start sending the data. Data is transported inside data packets (DPs). A DP consists of a generic header and the QoS map obtained from an AP, followed by application data. DPs using the same FlwID are said to form a *flow*.

Each node maintains a number of flow relays (FRs). An FR is an abstract data type associated with a single flow or a group of flows (*flow bundle*) sharing common characteristics, e.g., the same QoS constraints. The information in the FRs is updated by CPs and APs associated with the flows and by QoS measurements performed by the node in question.

At each node a list of active CPs is maintained. A CP is active from the time it is forwarded towards the destination until a corresponding AP is received or a timeout occurs. Each list entry contains information uniquely identifying a CP: a copy of the SrcID, the DstID and the FlwID. Further, a timer T_{out} is associated with every CP in the list. When the timer expires the corresponding CP is removed from the list.

15.3 Path Discovery

When a node in the overlay wants to open a route to another overlay node it assembles a CP with the desired QoS constraints. The requesting node, also called *source node*, sends the CP to all adjacent nodes connected by links satisfying the QoS constraints. If at least one feasible link is found, the CP is added to the list of active CPs and a timer is started accordingly. If after T_{out} seconds no information is received the CP is considered lost and it is removed from the active CP list.

We compute the value of T_{out} by the following formula:

$$T_{out} = 0.2 \times (\text{TTL} - \text{Hops}).$$

Initially, (TTL − Hops) was multiplied by 2 instead of 0.2 in order to obtain a conservative estimate of the round-trip time. However, it was observed that the T_{out} values were excessively large, keeping links blocked for unnecessary

---

[2] Traveling on the reverse feasible path means traveling in the opposite direction on the feasible path (i.e., over hops $N, (N − 1), \ldots, 1$).

long durations of time. Based on empirical evidence, it was decided to scale down the T_{out} values by a factor of $1/10$.

Each node receiving a CP checks whether its node UUID is matching an entry in the feasible path of the CP or not. A matching entry means that the CP has entered a loop and causes the CP to be disregarded. In this case however, the CP entry remains in the active CP list.

If no matching node UUID entry is found in the feasible path of the CP and at least one feasible link exists, then the received CP is added to the list of active CPs. For each feasible link found, the adjacent node UUID (denoted by Hop UUID in Figure 15.3) and the QoS statistics of the link are appended to a copy of the received CP. The modified CP is then forwarded over the link in question. This process is performed for each link, except for the one on which the packet arrived at the current node.

If no feasible link exists, the CP is dropped and no further actions are taken. The receiving and forwarding process is repeated at several nodes until one or more CPs reach the *destination node* or, alternatively, all CPs are dropped by intermediate nodes.

If all CPs are lost the nodes on the feasible path will eventually experience T_{out} timeouts and will thus be able to free any reserved resources.

The first CP that arrives to the destination node is used to obtain the feasible path between source and destination. The destination node will then create a FR for packets corresponding to the FlwID in the CP. The destination node sends an AP back to the source node over the reverse feasible path. If the received CP indicates that the source node wishes bidirectional communication, then the destination node begins immediately a route discovery process towards the source node, using the same QoS constraints specified in the CP.

All subsequent CPs that arrive to the destination node are used to construct corresponding APs. These APs are marked as redundant and then forwarded to the source on the reverse feasible path.

Each node receiving a AP checks whether the triple (SrcId, DstId, FlwID) is matching an entry in the list of active CPs or not. If a matching entry is found, the node either creates an FR or adds the flow to an existing flow bundle corresponding to an FR. Further, the CP entry is removed from the active CP list and the AP is forwarded to the next node on the reverse feasible path. If no matching entry is found, the AP is dropped silently. The manner in which the redundant APs are treated depends on the overlay policies. If the overlay policies favor backup paths or multipath routing, the redundant APs are treated just as regular APs. Otherwise, redundant APs are dropped.

The first AP to arrive at the source node signals that a feasible path has been set up and the application can begin sending DPs. A feasible path can be torn down by a CP with the delete (D) flag set.

15.4 Implementation

To evaluate the performance of RDP we use the public-source simulation environment *OMNeT++* [24]. OMNeT++ is an object-oriented, modular and open-architecture discrete event simulation environment with an embeddable simulation kernel.

An OMNeT++ simulation is build out of hierarchically nested modules, which is ideal for an object-oriented approach. Modules communicate with each other by means of messages and these messages may contain data of arbitrary length. Messages are transported through gates and over channels. A node maintains an arbitrary amount of gates and different gates are connected with channels. The topology of a network, in terms of gates, channels and modules, is defined in the Network Description (NED) language [24].

Our simulator includes two different modules: the ORP module and the DATACENTER module. The ORP module implements the RDP protocol and the DATACENTER module collects the simulation statistics. These statistics can easily be written to files with help of dedicated classes provided by the OMNeT++ framework.

As RDP is designed to run on top of the Internet we have attempted to use realistic Internet topologies in our simulations. There are several challenges in modeling the Internet topology, such as mapping the actual topology, characterizing it, and developing models that capture fundamental properties [25]. Several topology generators are available [25–27], but the generated topologies differ significantly according to the characteristics of the network models used.

We have used the BRITE [25] software to generate network models according to the Barabási–Albert model. BRITE is a universal topology generator developed at Boston University. It is designed to be a flexible, extensible, interoperable, portable and user friendly topology generator. We have chosen BRITE because:

(i) it has supports for realistic topology models based on power-law distributions,
(ii) it can generate router level topologies,
(iii) it is supported under OMNeT++, and

(iv) the source code is freely available.

OMNeT++ allows arbitrary parameters to be defined in an external initialization file that can be loaded in the simulation at any time. This allows the user to control the behaviour of the simulation without having to recompile the source code. The parameters available in our initialization file are:

- TTL value of the packets,
- destination node to which a route will be opened,
- delay and bandwidth QoS constraints used for route requests, and
- session arrival rate and session duration.

The destination node parameter can be a node identifier or a discrete probability distribution used to randomly select a node.

The RDP simulator currently supports Poisson arrivals and exponentially distributed session durations. Thus, the session arrival rate parameter describes the mean value λ of the Poisson distribution and the session duration parameter denotes the mean value $E[X] = 1/\lambda$ of the exponential distribution. More sophisticated distributions are planned to be used for future work.

In our simulations each node in the network attempts to establish a route at a time instant described by the arrival rate process. If RDP establishes a route, then that specific route will last for the duration of the session. The simulations are terminated when all established sessions end.

15.5 Simulation Testbed

We used the Barabási–Albert model for generating the network topologies in our simulations. This model is based on the idea that self-organization in large networks leads to a state described by a scale-free power-law distribution [28].

Power-laws are expressions of the form of

$$y \propto x^a$$

where $\propto$ means "proportional to", a is a constant and x and y are arbitrary measures. Besides characterizing the Internet, power-laws also appear to describe natural networks such as human respiratory systems and automobile networks [29].

The scale-free distribution in the Barabási–Albert model can be explained by two mechanisms: *incremental growth*, which refers to the gradual increase

Table 15.1 Parameter settings for the experiment.

Parameter	Value
TTL	7
Receiver	intuniform(0, number of nodes)
Delay	1000 ms
Bandwidth	intuniform(64, Y)
Session arrival rate	10
Session duration	15

in size of the network, and *preferential connectivity* referring to the tendency of new nodes joining a network to connect to nodes that are highly connected or popular.

The Barabási–Albert router model in BRITE interconnects the nodes following the incremental growth idea. The probability P that a node i wants to connect to another node j in the network is given by

$$P(i, j) = \frac{d_j}{\sum_{k \in V} d_k}$$

where d_j denotes the outdegree of node j, V is the set of nodes that joined the network and $\sum_{k \in V} d_k$ denotes the sum of outdegrees of all nodes that previously joined the network [25]. When we talk about the outdegree we refer to the amount of edges incident to a node.

The parameter settings for our simulations are provided in Table 15.1. The *intuniform* in the value field denotes a discrete uniform distribution.

The difference between each simulation run is determined by the amount of nodes and by the size of the interval from which the bandwidth value is chosen, i.e., by the variable Y in Table 15.1. Initially, there are 50 simulated nodes and this number is incremented each simulation run by 50, until the number of nodes reaches 950. The value of the bandwidth constraint assigned to a route request is drawn from a discrete uniform distribution. We use three intervals for the uniform distribution: 64–1024, 64–2048 and 64–5120 KB/s, respectively. These intervals were selected in order to study the effect of bandwidth reservation on the scalability of the algorithm. In each case, the upper bound of the interval corresponds to the Y variable in Table 15.1. This results in three curves on each graph, each curve having a total of 19 simulation points. Each simulation point is simulated 10 times and the results are used to compute the average. Furthermore, each node runs in "idle"-mode, which means that if a source node does not receive an AP in time it will make no further attempt to try to open a new connection.

The TTL value was selected empirically to obtain a reasonable trade-off between success in finding a feasible path and the amount of flooding that occurs during the process.

We have instructed BRITE to generate flat "ROUTER (IP) ONLY" topologies with nodes randomly placed on a plane of size 1000×1000 points. We have increased the number of nodes by 50 for each generated topology, starting from a size of 50 to 950 nodes.

The router Barabási–Albert model does not handle delays, but BRITE still assigns propagation delay mapped to the distance between nodes in the plane. The delay constraint is set to the opportunistic value of 1 second. This exceeds by far the link delays assigned by BRITE, which are in the range of milliseconds. As a consequence, the QoS delay constraint will always be satisfied. Furthermore, we have assumed in our simulations loss-free links. Therefore our current experiments with the Barabási–Albert router model analyse only the bandwidth performance of the RDP.

We generate session arrival rates following the Poisson distribution with $\lambda = 10$ and session duration times following the exponential distribution with expected value $E[X] = 15$. These values were selected in order to generate a fair amount of session churn. Each time a node is scheduled to open a session, as decided by the session arrival distribution, it select the destination node (i.e., the receiver) from a discrete uniform distribution

15.6 Performance Analysis

We evaluate the RDP performance in terms of protocol overhead, which is a function of the number of nodes in the overlay, and other parameters. We use the following metrics to determine the protocol overhead:

- *route establishment ratio* (%), computed as

$$\frac{\text{total number of established routes}}{\text{total number of route requests}}$$

- *average bandwidth utilization* (B/s), obtained by

$$\frac{\text{total number of bytes sent}}{T} \tag{15.1}$$

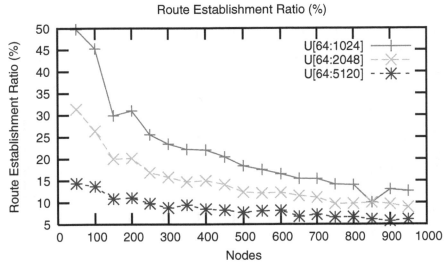

Figure 15.4 Route establishment ratio (%).

- *link load ratio*, defined as

$$\frac{1}{i}\sum_{j=1}^{i}\frac{n_j}{T \times b_j}$$

where i denotes the total number of links, n_j denotes the number of bytes sent over link j, T is the time duration of the simulation and b_j represents the bandwidth of link j.

It is observed that the route establishment ratio, shown in Figure 15.4, has an overall decreasing slope. This indicates that finding feasible paths becomes more difficult with an increasing number of nodes. There are several reasons for why a route cannot be established:

(i) no feasible path exists between source and destination,
(ii) insufficient free bandwidth on feasible links due to previous RDP requests,
(iii) too small TTL value in the CP (henceforth referred to as the *TTL problem*).

In the case when there is no feasible path between source and destination, there is not much that can be done. This case occurs when there is no path

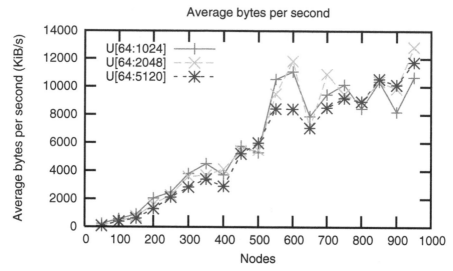

Figure 15.5 Average bandwidth utilization (KB/s).

connecting the source node to the destination node or when there is no feasible path satisfying the combination of QoS constraints.

The second case for failing to establish a route is when previous RDP requests on a feasible link have allocated so much bandwidth that the amount of remaining free bandwidth is too small to satisfy the current constraint. In this case the application that uses RDP can use a lower bandwidth constraint than this, which may allow it to establish a feasible path.

The TTL problem is the particular reason for the decay of the curves in Figure 15.4. This problem can be solved by increasing the TTL value. However, an increase of the TTL may also increase the timeout T_{out}, which will result in longer durations for bandwidth reservation as well as additional flooding of the network by CPs.

Figure 15.5 shows the number of average number of bytes sent per second due to RDP messages. It can be observed that the curves follow roughly a linear growth. This is an indication that from the perspective of this metric the overlay network can scale to a large number of nodes. The fluctuations in the tail of the curve (i.e., in the region of large number of nodes) can be explained by differences in the topology and simulation time between each simulation run.

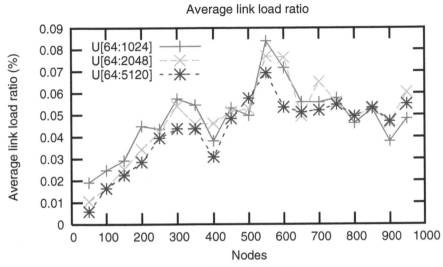
Figure 15.6 Link load ratio (%).

The average link load ratio curves shown in Figure 15.6 appear to level out when the number of nodes increases. This is a manifestation of the TTL-problem. When the network radius grows source nodes will not be able to reach all destinations in the network due to small TTL value in the CPs. This means that CPs will traverse only a fraction of the links available in the network. Since the average link load ratio is computed over all links in the network, this value will decrease while the network radius increases, provided that the TTL is kept the same.

Furthermore, during the RDP simulations we observed that if, for a given request there is a high number of feasible links, this will create a large number of CPs that are duplicated and forwarded. Often, different CPs from the same flow arrive at the same intermediate node by travelling over different routes.

Therefore, it can be useful to introduce a mechanism that alleviates this behavior. Such a mechanism should block each CP that arrives at a node if a CP belonging to the same flow has already passed that node. This solution is expected to lower the protocol overhead at the cost of increased computational overhead per node. We are currently in the process of extending RDP to include this feature.

Throughout our simulation runs we used the following formula to calculate the timeout value of the timers:

$$T_{\text{out}} = 0.2 \cdot (\text{TTL} - \text{Hops}).$$

This is a simple formula but has the downside of yielding values that are larger than necessary. A better approach would be that the timer value should depend on the QoS delay constraint, which is available in every CP packet. Since the CP will be dropped if the overall path delay exceeds the QoS delay constraint, the delay constraint can act as an upper bound. We are therefore considering the following new solution to compute the timeout:

$$
\begin{aligned}
T_{\text{out}} &= 2\frac{D_{\text{QoS}}}{\text{TTL}} \times \text{SSF} \times (\text{TTL} - \text{Hops}) \\
&= 2D_{\text{QoS}} \times \text{SSF} \times \left(1 - \frac{\text{Hops}}{\text{TTL}}\right)
\end{aligned}
$$

D_{QoS} is the QoS delay constraint, the factor 2 is used to allow an AP to be sent in response to a CP, SSF is a safety scaling factor used to overcome problems when for example the last links on a path to a destination has a significantly greater value then the first links, TTL is the time-to-live value and $Hops$ the amount of hops that the CP already has passed. All these parameters are available in CPs. It is expected that the introduction of the D_{QoS} parameter will solve the problem with bandwidth being reserved for too long time by one flow when the TTL-value is increased.

15.7 Conclusions

The paper has reported simulation results for RDP, which is a protocol used for constrained-path selection. The results indicate that the RDP scales almost linearly with the number of nodes in the network. The system's ability to find feasible paths is intimately related to the TTL value used in RDP messages: a large TTL value increases the chance of finding a feasible path at the cost of a higher volume of RDP traffic.

Our future work will focus on issues regarding performance improvement of the current version of RDP. In particular, we plan to address the TTL-problem, extend the protocol to alleviate the issue with multiple CPs being routed over the same link and to test the new timeout formula. Another important issue is the behavior of RDP in the presence of churn.

Additionally, we intend to run the simulation on different types of topologies (e.g., Waxman and hierarchical topologies). Also, we plan to observe the protocol behavior in the presence of session arrivals and session durations generated by long-range dependence processes. We expect that these results will provide additional clues on how to improve RDP's performance. When the protocol reaches maturity we plan to test it in a live environment such as PlanetLab.

Acknowledgements

We would like to thank the Swedish Internet Infrastructure Foundation (IIS) and Euro-NGI for granting and supporting the ROVER project during 2006 and 2007.

References

[1] R. Braden, D. D. Clark and S. Shenker, *Integrated Services in the Internet Architecture: An Overview*, RFC1633, IETF, Category: Informational, June 1994.

[2] S. Blake, D. L. Black, M. A. Carlson, E. Davies, Z. Wang and W. Weiss, *An Architecture for Differentiated Services*, RFC 2475, IETF, Category: Informational, December 1998.

[3] G. J. Armitage, Revisiting IP QOS, *ACM SIGCOMM Computer Communications Review*, vol. 33, no. 5, pp. 81–88, October 2003.

[4] G. Bell, Failure to thrive: QoS and the culture of operational networking, in *Proceedings of the ACM SIGCOMM Workshops*, Karlsruhe, Germany, August, pp. 115–120, 2003.

[5] L. Burgsthaler, K. Dolzer, C. Hauser, J. Jähnert, S. Junghans, C. Macián and W. Payer, Beyond technology: The missing pieces for QoS success, in *Proceedings of the ACM SIGCOMM Workshops*, Karlsruhe, Germany, August, pp. 121–130, 2003.

[6] The ATM Forum, *Private Network-Network Interface Specification Version 1.0 (PNNI 1.0)*. The ATM Forum, af-pnni-0055.000, March 1996.

[7] O. C. Ibe, *Essentials of ATM Networks and Services*, Addison Wesley, Boston, MA, 1997.

[8] D. G. Andersen, Resilient overlay networks, Master's Thesis, Department of Electrical Engineering and Computer Science, Massachusetts Institute of Technology, May 2001.

[9] M. Castro, P. Druschel, A.-M. Kermarrec, A. Nandi, A. Rowstron and A. Singh, Splitstream: High-bandwidth multicast in a cooperative environment, in *Proceeding of IPTPS'03*, Berkeley, CA, February, 2003.

[10] Y. Cui, B. Li and K. Nahrstedt, oStream: Asynchronous streaming multicast in application-layer overlay networks, *IEEE Journal on Selected Areas in Communications*, vol. 22, no. 1, pp. 91–106, January 2004.

[11] Z. Li and P. Mohapatra, QRON: QoS-aware routing in overlay networks, *IEEE Journal on Selected Areas in Communications*, vol. 22, no. 1, pp. 29–40, January 2004.

[12] L. Subramanian, I. Stoica, H. Balakrishnan and R. Katz, OverQoS: An overlay based architecture for enhancing Internet QoS, in *Proceedings of NSDI*, San Francisco, CA, March, pp. 71–84, 2004.

[13] D. Ilie and A. Popescu, A framework for overlay QoS routing, in *Proceedings of 4th Euro-FGI Workshop*, Ghent, Belgium, May 2007.

[14] K. De Vogeleer, QoS routing in overlay networks, Master's Thesis, Blekinge Institute of Technology (BTH), Karlskrona, Sweden, June 2007.

[15] D. Ilie, Overlay routing protocol (ORP), Unpublished architecture and design document, December 2004.

[16] S. Chen and K. Nahrstedt, Distributed quality-of-service routing in high-speed networks based on selective probing, in *Proceedings of LCN*, Lowell, MA, October, pp. 80–89, 1998.

[17] E. Gelenbe, M. Gellman, R. Lent, P. Lei and P. Su, Autonoumous smart routing for network QoS, in *Proceedings of ICAC*, New York, NY, pp. 232–239, May 2004.

[18] E. Gelenbe, R. Lent, A. Montuori and Z. Xu, Cognitive packet networks: QoS and performance, in *Proceedings of IEEE MASCOTS*, Ft. Worth, TX, October, pp. 3–12, 2002.

[19] D. Ilie, Optimization algorithms with applications to unicast QoS routing in overlay networks, Research Report 2007:09, Blekinge Institute of Technology, Karlskrona, Sweden, September 2007.

[20] J. Behrens and J. J. Garcia-Luna-Aceves, Distributed, scalable routing based on link-state vectors, in *Proceedings of SIGCOMM*, London, UK, August, pp. 136–147, 1994.

[21] J. J. Garcia-Luna-Aceves, Loop-free routing using diffusing computations, *IEEE/ACM Transactions on Networking*, vol. 1, no. 1, pp. 130–141, February 1993.

[22] D. Ilie, *On unicast QoS routing in overlay networks*, PhD Thesis, Blekinge Institute of Technology (BTH), Karlskrona, Sweden, October 2008.

[23] P. Leach, M. Mealling and R. Salz, *A Universally Unique IDentifier (UUID) URN Namespace*, RFC 4122, Category: Standards Track, July 2005.

[24] A. Varga, OMNeT++, http://www.omnetpp.org, March 2006.

[25] A. Medina, A. Lakhina, I. Matta and J. Byers, BRITE: Universal topology generation from a user's perspective, Technical Report BUCS-TR-20001-03, Boston University, Boston, MA, April 2001.

[26] J. Winick and S. Jamin, Inet-3.0: Internet topology generator, Technical Report CSE-TR-456-02, University of Michigan, Ann Arbor, MI, 2002.

[27] E. W. Zegura, K. L. Calvert and S. Bhattacharjee, How to model an internetwork, in *Proceedings of IEEE Infocom*, San Francisco, CA, March, vol. 2, pp. 594–602, 1996.

[28] A.-L. Barabási and R. Albert, Emergence of scaling in random networks, *Science*, vol. 286, pp. 509–512, 1999.

[29] M. Faloutsos, P. Faloutsos and C. Faloutsos, On power-law relationships of the internet topology, in *Proceedings of SIGCOMM*, Cambridge, MA, August, pp. 251–262, 1999.

16

Virtual Paths Networks Fast Performance Analysis

P. Belzarena, P. Bermolen, P. Casas and M. Simon

ARTES*, Facultad de Ingeniería, Universidad de la República, 11300 Montevideo, Uruguay; e-mail: artes@fing.edu.uy

Abstract

The performance analysis of a link is a well studied problem. However, for a service provider the most interesting issue is the end-to-end quality of service (QoS) evaluation. The focus of this work is to go from the link to the network analysis. This can be done in a complete but complex way or using an approximation to speedup the calculations. We analyze and compare both methods.

Large Deviations Theory applications to Data Networks are mainly based on the many sources asymptotic. This asymptotic is adequate for networks like Internet backbones, where the assumption that the network is fed by a large number of sources is reasonable.

Recently, Ozturk et al. have proposed a slightly different model called many sources and small buffer asymptotic. They give a formula to calculate the link overflow probability and the end-to-end Loss Ratio of traffic streams in a virtual path feed forward network of general topology. They also define the *fictitious network* concept. The fictitious network has the same topology than the real one, but each traffic stream goes across a link on its path without being affected by the upstream links until that one. So, in the fictitious network each internal link can be analyzed as an external one. Therefore,

---

* ARTES: Joint Research Group of the Electrical Engineering and Mathematics and Statistics Departments.

D. D. Kouvatsos (ed.), Mobility Management and Quality-of-Service for Heterogeneous Networks, 359–385.

the fictitious network usage simplifies dramatically the network performance analysis.

Our main motivation to simplify this task is to allow on-line performance analysis and traffic engineering algorithms in virtual path networks as MPLS or ATM.

Ozturk et al. show that the fictitious network overestimates the overflow probability and the end to end Loss Ratio. Therefore, decisions based on the fictitious network analysis are safe. However, this overestimation leads to network resources under-utilization.

Under certain conditions the real and the fictitious network analysis give the same results (there is no overestimation). In this work we establish sufficient conditions to assure that this coincidence arises. Those conditions are not necessary, and we give an easy way to check if exact results may be obtained even though sufficient conditions are not met. When the real and fictitious networks analysis give different results, we find a method to bound the overestimation.

Finally, we show some numerical examples to compare the performance analysis in the real vs. the fictitious network, and to validate our main results.

Keywords: Network performance analysis, many sources and small buffer asymptotic, large deviations principle.

16.1 Introduction

MPLS (MultiProtocol Label Swiching) [10] is an architecture that enables to perform traffic engineering in IP networks. MPLS introduces the notion of Forwarding Equivalence Class (FEC), giving the network operator the possibility to split the traffic in aggregated flows according to the service model adopted by the Internet Service Provider (ISP).

The edge routers in a MPLS network (or LER for Label Edge Router) are responsible for establishing MPLS tunnels named LSPs (Label Switched Path) between the endpoints of the MPLS domain, and to send each arriving packet to the corresponding LSP.

Explicit Routing (ER), a typical MPLS feature, is the main function that enables Traffic Engineering (TE). Using ER the network operator can establish for each FEC one or more LSPs.

In order to satisfy end to end QoS guarantees in a MPLS network, a performance model is required. We are interested in a performance evaluation model simple enough to be used in on-line algorithms. In this work we will analyze the model called the "fictitious network model" in the context of many sources and small buffer asymptotic.

Ozturk et al. [9] find a useful way to analyze the overflow probability in a network interior link, in the many sources and small buffer asymptotic. Applying their model, the end-to-end Loss Ratio can also be evaluated. This model can be used for off-line performance evaluation and traffic engineering. They also introduce the fictitious network model and show that this latter overestimates the link overflow probability. The fictitious network is a simplified network model that can be used for on-line performance evaluation and traffic engineering. In this work we study in detail the fictitious network model and we find conditions to assure that the fictitious network analysis in an interior link gives the same overflow probability than the real network analysis, being much simpler. We also find a method to bound the overestimation when these conditions are not met.

In Section 16.2 we summarize some related works in order to give a brief description of the problem's context. The model introduced by Ozturk et al. is explained in Section 16.3 where we summarize their main results. Our main results are introduced in Section 16.4. In Section 16.5 we show some numerical examples, comparing the real and the fictitious network based performance analysis and evaluate our results. Finally, we summarize and comment the main results in Section 16.6.

16.2 Related Works

Network designers and operators need a model to evaluate the end to end network performance in an Internet core backbone. Since losses are "rare" events, some researchers have proposed the use of Large Deviations methods for network performance evaluation.

In this context, the effective bandwidth notion was introduced some years ago. The notion of equivalent bandwidth was formerly used to study the access control problem for some networks, as ATM. Many contributions following this approach were done during the 90's to analyze the access control in some networks based on the IntServ model or others. In that situation, the access node receives a connection request and has to estimate the resources it requires, in order to allow or deny the new connection. Kelly's Effective Bandwidth (EB) [7] may be used in such situations as the "equivalent capacity" needed by the new connection. In this context, the flow to be statistically characterized is an individual flow, and may be directly related with the data source (for instance, voice or video codecs). This situation was studied using the so called large buffer asymptotic, in which the link buffer grows to infinity, and its filling above some threshold is analysed. This approach cannot be used in backbone links, where buffers are not devised to store bursts but to resolve simultaneous packet arrival, being consequently small. The application of Large Deviations Theory to the analysis of the MPLS backbone must be performed on the basis of the many sources asymptotic. In this regime we take buffer size $B = Nb$ (with N the number of sources), output capacity $C = Nc$ and make N go to infinity. Results about loss probability in this regime can be found in [2, 5, 11, 13].

Using Large Deviations, Wischik [13] proves the following formula (called *inf sup* formula) for the overflow probability:

$$\log \mathbf{P}(Q_N > B) \approx - \inf_{t \geq 0} \sup_{s \geq 0} ((B + Ct)s - Nst\alpha(s, t))$$

where Q_N represents the stationary amount of work in the queue, C is the link capacity, B is the buffer size and N is the number of incoming multiplexed sources of effective bandwidth $\alpha(s, t)$.

Wischik also shows in [12] that in the many sources asymptotic regime the aggregation of independent copies of a traffic source at the link output and the aggregation of similar characteristics at the link input, have the same effective bandwidth in the limit when the number of sources goes to infinity. This result allows to evaluate the end to end performance of some kind of networks like "in-tree" networks. Unfortunately this analysis cannot be extended to networks like a MPLS backbone.

Eun and Shroff [6], have recently shown that in the many sources asymptotic regime, the probability of the buffer size to be grater than zero goes to zero when the number of sources goes to infinity. This result is valid for a discrete time queue and is valid for a continuous time queue if the source is bounded or can be expressed as an integral of a stationary stochastic process.

Recently, a slightly different asymptotic with many sources and small buffer characteristics was proposed by Ozturk, Mazumdar and Likhanov in [9]. They consider an asymptotic regime defined by N traffic sources, link capacity increasing proportionally with N but buffer size such $\lim B(N)/N \to 0$. In their work they calculate the rate function for the buffer overflow probability and also for the end to end Loss Ratio. This last result can be used to evaluate the end to end QoS performance in a MPLS backbone in contrast with the Wischick result explained before, where it is necessary to aggregate at each link N i.i.d. copies of the previous output link.

Ozturk et al. also introduce the "fictitious network" model. The fictitious network is a network with the same topology than the real one, but where each flow aggregate goes to a link on its path without being affected by the upstream links until that link. The fictitious network analysis is simpler and so, more adequate to on-line performance evaluation and traffic engineering. Ozturk et al. show that the fictitious network analysis overestimates the overflow probability. In this work we analyze when, for an interior network link, the overflow probability calculated using the fictitious network is equal to the overflow probability of the real network. Ramon Casellas [1] has also studied the overestimation problem in the fictitious network. He found a condition to assure that there is no overestimation. This condition is a particular case of the sufficient condition proven in this work.

In the next section we introduce some concepts of the Large Deviations Theory and summarize Ozturk et al.'s work.

16.3 Many Sources and Small Buffer Asymptotic Performance Model

16.3.1 Large Deviations Principle [4]

Definition 3.1.

(1) $I : \chi \to [0, \infty)$, with (χ, B) a measure space is a rate function (RF) if it is a lower semicontinous function.
(2) $I : \chi \to [0, \infty)$, with (χ, B) a measure space is a good rate function (GRF) if all the level set of I ($\Psi_\alpha(I) = \{x : I(x) \leqslant \alpha\}$) are compact sets.

Definition 3.2. A probability measure family μ_n in (χ, B) satisfies a large deviation principle (LDP) with rate function I if

$$- \inf_{x \in \Gamma^o} I(x) \leq \liminf_{n \to \infty} \frac{1}{n} \log \mu_n(\Gamma) \leq \limsup_{n \to \infty} \frac{1}{n} \log \mu_n(\Gamma) \leq - \inf_{x \in \bar{\Gamma}} I(x)$$
(16.1)

$\forall \Gamma \in B$ with interior Γ^o and closure $\bar{\Gamma}$.

16.3.2 Ozturk, Mazumdar and Likhanov's Work

Consider a network of L links which is accessed by M types of independent traffic. Consider a discrete time fluid FIFO model where traffic arrives at time $t \in Z$ and is served immediately if buffer is empty and is buffered otherwise.

Each link k has capacity NC_k and buffer size $B_k(N)$ where $B_k(N)/N \to 0$ with $N \to \infty$. Input traffic of type $m = 1, \ldots, M$, denoted $X^{m,N}$ is stationary and ergodic and has rate $X_t^{m,N}$ at time t (workload at time t of N sources of type m).

Let $\mu_m^N = \mathbb{E}(X_0^{m,N})/N$ and $X^{m,N}(t_1, t_2) = \sum_{t=t_1}^{t_2} X_t^{m,N}$. We assume that $\mu_m^N \xrightarrow[N \to \infty]{} \mu_m$ and $X^{m,N}(0, t)/N$ satisfies the following Large Deviation Principle (LDP) with *good rate function* $I_t^{X^m}(x)$:

$$- \inf_{x \in \Gamma^o} I_t^{X^m}(x) \le \liminf_{N \to \infty} \frac{1}{N} \log \mathbb{P} \left(\frac{X^{m,N}(0, t)}{N} \in \Gamma \right) \qquad (16.2)$$

$$\le \limsup_{N \to \infty} \frac{1}{N} \log \mathbb{P} \left(\frac{X^{m,N}(0, t)}{N} \in \Gamma \right) \le - \inf_{x \in \overline{\Gamma}} I_t^{X^m}(x) \qquad (16.3)$$

where $\Gamma \subset \mathbb{R}$ is a Borel set with interior Γ^o and closure $\overline{\Gamma}$ and $I_t^{X^m}(x) : \mathbb{R} \to [0, \infty)$ is a continuous mapping with compact level sets. We also assume the following technical condition: $\forall m$ and $a > \mu_m$,

$$\liminf_{t \to \infty} \frac{I_t^{X^m}(at)}{\log t} > 0$$

Type m traffic has a fixed route without loops (as in MPLS and ATM networks) and its path is represented by the vector $\mathbf{k}^m = (k_1^m, \ldots, k_{l_m}^m)$, where $k_i^m \in (1, \ldots, L)$. The set $\mathcal{M}_k = \{m : k_i^m = k, 1 \le i \le l_m\}$ denotes the types of traffic that goes through link k. To guarantee system stability it is assumed that

$$\sum_{m \in \mathcal{M}_k} \mu_m < C_k \qquad (16.4)$$

The main result of Ozturk et al.'s work is the following theorem:

Theorem 3.3. *Let $X_{k,t}^{m,N}$ be the rate of type m traffic at link k at time t. There exist a continuous function $g_k^m : \mathbb{R}^M \to \mathbb{R}$ relating the instantaneous input rate at link k for traffic m to all of the instantaneous external input traffic rates such that:*

$$\frac{X_{k,0}^{m,N}}{N} = g_k^m \left(\frac{X_0^{1,N}}{N}, \ldots, \frac{X_0^{M,N}}{N} \right) + o(1) \qquad (16.5)$$

The buffer overflow probabilities are given by

$$\lim_{N \to \infty} \frac{1}{N} \log P(\text{overflow in link k}) = -\mathbf{I}_k$$

$$= -\inf \left\{ \sum_{m=1}^{M} I_1^{X^m}(x_m) : x = (x_m) \in \mathbb{R}^M, \sum_{m=1}^{M} g_k^m(x) > C_k \right\} \quad (16.6)$$

In (16.5), $o(1)$ verifies that $\lim_{N \to \infty} o(1) = 0$ since $B_k(N)/N \to_{N \to \infty} 0$. The function $g_k^m(x)$ is constructed in the proof of the theorem. Ozturk et al. prove that the continuous function relating the instantaneous input rate at link i for traffic m to all of the instantaneous external input traffic rates is the same function relating these variables in a no buffers network. The function relating the instantaneous output rate at link i for traffic m to all of the instantaneous input traffic rates at this link is:

$$f_i^m(x, C_i) = \frac{x_m C_i}{\max \left(\sum_{j \in \mathcal{M}_i} x_j, C_i \right)} \quad (16.7)$$

In a feed-forward network the function $g_k^m(x)$ can be written as composition of the functions of type (16.7) in a recursive way. Using equation (16.7) the buffer overflow probability can be calculated for any network link, by solving the optimization problem of equation (16.6). We need to know the network topology, the link's capacities and, for each arrival traffic type m, the rate functions $I_1^{X^m}$.

Ozturk et al. define also the total (end to end) Loss Ratio as the ratio between the expected value of lost bits at all links along a route and the mean of input traffic in bits, for stream m identified by X^m. With the previous definition they find the following asymptotic for the Loss Ratio $\mathbf{L}^{m,N}$:

$$\lim_{N \to \infty} \frac{1}{N} \log \mathbf{L}^{m,N} = - \min_{k \in k^m} \mathbf{I}_k \quad (16.8)$$

However, in a big network, the optimization problem of equation (16.6) could be very hard to solve. The calculation of the function $g_k^m(x)$ is recursive and so, when there are many links it becomes complex. In addition, the virtual paths can change during the network operation. Therefore, it is necessary to recalculate on-line the function $g_k^m(x)$. To solve equation (16.6), it is also necessary to optimize a nonlinear function under nonlinear constraints. In order to simplify this problem, Ozturk et al. introduce the "fictitious network" concept, that is simpler and gives conservative results.

In the next section we find conditions to assure that there is no overestimation in the calculus of the link overflow probability in the fictitious network analysis. We also find a bound for the error (overestimation) in those cases where the previous condition is not satisfied.

16.4 Fictitious Network Analysis

We analyze an interior network link k under the same assumptions that in Ozturk et al. work. $\mathcal{M}$ is the set of traffic types that access the network and $\mathcal{M}_i$ is the set of traffic types that go through link i. We suppose that the network is feed-forward, this means that each traffic type has a fixed route without loops. In the real network, the link k overflow probability large deviation function (or rate function) is given by

$$I_k^R = \inf \left\{ \sum_{i \in \mathcal{M}} I_1^{X^i}(x_i) : x = (x_i)_{i \in \mathcal{M}}, \sum_{i \in \mathcal{M}} g_k^i(x) > C_k \right\} \qquad (16.9)$$

In the fictitious network this function is given by

$$I_k^F = \inf \left\{ \sum_{i \in \mathcal{M}_k} I_1^{X^i}(x_i) : x = (x_i)_{i \in \mathcal{M}_k}, \sum_{i \in \mathcal{M}_k} x_i > C_k \right\} \qquad (16.10)$$

In the following it is assumed that each traffic type is an aggregate of N i.i.d sources. This implies that each rate function $I_1^{X^i}$ is convex and $I_1^{X^i}(\mu_i) = 0$ for all i. Then, (16.9) and (16.10) are convex optimization problems under constraints. The second one has the advantage that the constraints are linear and there are well known fast methods to solve it. The functions $I_1^{X^i}$ are continuous, so we solve the following problems corresponding to the real and fictitious network respectively.

$$P_R \begin{cases} \min \sum\limits_{i \in \mathcal{M}} I_1^{X^i}(x_i) \\ \\ \sum\limits_{i \in \mathcal{M}} g_k^i(x) \geq C_k \end{cases} \qquad P_F \begin{cases} \min \sum\limits_{i \in \mathcal{M}_k} I_1^{X^i}(x_i) \\ \\ \sum\limits_{i \in \mathcal{M}_k} x_i \geq C_k \end{cases}$$

Definition 4.1. *Consider two optimization problems*

$$P_1 \begin{cases} \min f_1(x) \\ x \in D_1 \end{cases} \quad \text{and} \quad P_2 \begin{cases} \min f_2(x) \\ x \in D_2 \end{cases}$$

P_2 *is called a relaxation of* P_1 *if* $D_1 \subseteq D_2$ *and* $f_2(x) \leq f_1(x), \forall x \in D_1.$

Proposition 4.2. *If* P_2 *is a relaxation of* P_1 *and* x_2 *is optimum for* P_2 *such* $x_2 \in D_1$ *and* $f_2(x_2) = f_1(x_2)$, *then* x_2 *is optimum for* P_1.
    Proof. $f_1(x_2) = f_2(x_2) \leq f_2(x) \leq f_1(x), \forall x \in D_1 \subseteq D_2$, so x_2 is optimum for P_1 because it minimizes f_1 and belongs to D_1. $\square$

Proposition 4.3. P_F *is a relaxation of* P_R.
    Proof. Since the functions $I_1^{X^i}$ are nonnegatives, it is clear that

$$\sum_{i \in \mathcal{M}_k} I_1^{X^i}(x_i) \leq \sum_{i \in \mathcal{M}_r} I_1^{X^i}(x_i) \qquad \forall x = (x_i)_{i \in \mathcal{M}_r}.$$

Then, we have to prove that

$$\left\{ x : \sum_{i \in \mathcal{M}_r} g_k^i(x) \geq C_k \right\} \subseteq \left\{ x : \sum_{i \in \mathcal{M}_k} x_i \geq C_k \right\}$$

By definition, $g_k^i(x) = 0, \forall i \notin \mathcal{M}_k$ and $g_k^i(x) \leq x_i, \forall i \in \mathcal{M}_k$ (since g_k^i can be written as composition of functions of type (16.7)) then

$$\sum_{i \in \mathcal{M}_r} g_k^i(x) = \sum_{i \in \mathcal{M}_k} g_k^i(x) \leq \sum_{i \in \mathcal{M}_k} x_i$$

and therefore $\sum_{i \in \mathcal{M}_k} g_k^i(x) \geq C_k$, implies $\sum_{i \in \mathcal{M}_k} x_i \geq C_k$. $\square$

Remark 4.4. *If an optimum of the fictitious problem* P_F *verifies the real problem* P_R *constraints and the objective functions take the same value at this point, then it is an optimum of the real problem too. In the next remark we find the optimality conditions for* P_F.

Remark 4.5. *The optimality conditions (KKT* [1]*) for the fictitious problem*

$$P_F \begin{cases} \min \sum_{i \in \mathcal{M}_k} I_1^{X^i}(x_i) \\ \\ \sum_{i \in \mathcal{M}_k} x_i \geq C_k \end{cases}$$

are the following:

---

[1] Karush–Khum–Tucker [8].

(1) $\nabla \left(\sum_{i \in \mathcal{M}_k} I_1^{X^i}(x_i) + \lambda(C_k - \sum_{i \in \mathcal{M}_k} x_i) \right) = 0$, with λ *Lagrange multiplier.*

(2) $\lambda \geq 0.$

(3) $\sum_{i \in \mathcal{M}_k} x_i \geq C_k.$

(4) $\lambda \left(C_k - \sum_{i \in \mathcal{M}_k} x_i \right) = 0.$

The first condition implies that

$$\frac{\partial I_1^{X^i}}{\partial x_i}(x_i) = \lambda \qquad \forall i \in \mathcal{M}_k$$

and by the second one,

$$\frac{\partial I_1^{X^i}}{\partial x_i}(x_i) = \lambda \geq 0$$

If $\lambda = 0$, $x_i = \mu_i$, $\forall i$. *In this case* $\sum_{i \in \mathcal{M}_k} I_1^{X^i}(\mu_i) = 0$ *and it is not considered.*
Then we suppose that $\frac{\partial I_1^{X^i}}{\partial x_i}(x_i) > 0$, *which implies* $x_i > \mu_i$. *Finally, since*
$\lambda \neq 0$, *the last condition implies that*

$$C_k - \sum_{i \in \mathcal{M}_k} x_i = 0$$

Then, $\widetilde{x} = (\widetilde{x}_i)_{i \in \mathcal{M}_k}$ *optimum for* P_F *verifies:*

$$\begin{cases} \widetilde{x}_i > \mu_i \quad \forall i \in \mathcal{M}_k \\ \sum_{i \in \mathcal{M}_k} \widetilde{x}_i = C_k \end{cases} \qquad (16.11)$$

The following theorem gives conditions over the network to assure that the link k overflow probability rate function for the real and for the fictitious network are equal ($E = I_k^R - I_k^F = 0$). Since the network is feed forward, it is possible to establish an order between the links. We say that link i is "previous to" or "less than" link j if for one path, link i is found before than link j in the flow direction.

Theorem 4.6. *If* $\tilde{x} = (\tilde{x}_i)_{i \in \mathcal{M}_k}$ *is optimum for* P_F, *and the following condition is verified for all links i less than k*:

$$C_k - \sum_{j \in \mathcal{M}_k \setminus \mathcal{M}_i} \mu_j \leq C_i - \sum_{j \in \mathcal{M}_i \setminus \mathcal{M}_k} \mu_j \qquad \forall i < k \qquad (16.12)$$

then x^* *defined by*

$$(x^*)_i = \begin{cases} \tilde{x}_i & \text{if } i \in \mathcal{M}_k \\ \mu_i & \text{if } i \notin \mathcal{M}_k \end{cases}$$

is optimum for P_R.

Proof. The objective functions of the optimization problems (16.9) and (16.10) take the same values at x^* because $I_1^{X^i}(\mu_i) = 0, \forall i$:

$$\sum_{i \in \mathcal{M}_r} I_1^{X^i}(x_i^*) = \sum_{i \in \mathcal{M}_k} I_1^{X^i}(\tilde{x}_i) + \sum_{i \in \mathcal{M}_r \setminus \mathcal{M}_k} I_1^{X^i}(\mu_i) = \sum_{i \in \mathcal{M}_k} I_1^{X^i}(\tilde{x}_i)$$

Considering Proposition 4.3, it is enough to prove that x^* satisfy the real problem constraints:

$$\sum_{i \in \mathcal{M}_r} g_k^i(x_i^*) \geq C_k$$

By definition $g_k^i(x^*) = 0, \forall i \notin \mathcal{M}_k$. Moreover the function g_k^i can be written as composition of function of type (16.7), so if $\sum_{j \in \mathcal{M}_i}(x_j^*) \leq C_i, \forall i$, then $g_k^i(x^*) = (x^*)_i, \forall i \in \mathcal{M}_k$ and

$$\sum_{i \in \mathcal{M}} g_k^i(x^*) = \sum_{i \in \mathcal{M}_k}(x^*)_i = \sum_{i \in \mathcal{M}_k} \tilde{x}_i = C_k$$

proving the theorem. In the last equality we use that $\tilde{x}$ verifies (16.11), since it is optimum for P_F. Then, it is sufficient to prove that $\sum_{j \in \mathcal{M}_i}(x^*)_j \leq C_i$, $\forall i < k$. Separating the sum,

$$\sum_{j \in \mathcal{M}_i}(x^*)_j = \sum_{j \in \mathcal{M}_i \cap \mathcal{M}_k} \tilde{x}_j + \sum_{j \in \mathcal{M}_i \setminus \mathcal{M}_k} \mu_j \leq C_i \qquad \forall i < k \qquad (16.13)$$

and then we have to guarantee that

$$\sum_{j \in \mathcal{M}_i \cap \mathcal{M}_k} \tilde{x}_j \leq C_i - \sum_{j \in \mathcal{M}_i \setminus \mathcal{M}_k} \mu_j \qquad \forall i < k$$

Since $\tilde{x}$ is optimum for P_F, it satisfy $C_k = \sum_{j \in \mathcal{M}_k} \tilde{x}_j$, and therefore

$$\sum_{j \in \mathcal{M}_k \cap \mathcal{M}_i} \tilde{x}_j = C_k - \sum_{j \in \mathcal{M}_k \setminus \mathcal{M}_i} \tilde{x}_j$$

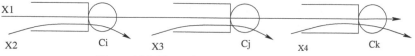

Figure 16.1 Network topology of Examle 4.7.

Also, $\tilde{x}_j > \mu_j, \forall j \in \mathcal{M}_k$

$$\sum_{j \in \mathcal{M}_k \cap \mathcal{M}_i} \tilde{x}_j \le C_k - \sum_{j \in \mathcal{M}_k \setminus \mathcal{M}_i} \mu_j$$

Using the hypothesis, we have that

$$\sum_{j \in \mathcal{M}_k \cap \mathcal{M}_i} \tilde{x}_j = C_k - \sum_{j \in \mathcal{M}_k \setminus \mathcal{M}_i} \mu_j \le C_i - \sum_{j \in \mathcal{M}_i \setminus \mathcal{M}_k} \mu_j \qquad \forall i < k$$

which proves (16.13) and the theorem.   □

Example 4.7. *Consider a network like in Figure 16.1. We analyze the overflow probability at link k.*

If condition (17.12) is attained for link k, then $E = I_k^R - I_k^F = 0$. This condition is:

$$\begin{cases} C_k - \mu_4 \le C_i - \mu_2 \\ C_k - \mu_4 \le C_j - \mu_3 \end{cases}$$

16.4.1 Theorem 4.6 in Terms of Available Bandwidth

Definition 4.8. *For a traffic type m in a link j, it is defined the available bandwidth ABW_j^m as the difference between the link j capacity and the mean value of the transmission rate of the other traffic types in j.*

In terms of the previous definition, the theorem condition (17.12) assures that the overflow probability rate function at link k on real and fictitious network are the same if for all link $j < k$, and for all m traffic type in $\mathcal{M}_j \cap \mathcal{M}_k$, $ABW_j^m > ABW_k^m$. This condition is represented in Figure 16.2 for a simple network with two links.

16.4.2 Sufficient But Not Necessary Condition

The theorem condition (17.12) is sufficient to assure that the overflow probability rate function at link k on real and fictitious networks are the same,

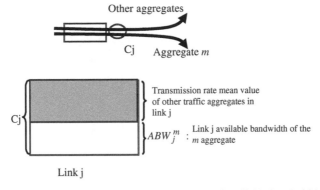

Figure 16.2 Theorem 4.6 condition in terms of available bandwidth.

but it is not a necessary condition. In fact, if $\tilde{x}$ is optimum for the fictitious problem, and if x^* defined as

$$(x^*)_i = \begin{cases} \tilde{x}_i & \text{if } i \in \mathcal{M}_k \\ \mu_i & \text{if } i \notin \mathcal{M}_k \end{cases} \qquad (16.14)$$

satisfies the real problem constraints, then x^* is optimum for the real problem. If x^* verifies the following condition

$$\sum_{j \in \mathcal{M}_i} (x^*)_j \leq C_i \qquad \forall i < k \qquad (16.15)$$

it also verifies the real problem constraints and therefore is optimum for the real problem.

Therefore, in the case that the theorem condition is not fulfilled, if we found $\tilde{x}$ optimum for the fictitious problem, then is easy to check if the rate functions are equal or no. It is enough to check (16.15), where x^* is defined in (16.14).

16.4.3 Error Bound

Since the functions $I_1^{X^i}$ are nonnegatives, it is clear that the rate function for the real problem is always greater than the fictitious one. Then the error $E = I_k^R - I_k^F$ is always nonnegative. This implies that the fictitious network overestimates the overflow probability. We are interested in finding an error bound for the overestimation of the fictitious analysis when conditions (16.12) and (16.15) are not satisfied. A simple way to get this bound is to find a point x which verifies the real problem constraints. In this case, we have that:

$$E = I_k^R - I_k^F \leq \sum_{i \in \mathcal{M}} I_1^{X^i}(x_i) - \sum_{i \in \mathcal{M}_k} I_1^{X^i}(\tilde{x}_i)$$

To assure that x verifies the real problem constraints, we have already seen that it is enough to show that

$$\begin{cases} \displaystyle\sum_{j \in \mathcal{M}_i} x_j \leq C_i & \forall i < k \\ \displaystyle\sum_{j \in \mathcal{M}_k} x_j \geq C_k \end{cases}$$

Therefore, we have to solve this inequalities system. From Remark 4.5, it can be seen that the optimum of the fictitious problem is in the boundary of the feasible region ($\sum_{i \in \mathcal{M}_k} \tilde{x}_i = C_k$). Since we are looking for a point near the optimum of the fictitious problem in the sense that the error bound be as small as possible, we solve the following system:

$$\begin{cases} \displaystyle\sum_{j \in \mathcal{M}_i} x_j \leq C_i & \forall i < k \\ \\ \displaystyle\sum_{j \in \mathcal{M}_k} x_j = C_k \end{cases} \tag{16.16}$$

For the interesting cases, where there are losses at link k, this system always has a solution. In the following, an algorithm to find a solution of this system is defined. We define the following point:

$$(x^*)_j = \begin{cases} \tilde{x}_j & \text{if } j \in \mathcal{M}_k \\ 0 & \text{if } j \notin \mathcal{M}_k \end{cases}$$

If x^* verifies the conditions (16.16), we find a point that verifies the real problem constraints. In some cases this is not useful because $I_1^{X^j}(0) = \infty$ and we have that the error bound is infinite. If $P(X_1^{j,N} \leq 0) \neq 0$, the function $I_1^{X^j}(0) < \infty$ and a finite error bound is obtained. If x^* is not solution for system (16.16), then we redefine (by some small value) the coordinates where $\sum_{j \in \mathcal{M}_i} x_j > C_i$ in such a way that $\sum_{j \in \mathcal{M}_i} x_j = C_i$. The second equation must be verified too and, since some coordinates were reduced, others co-ordinates have to increase to get the total sum equal to C_k. Since the system is compatible, following this method, a solution is always found. There is no guarantee that the solution given by this method minimizes the error bound.

However, this method has a very simple implementation and gives reasonable error bounds as we can see in the numerical examples of the last section.

16.4.4 Error Bound in a Particular Case

We analyze a particular case in which the following conditions are verified:

(1) $\mathcal{M}_i \setminus \mathcal{M}_k \neq \emptyset, \forall i < k$ this means that for all link i less than k, there exists at least one traffic type going through link i and not arriving at link k.

(2) $C_i - \left(C_k - \sum_{j \in \mathcal{M}_k \setminus \mathcal{M}_i} \mu_j \right) \geq 0, \forall i.$

Consider $\widetilde{x} = (\widetilde{x}_i)_{i \in \mathcal{M}_k}$ optimum for (P_F) and x^* defined by

$$(x^*)_j = \begin{cases} \widetilde{x}_j & \text{if } j \in \mathcal{M}_k \\ x_j^* & \text{if } j \notin \mathcal{M}_k \end{cases} \tag{16.17}$$

If x^* verify the real problem constraints, the following error bound is obtained:

$$E \leq \sum_{i \in \mathcal{M}} I_1^{X^i}((x^*)_i) - \sum_{i \in \mathcal{M}_k} I_1^{X^i}(\widetilde{x}_i) \tag{16.18}$$

$$= \sum_{i \in \mathcal{M}_k} I_1^{X^i}(\widetilde{x}_i) + \sum_{i \in \mathcal{M} \setminus \mathcal{M}_k} I_1^{X^i}(x_i^*) - \sum_{i \in \mathcal{M}_k} I_1^{X^i}(\widetilde{x}_i) \tag{16.19}$$

$$= \sum_{i \in \mathcal{M} \setminus \mathcal{M}_k} I_1^{X^i}(x_i^*) \tag{16.20}$$

By definition of x^* and the optimality conditions for the fictitious problem, it follows that:

$$\sum_{j \in \mathcal{M}_k} (x^*)_j = \sum_{j \in \mathcal{M}_k} (\widetilde{x})_j = C_k$$

Therefore, to prove that x^* verify the real problem constraints (16.16), it is enough to show that x^*, verify:

$$\sum_{j \in \mathcal{M}_i} (x^*)_j = \sum_{j \in \mathcal{M}_i \cap \mathcal{M}_k} \widetilde{x}_j + \sum_{j \in \mathcal{M}_i \setminus \mathcal{M}_k} x_j^* \leq C_i \qquad \forall i \tag{16.21}$$

In this particular case, by the second condition, it is possible to define x_j^* for $j \in \mathcal{M}_i \backslash \mathcal{M}_k$ such that

$$\sum_{j \in \mathcal{M}_i \backslash \mathcal{M}_k} x_j^* \leq C_i - \left(C_k - \sum_{j \in \mathcal{M}_k \backslash \mathcal{M}_i} \mu_j \right) \qquad \forall i$$

and therefore

$$\sum_{j \in \mathcal{M}_i} (x^*)_j \leq \sum_{j \in \mathcal{M}_i \cap \mathcal{M}_k} \tilde{x}_j + C_i - C_k + \sum_{j \in \mathcal{M}_k \backslash \mathcal{M}_i} \mu_j$$

On the other hand, since $\tilde{x}$ is optimum for P_F, $\sum_{j \in \mathcal{M}_k} \tilde{x}_j = C_k$ and

$$\sum_{j \in \mathcal{M}_i \cap \mathcal{M}_k} \tilde{x}_j = C_k - \sum_{j \in \mathcal{M}_k \backslash \mathcal{M}_i} \tilde{x}_j$$

Replacing in the previous equation results

$$\sum_{j \in \mathcal{M}_i} (x^*)_j \leq C_k - \sum_{j \in \mathcal{M}_k \backslash \mathcal{M}_i} \tilde{x}_j + C_i - C_k + \sum_{j \in \mathcal{M}_k \backslash \mathcal{M}_i} \mu_j$$

$$= C_i + \sum_{j \in \mathcal{M}_k \backslash \mathcal{M}_i} (-\tilde{x}_j + \mu_j) < C_i$$

since from (16.11), $\tilde{x}_j > \mu_j, \forall j \in \mathcal{M}_k$.

Then x^* verifies (16.16) and therefore is optimum for the real problem. The error bound obtained is (16.20). We can found $(x^*)_{j \in \mathcal{M}_r \backslash \mathcal{M}_k}$ such that, the error bound (16.20) be minimum in the set of $(x^*)_{j \in \mathcal{M}_r}$ defined in (16.17) that verifies the real problem constraints. It is necessary to solve the following convex optimization problem:

$$\begin{cases} \min \sum_{i \in \mathcal{M}_r \backslash \mathcal{M}_k} I_1^{X^i}(x_i) \\ \\ \sum_{j \in \mathcal{M}_i \backslash \mathcal{M}_k} x_j \geq C_i - \sum_{j \in \mathcal{M}_i \cap \mathcal{M}_k} \tilde{x}_j \qquad \forall i = 1, \dots, L \end{cases}$$

Once again it is sufficient to find $(x^*)_{j \in \mathcal{M}_r}$ that verifies the KKT optimality conditions:

(1) $\dfrac{\partial}{\partial x_j} I_1^{X^j}(x_j) + \sum_{i \in k^j} \lambda_i = 0, \forall j \in \mathcal{M}_r \backslash \mathcal{M}_k$.

(2) $\lambda_i \geq 0 \qquad \forall i.$

(3) $\displaystyle\sum_{j \in \mathcal{M}_i \setminus \mathcal{M}_k} x_j \geq C_i - \sum_{j \in \mathcal{M}_i \cap \mathcal{M}_k} \tilde{x}_j \qquad \forall i.$

(4) $\lambda_i \left(\displaystyle\sum_{j \in \mathcal{M}_i \setminus \mathcal{M}_k} x_j - \left(C_i - \sum_{j \in \mathcal{M}_i \cap \mathcal{M}_k} \tilde{x}_j \right) \right) = 0 \qquad \forall i.$

We will define an algorithm to find such point. If for $j \in \mathcal{M}_r \setminus \mathcal{M}_k$, there is a link $i \in k^j$ that verifies

$$C_i - \sum_{h \in \mathcal{M}_i \cap \mathcal{M}_k} \tilde{x}_h \leq \sum_{h \in \mathcal{M}_i \setminus \mathcal{M}_k} \mu_h$$

we define x^* as follows:

$$\sum_{h \in \mathcal{M}_i \setminus \mathcal{M}_k} x_h = C_i - \sum_{j \in \mathcal{M}_i \cap \mathcal{M}_k} \tilde{x}_j$$

This determines a linear equations system that always has a solution, but it can be undetermined. The choice in that case it is not important because the optimum obtained is global. For the coordinates $j \in \mathcal{M}_r \setminus \mathcal{M}_k$, that are not determined with the previous equations, we define $x_j = \mu_j$. It is easy to check that $(x^*)_{j \in \mathcal{M}_r \setminus \mathcal{M}_k}$ defined by this algorithm verifies the KKT optimality conditions. Then we have defined an algorithm that gives the minimum error bound for this particular case.

16.5 Numerical Examples

Example 5.1. *Consider a network like in Figure 16.3. We analyze the overflow probability at link k, assuming that $C_i > C_k$.*

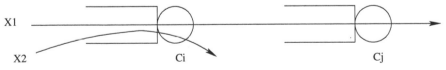

Figure 16.3 Network topology of Example 5.1.

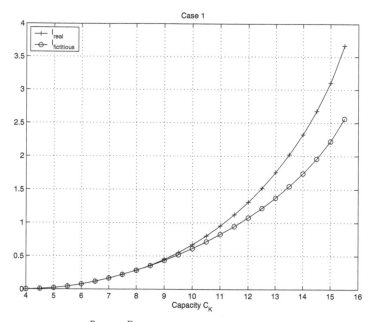

Figure 16.4 I^R and I^F as a function of link capacity of Example 5.1.

If condition (16.12) is attained for link k, then $E = I_k^R - I_k^F = 0$. This condition is:

$$C_k \leq C_i - \mu_2$$

If this condition is not satisfied, since $\tilde{x} = C_k$ is optimum for P_F, we first verify if $x^* = (C_k, \mu_2)$ is optimum for (P_R). It is sufficient to show that x^* verifies the real problem constraints, i.e:

$$\begin{cases} C_k + \mu_2 \leq C_i \\ C_k = C_k \end{cases}$$

If $C_k + \mu_2 > C_i$, we look for $x^* = (x_1^*, x_2^*)$ that verifies

$$\begin{cases} x_1^* + x_2^* \leq C_i \\ x_1^* = C_k \end{cases}$$

It is possible to choose $x_1^* = C_k$ and $x_2^* = C_i - C_k > 0$ resulting in the following error bound:

$$E \leq I_1(C_k) + I_2(C_i - C_k) - I_1(\tilde{x}_1) = I_2(C_i - C_k) \qquad (16.22)$$

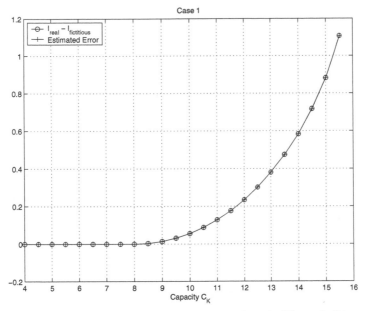

Figure 16.5 Overestimation error and an upper bound of Example 5.1.

In the following numerical example, we calculate the overflow probability rate function for the real and fictitious network. Let $C_i = 16$ kb/s per source and C_k increasing from 4 to 15.5 kb/s per source. All traffic sources are on-off Markov processes. For X_1, the bit rate in the on-state is 16 kb/s, and average times are 0.5 s in the on-state and 1.5 s in the off-state. For X_2, the bit rate in the on-state is 16 kb/s, and average times are 1 s in the on-state and 1 s in the off-state. Since $\mu_1 = 4$ kb/s the stability condition is $C_k > \mu_1 = 4$ kb/s. Using these values, the sufficient condition (16.12) is, $C_k \leq 8$ kb/s. Figures 16.4 and 16.5 show that while this condition is satisfied both functions match, but after $C_k \geq 8$ kb/s they separate. Figure 16.5 also shows the overestimation error ($E = I_k^R - I_k^F$) and the error bound (16.22) described before. In this case, the error bound is exactly the error.

Example 5.2. *Consider a network like in Figure 16.6. We analyze the overflow probability at link k.*

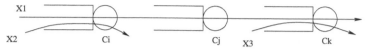

Figure 16.6 Network topology of Example 5.2.

If condition (16.12) is attained for link k, then $E = I_k^R - I_k^F = 0$. This condition is:

$$\begin{cases} C_k - \mu_3 \le C_i - \mu_2 \\ C_k - \mu_3 \le C_j \end{cases}$$

If this condition is not satisfied, and $\tilde{x} = (\tilde{x}_1, \tilde{x}_3)$ is optimum for P_F, we first verify if $x^* = (\tilde{x}_1, \mu_2, \tilde{x}_3)$ is optimum for P_R. It is sufficient to show that x^* verifies the real problem, i.e:

$$\begin{cases} \tilde{x}_1 + \mu_2 \le C_i \\ \tilde{x}_1 \le C_j \\ \tilde{x}_1 + \tilde{x}_3 = C_k \end{cases}$$

If these conditions are not satisfied, we look for $x^* = (x_1^*, x_2^*, x_3^*)$ that satisfies:

$$\begin{cases} x_1^* + x_2^* \le C_i \\ x_1^* \le C_j \\ x_1^* + x_3^* = C_k \end{cases}$$

We choose $x_1^* = \min(\tilde{x}_1, C_i, C_j)$. Three different cases are identified. For the first case $x_1^* = \tilde{x}_1$, we choose:

$$\begin{cases} x_1^* = \tilde{x}_1 \\ x_2^* = C_i - \tilde{x}_1 \\ x_3^* = C_k - \tilde{x}_1 \end{cases} \qquad (16.23)$$

In this case the error bound is:

$$\begin{aligned} E &\le I_1(\tilde{x}_1) + I_2(C_i - \tilde{x}_1) + I_3(C_k - \tilde{x}_1) - I_1(\tilde{x}_1) - I_3(\tilde{x}_3) \\ &= I_2(C_i - \tilde{x}_1) + I_3(C_k - \tilde{x}_1) - I_3(\tilde{x}_3) \qquad (16.24) \end{aligned}$$

Using that $\tilde{x}_1 + \tilde{x}_3 = C_k$, we have another possibility for determining an error bound. We can rewrite the first equation as $C_k - \tilde{x}_3 + x_2^* \le C_i$. And, since

$\tilde{x}_3 > \mu_3$ we can choose $x_2^* = C_i - (C_k - \mu_3)$ (or any lower value). In this case the error bound is

$$E \ \leq \ I_1(\tilde{x}_1) + I_2(C_i - (C_k - \mu_3)) + I_3(C_k - \tilde{x}_1) - I_1(\tilde{x}_1) - I_3(\tilde{x}_3)$$

$$= \ I_2(C_i - (C_k - \mu_3)) + I_3(C_k - \tilde{x}_1) - I_3(\tilde{x}_3) \tag{16.25}$$

The best error bound depends on the relative position of the points $C_i - (C_k - \mu_3)$ and $C_i - \tilde{x}_1$. Since $C_i - \tilde{x}_1 \leq C_i - (C_k - \mu_3)$, if both are less than μ_2 then the best error bound is (16.24).

For the second case $x_1^* = C_i$ $(C_i \leq C_j)$, we choose:

$$\begin{cases} x_1^* = C_i \\ x_2^* = 0 \\ x_3^* = C_k - C_i \end{cases} \tag{16.26}$$

In this case, the error bound is

$$E \leq I_1(C_i) + I_2(0) + I_3(C_k - C_i) - I_1(\tilde{x}_1) - I_3(\tilde{x}_3) \tag{16.27}$$

For the last case $x_1^* = C_j$ $(C_j \leq C_i)$, we choose:

$$\begin{cases} x_1^* = C_j \\ x_2^* = C_i - C_j \\ x_3^* = C_k - C_j \end{cases} \tag{16.28}$$

In this case, the error bound is

$$E \leq I_1(C_j) + I_2(C_i - C_j) + I_3(C_k - C_j) - I_1(\tilde{x}_1) - I_3(\tilde{x}_3) \tag{16.29}$$

For the following numerical example, we calculate the overflow probability rate function for the real and the fictitious network. Let $C_i = 5.5$ kb/s, $C_j = 7$ kb/s per source and C_k ranging from 7 to 25 kb/s per source. All traffic sources are on-off Markov processes. For X_1, the bit rate in the on-state is 8 kb/s. For X_2, the bit rate in the on-state is 10 kb/s. For X_3 the bit rate in the on-state is 20 kb/s. The average time for all traffic types are 0.5 s on the on-state and 1.5 s in the off-state. Since $\mu_1 = 2$ kb/s and $\mu_3 = 5$ kb/s the stability condition is $C_k > \mu_1 + \mu_3 = 7$ kb/s. Using these values, the sufficient conditions (16.12) are:

$$\begin{cases} C_k \leq C_i - \mu_2 + \mu_3 = 8 \text{ kb/s} \\ C_k \leq C_j + \mu_3 = 12 \text{ kb/s} \end{cases}$$

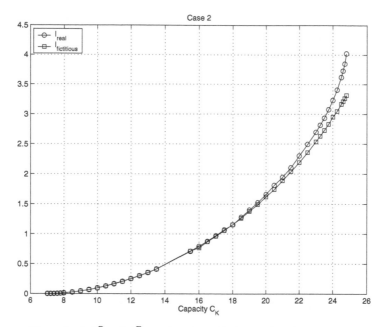

Figure 16.7 I^R and I^F as a function of link capacity of Example 5.2.

The conditions are satisfied for values of C_k less than 8 kb/s. Figure 16.7 shows that both functions match even after the condition is not satisfied and up to $C_k \simeq 15$ kb/s. The reason is that $x^* = (\tilde{x}_1, \mu_2, \tilde{x}_3)$ is optimum for P_R. From this point the functions begin to separate. Figures 16.7 and 16.8 also show the functions I_k^R, I_k^F and $I_k^F + E'$, where E' is the error bound. Until $C_k = 24$ kb/s, E' is calculated using (16.23), and then using (16.26). It is important to note that when $C_k > 14$ kb/s per source, the link utilization falls to less than 50% and therefore, as can be seen in Figure 16.8, the rate function $I_1^{X^k}$ takes values bigger than 0.5. If for example the number of sources feeding the network is $N = 100$, the losses are near 10^{-22}. Finally, we have seen that the estimated error bound is tight and when the error is big, the link overflow probability is small and, therefore, these links are not relevant for the QoS evaluation.

Example 5.3. *Consider a network like in Figure 16.9. We analyse the overflow probability at link k, assuming that $C_i + C_j > C_k$.*

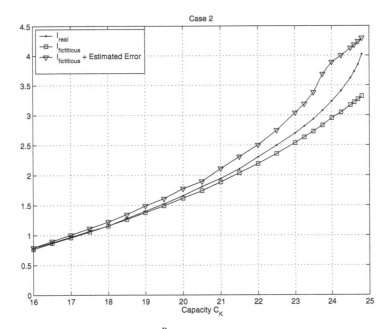

Figure 16.8 Upper bound for I^R as a function of link capacity of Example 5.2.

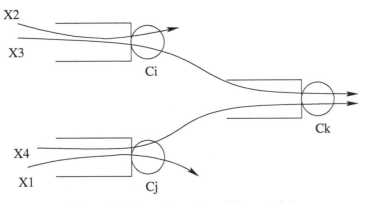

Figure 16.9 Network topology of Example 5.3.

If condition (16.12) is attained for link k, then $E = I_k^R - I_k^F = 0$. This condition is

$$\begin{cases} C_k - \mu_3 \le C_i - \mu_1 \\ C_k - \mu_2 \le C_j - \mu_4 \end{cases}$$

If this condition is not satisfied, since $\tilde{x} = (\tilde{x}_2, \tilde{x}_3)$ is optimum for P_F, we first verify if $x^* = (\mu_1, \tilde{x}_2, \tilde{x}_3, \mu_4)$ is optimum for P_R. It sufficient to show that x^* verifies the real problem constraints, i.e.:

$$\begin{cases} \tilde{x}_2 + \mu_1 \le C_i \\ \tilde{x}_3 + \mu_4 \le C_j \\ \tilde{x}_2 + \tilde{x}_3 = C_k \end{cases} \tag{16.30}$$

If these conditions are not satisfied, we look at first for $x^* = (x_1^*, \tilde{x}_2, \tilde{x}_3, x_4^*)$ that satisfies

$$\begin{cases} \tilde{x}_2 + x_1^* \le C_i \\ \tilde{x}_3 + x_4^* \le C_j \\ \tilde{x}_2 + \tilde{x}_3 = C_k \end{cases}$$

If $\tilde{x}_2 > C_i$ or $\tilde{x}_3 > C_j$ then it is not possible to choose such point. So, we look for $x^* = (x_1^*, x_2^*, x_3^*, x_4^*)$ that verifies

$$\begin{cases} x_1^* + x_2^* = C_i \\ x_3^* + x_4^* = C_j \\ x_1^* + x_3^* = C_k \end{cases}$$

One possible choice is

$$\begin{cases} x_1^* = C_i \\ x_2^* = 0 \\ x_3^* = C_k - C_i \\ x_4^* = C_j - (C_k - C_i) \end{cases} \tag{16.31}$$

For the following numerical example, we calculate the overflow probability rate function for the real and fictitious network. Let $C_i = 12$ kb/s, $C_j = 14$ kb/s per source and C_k increasing from 8 to 25.5 kb/s per source. All traffic sources are on-off Markov processes. For X_1, the bit rate in the on-state is 20 kb/s. For X_2, the bit rate in the on-state is 16 kb/s. For X_3, the bit rate in the on-state is 16 kb/s. For X_4, the bit rate in the on-state is 12 kb/s. The average times for all traffic types are 0.5 s in the on-state and 1.5 s in the off-state. Since $\mu_2 = 4$ kb/s and $\mu_3 = 4$ kb/s the stability condition is $C_k > \mu_1 + \mu_3 = 8$ kb/s. Using these values, the sufficient condition (16.12) are:

$$\begin{cases} C_k < C_i - \mu_1 + \mu_3 = 11 \text{ kb/s} \\ C_k < C_j - \mu_4 + \mu_2 = 15 \text{ kb/s} \end{cases}$$

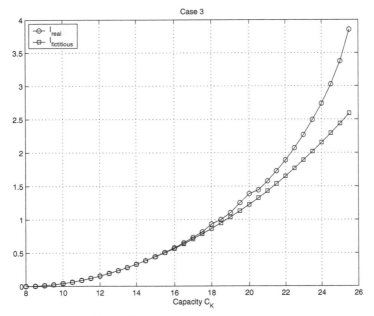

Figure 16.10 I^R and I^F as a function of link capacity of Example 5.3.

Figures 16.10 and 16.11 show that both functions match even after the condition is not satisfied and up to $C_k \simeq 15$ kb/s. The reason is that $x^* = (\mu_1, \tilde{x}_2, \tilde{x}_3, \mu_4)$ is optimum for the real problem. From this point the functions begin to separate.

Figure 16.11 also shows the functions I_k^R, I_k^F and $I_k^F + E'$, where E' is the error bound. Until $C_k = 24$ kb/s, E' is calculated using (16.30), and then using (16.31). As in the previous example, when $C_k > 16$ kb/s, the link utilization is less than 50% and that the error bound is tight for the relevant cases in the QoS evaluation.

All calculations of overflow probability rate functions were done with a software package developed by our group available in the web [3].

16.6 Conclusions

In this paper we have explained the fictitious network analysis. We have seen that the calculus of the overflow probability rate function of an interior network link is simpler and faster than the equivalent task in the real network. For this reason, on-line performance analysis and on-line traffic engineering will

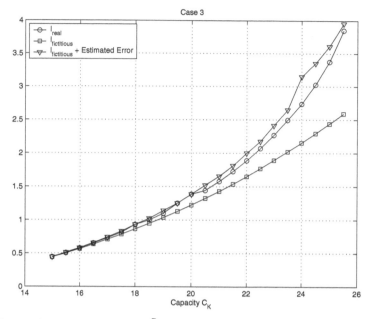

Figure 16.11 Upper bound for I^R as a function of link capacity of Example 5.3.

be easier (or even feasible instead of impossible) using the fictitious network analysis.

Generally, the fictitious network analysis overestimates the overflow probability and the end-to-end Loss Ratio. Therefore, this approach is safe but network resources can be under-utilized. To solve this problem we have found a condition that depends only on link's capacities and mean traffic rates. If this condition is satisfied, the overflow probability calculated using the fictitious network has the same value that the one calculated in the real one. When this condition is not satisfied, the rate function of the link overflow probability calculated in the fictitious network can be smaller or equal than the same rate function calculated in the real network. We have shown that once the fictitious rate function is calculated, it is very simple to verify if both rate functions are equal or not. If they are not equal, we have found a simple algorithm to find an error bound.

In the numerical examples we have found that the error bound is tight. In these examples, it can be seen that when the error is big, the link overflow probability is very small and, therefore, these links are not relevant for the QoS evaluation. In spite of this, we can affirm that when the overflow probability at link k in the fictitious network is very small, even if the error is big, this link is not considered for the QoS evaluation.

Acknowledgements

This research was partially supported by PDT (Programa de Desarrollo Tecnológico, Préstamo 1293/OC-UR): S/C/OP/17/02, S/C/OP/17/03, CSIC and program FCE (Fondo Clemente Estable) 8079. The authors wish to thank the referees for their careful review and helpful comments of this work.

References

[1] R. Casellas, *MPLS Traffic Engineering*, ENST, Paris, 2002.
[2] C. Courcoubetis and R. Weber, Buffer overflow asymptotics for a switch handling many traffic sources, *Journal of Applied Probability*, vol. 33, pp. 886–903, 1996.
[3] D. Buschiazzo, A. Ferragut, A. Vazquez and P. Belzarena, ARCA, http://iie.fing.edu.uy/investigacion/grupos/artes/arca, 2004.
[4] A. Dembo and O. Zeitouni, *Large Deviations Techniques and Applications*, Jones and Barlett Publishers, 1993.
[5] N. G. Duffield and N. O'Connell, Large deviations and overflow probabilities for the general single server queue, with applications. *Mathematical Proceedings of the Cambridge Philosophical Society*, vol. 118, pp. 363–374, 1995.
[6] D. Y. Eun and N. Shroff, Network decomposition in the many-sources regime, *Advances in Applied Probability*, vol. 36, no. 3, 893–918, 2004.
[7] F. P. Kelly, Notes on effective bandwidth, in *Stochastic Networks: Theory and Applications*, F. P. Kelly, S. Zachary and I. Ziedins (Eds.), Oxford University Press, pp. 141–168, 1996.
[8] S. Mokhtar, D. Bazaraa Hanif, C. Sherali and M. Shetty, *Nonlinear Programming Theory and Algorithms*, John Wiley & Sons, 1993.
[9] O. Ozturk, R. Mazumdar and N. Likhanov, Many sources asymptotics in networks with small buffers. *Queueing Systems (QUESTA)*, vol. 46, nos. 1–2, pp. 129–147, 2004.
[10] E. Rosen and A. Viswanathan, Multiprotocol label switching architecture, RFC3031, January 2001.
[11] A. Simonian and J. Guibert, Large deviations approximations for fluid queues fed by a large number of on/off sources, *IEEE Journal on Selected Areas in Communications*, vol. 13, no. 7, pp. 1017–1027, 1995.
[12] D. Wischik, The output of a switch or effective bandwidths for networks, *Queueing Systems*, vol. 32, pp. 383–396, 1999.
[13] D. Wischik, Sample path large deviations for queues with many inputs, *Annals of Applied Probability*, vol. 11, no. 2, pp. 379–404, 2001.

17

QoS Guarantee in a Multirate Loss Model of Batched Poisson Arrival Processes

Ioannis D. Moscholios and Michael D. Logothetis*

WCL, Department. of Electrical & Computer Engineering, University of Patras, 265 04 Patras, Greece; e-mail: m-logo@wcl.ee.upatras.gr

Abstract

Bandwidth Reservation (BR) is a sine qua non of multi-service loss systems, in order to guarantee a certain Quality of Service (QoS) for each service accommodated in the system/link. First, we review an extension of the Erlang Multirate Loss Model (EMLM), namely the Batched Poisson Erlang Multirate Loss Model (BP-EMLM), in which calls of each service arrive in a link of certain capacity, following a batch Poisson process and compete for the available link bandwidth under the Complete Sharing (CS) policy. The batch size is generally distributed while the partial batch blocking discipline is applied, i.e. depending on the available link bandwidth a part of an arriving batch can be accepted while the rest of it is discarded. Second, we propose the BP-EMLM under BR policy. The importance of this proposal is not only that we can guarantee specific call-level QoS for each service, but also that we strengthen the applicability of the BP-EMLM to overflow traffic modeling of alternate route systems, where BR policy must be applied in the overflow routes to protect direct traffic from overflow traffic. For the application of the BR policy in the BP-EMLM, we study two methods: (1) the Roberts and (2) the Stasiak–Glabowski (S&G) method. The new model (with BR) does not have a product form solution and therefore we propose approximate but recursive formulas for the calculation of various performance measures,

---

* Author for correspondence.

D. D. Kouvatsos (ed.), Mobility Management and Quality-of-Service for Hetero-geneous Networks, 387–416.

such as time and call congestion probabilities and the link utilization. By comparing the analytical with simulation results we show that both methods achieve satisfactory results, but the S&G method performs better than the Roberts method under heavy traffic load, in the case of QoS equalization.

Keywords: Batched Poisson process, bandwidth reservation, QoS, time congestion, call congestion, recursive formula, Markov chain.

List of Abbreviations

b.u.	bandwidth units
BP-EMLM	Batched Poisson Erlang Multirate Loss Model
BR	Bandwidth Reservation
CBP	Call Blocking Probabilities
CC	Call Congestion
CS	Complete Sharing
EMLM	Erlang Multirate Loss Model
EMLM/BR	EMLM under the BR policy
MRAE	Mean Relative Approximation Error
PASTA	Poisson Arrivals See Time Averages
PFS	Product Form Solution
QoS	Quality of Service
RAE	Relative Approximation Error
RS	Reservation Space
S&G	Stasiak–Glabowski
TC	Time Congestion

17.1 Introduction

The classical Erlang Multirate Loss Model (EMLM) is the basis of teletraffic analysis of various traffic streams multiplexed in a link of certain capacity, both of wired and wireless networks [1–7]. In the EMLM, calls of different services (with different traffic and bandwidth requirements) arrive to a single link according to a Poisson process, require fixed bandwidth and compete for the available link bandwidth under the Complete Sharing (CS) policy (all call types compete for all bandwidth resources). An arriving call is accepted in the system if and only if its bandwidth requirement, not necessarily contiguous, is available (see [8] for contiguous bandwidth requirements); blocked calls are lost (do not further affect the system). The mean call holding (service) time is arbitrarily distributed [9, 10].

The importance of the EMLM in teletraffic comes from the following features:

- The existence of a Product Form Solution (PFS), as far as the steady-state probabilities is concerned. The PFS leads to an accurate calculation of Call Blocking Probabilities (CBP), the main QoS performance index [9].
- The macro-state probabilities (link occupancy distribution) are calculated recursively; a feature that facilitates the calculation of various performance indexes, including CBP, and broadens the EMLM's applicability range to links of large capacities [9–11].

Due to these features the EMLM has been extended to the study of other loss systems [12–17]. A quite interesting extension of the EMLM is a model in which calls of each service arrive at the link according to a batch Poisson process. In the batch Poisson arrival process, one basic principle of random arrivals, according to which no simultaneous arrivals occur, is abolished, while another basic principle of random arrivals, according to which arrivals (the batches) occur at time-points following a negative exponential distribution, is kept (Figure 17.1) [18–21]. As pointed out in these papers the batch Poisson process is important, not only because in several applications calls arrive as batches (groups), but also because it can represent, in an approximate way, arrival processes that are more "peaked" and "bursty" (expressed by the peakedness factor) than the Poisson process. The peakedness factor, z, is the ratio of the variance over the mean of the number of arrivals; if $z = 1$, the arrival process is Poisson; if $z < 1$, the arrival process is quasi-random [22]; if $z > 1$, the process is more peaked and bursty than Poisson. For example, batch Poisson processes can be used to model overflow traffic (where $z > 1$) [19].

In [18], the EMLM with Batch Poisson arrivals (BP-EMLM) is examined under the hypothesis of partial batch blocking and geometric batch size distribution, while in [19] a general batch size distribution is considered. Partial batch blocking means that when the available link bandwidth is not enough in order for the entire batch to be accepted in the system, then, a part of it is accepted (one or more calls), while the rest of it (one or more calls) is discarded. When the geometric batch size distribution is considered, the BP-EMLM coincides with the Delbrouck model [23]. The latter is a generalization of the EMLM since it allows services to have different peakedness

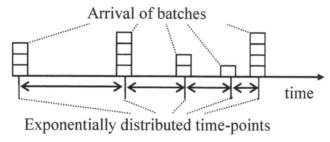

Figure 17.1 The arrival process of the multirate batched Poisson loss model.

factors. In both [18] and [19] the model has a PFS, and the link occupancy distribution is determined via an accurate recursive formula which is similar to that of the EMLM. In [20] an asymptotic analysis of the model of [19] is presented. In [21] the models of [18, 19] are extended to include: (a) a fixed routing network, (b) batches that arrive in a state-dependent manner, (c) the complete batch blocking discipline, according to which the whole batch is blocked, even if only one call of the batch cannot be accepted in the system, and (d) two more bandwidth sharing policies namely the upper limit and guaranteed minimum policies. Note that the results of [18–21] hold only for exponentially distributed holding times. As it is written in [19], the state distribution depends on the service time distribution rather than only on the mean service times.

In the BP-EMLM, it is important to note the distinction between time congestion (TC) and call congestion (CC) probabilities. These probabilities coincide in the case of Poisson arrivals (Poisson Arrivals See Time Averages, PASTA property [22]), but not in the case of batch Poisson arrivals; when coincide they are commonly referred as CBP. TC probability is determined by the proportion of time that the system is congested. An observer, who is not part of the system, can measure this probability. CC probability is determined by the proportion of arriving calls that find the system congested. An observer who is part of the system (i.e. an arriving call) can measure this probability.

In this article we focus on the BP-EMLM of [19] (partial batch blocking – generally distributed batch size) and examine the application of the Bandwidth Reservation (BR) policy on it. This policy is a sine qua non of multi-service loss systems, because we can guarantee a certain QoS for each service accommodated in the system/link. According to the BR policy, a fraction of the available link bandwidth (denoted as BR parameter) is reserved to benefit services with higher bandwidth per call requirements. This is necessary because such services receive worse call-level QoS than other services with less bandwidth per call requirements. Besides, since the BP-EMLM can be used to model overflow traffic (in alternate route systems), the BR policy must be applied in the overflow routes to protect direct traffic of these routes from overflow traffic [22]; in alternate route systems, the BR policy is usually referred to as trunk reservation scheme. Therefore, the study of the BP-EMLM under the BR policy is essential.

Contrary to the CS policy, the BR policy destroys the PFS of the BP-EMLM and, therefore, the calculation of various performance measures, such as TC and CC probabilities and link utilization, cannot be accurate. To this end, we propose approximate but still recursive formulas. In order to study the BP-EMLM under the BR policy, we examine and compare the applicability of two methods, namely the Roberts method [24] and the Stasiak–Glabowski (S&G) method [25], already proposed in the case of the EMLM under the BR policy (EMLM/BR). In [24], Roberts proposes a simple recursive approximation for the calculation of the link occupancy distribution. The simplicity of that approximation is based on the assumption that the population of calls of a service inside its Reservation Space (RS) is negligible; the RS of a service is the fraction of the system state space where the access of the calls (new arrivals) is denied. In general, the Roberts approximation is satisfactory, but it may become critical when one aims at equalizing CBP between services. In the latter usual case, the RS of a service may be quite large and, therefore, the Roberts assumption (calls of a service are negligible inside its RS) becomes problematic and may lead to unsatisfactory results compared to simulation results [25]. To increase the accuracy of the Roberts approximation when CBP equalization is required (in the EMLM/BR), Stasiak and Glabowski proposed a more realistic method [25], where the population of calls of a service inside its RS is not considered negligible and taken into account in the CBP calculation. The more accurate S&G method has the drawback of being more complex than the Roberts method.

In the BP-EMLM under the BR policy, we show that both methods provide satisfactory results, but when TC probabilities equalization is required and more than two services are accommodated in a link, then, the S&G method gives better results than the Roberts method, especially when the offered traffic-load is high. This comparison is based on simulation. TC probabilities equalization is achieved by a certain set of BR parameters (one BR parameter per service), which depend on the bandwidth requirements of calls. On the other hand, CC probability equalization cannot be achieved by a certain set of BR parameters, because it depends on the offered traffic load of each service.

The remainder of the article is organized as follows: in Section 17.2 we review the BP-EMLM [19]. In Section 17.3, we propose the BP-EMLM under the BR policy. In Section 17.3.1 we present the Roberts method, while in Section 17.3.2 we present the S&G method; in both sections, we propose the corresponding analytical formulas for the calculation of various performance measures. In Section 17.4 we present numerical results in order to evaluate the two methods by comparing analytical with simulation results. We conclude with Section 17.5.

17.2 Review of the BP-EMLM

Consider a single link of capacity C bandwidth units (b.u.) and K different services. Calls of each service k $(k = 1, \ldots, K)$ require b_k b.u., compete for the available link bandwidth under the CS policy and have an exponentially distributed service time with mean μ_k^{-1}. Calls arrive to the link according to a batch Poisson process with arrival rate λ_k and batch size distribution B_r^k, where B_r^k denotes the probability that there are r calls in an arriving batch of service k.

The batch blocking discipline adopted in the BP-EMLM is the partial batch blocking. As it is proved in [19], the steady-state probabilities in the BP-EMLM are given by a PFS. The latter is essential in order to derive an accurate recursive formula for the calculation of the link occupancy distribution. Indeed, if we denote by j the occupied link bandwidth then the link occupancy distribution, $G(j)$, which is the probability that j out of C b.u. are occupied, is given by the following accurate recursive formula [19]:

$$
G(j) = \begin{cases}
1 & \text{for } j = 0 \\
\dfrac{1}{j} \sum_{k=1}^{K} \alpha_k b_k \sum_{l=1}^{\lfloor j/b_k \rfloor} \hat{B}_{l-1}^k G(j - lb_k) & \text{for } j = 1, \ldots, C \\
0 & \text{otherwise}
\end{cases}
\tag{17.1}
$$

where $\alpha_k = \lambda_k / \mu_k$, $\lfloor j/b_k \rfloor$ is the largest integer less than or equal to j/b_k and $\hat{B}_l^k$ is the complementary batch size distribution given by $\hat{B}_l^k = \sum_{r=l+1}^{\infty} B_r^k$.

If $B_r^k = 1$ for $r = 1$ and $B_r^k = 0$ for $r > 1$, then the arrival process is Poisson and the EMLM results:

$$G(j) = \begin{cases} 1 & \text{for } j = 0 \\ \dfrac{1}{j} \displaystyle\sum_{k=1}^{K} \alpha_k b_k G(j - b_k) & \text{for } j = 1, \ldots, c \\ 0 & \text{otherwise} \end{cases} \tag{17.2}$$

Note: Equation (17.2) is usually referred to in the bibliography as the Kaufman–Roberts formula since it was independently proved by Kaufman in [9] and Roberts in [10].

Having recursively determined the $G(j)$'s in the BP-EMLM, various other performance measures can be calculated [19]:

- The average number of service k calls given that the system state is j, denoted as $E(n_k \mid j)$, can be determined as:

$$E(n_k \mid j) = \alpha_k \sum_{l=1}^{\lfloor j/b_k \rfloor} \hat{B}_{l-1}^k G(j - lb_k) \Big/ G(j) \tag{17.3}$$

Note: Equation (17.3) is an intermediate result of the proof of equation (17.1). Indeed, if we multiply both sides of equation (17.3) by b_k and sum over $k = 1, \ldots, K$ then equation (17.1) is the result (see [19]).

- The occupied link bandwidth in state j is given by

$$j = \sum_{k=1}^{K} b_k E(n_k \mid j) \tag{17.4}$$

- The average number of service k calls in the system, $\bar{n}_k$, is calculated via

$$\bar{n}_k = \sum_{j=1}^{c} E(n_k \mid j) G(j) \tag{17.5}$$

- CC probability of service k, C_{b_k}, which is the probability that an arriving call of service k cannot be accepted in the system, can be calculated by

$$C_{b_k} = \left(\alpha_k \hat{B}_k - \bar{n}_k \right) \Big/ \alpha_k \hat{B}_k \tag{17.6}$$

where $\hat{B}_k$ denotes the average size of service k arriving batches and is given by

$$\hat{B}_k = \sum_{r=1}^{\infty} r B_r^k.$$

- TC probability of service k, P_{b_k}, which is the probability that at least $C - b_k + 1$ b.u. are occupied is given by

$$P_{b_k} = \sum_{j=C-b_k+1}^{C} G^{-1}G(j) \tag{17.7}$$

  where $G = \sum_{j=0}^{C} G(j)$ is a normalization constant.
- The link utilization U can be determined via equation (17.1) as follows:

$$U = \sum_{j=1}^{C} jG(j) \tag{17.8}$$

If calls of service k arrive in batches of size s_k where s_k is given by the geometric distribution with parameter β_k, i.e. $Pr(s_k = r) = (1-\beta_k)\beta_k^{r-1}$ with $r \geq 1$, then the BP-EMLM coincides with the model proposed by Delbrouck in [23], as shown in [18], [19]. More precisely, since $\hat{B}_l^k = \beta^l$, equation (17.1) takes the form:

$$G(j) = \begin{cases} 1 & \text{for } j = 0 \\ \dfrac{1}{j}\displaystyle\sum_{k=1}^{K}\alpha_k b_k \sum_{l=1}^{\lfloor j/b_k \rfloor} \beta^{l-1}G(j - lb_k) & \text{for } j = 1, \dots, C \\ 0 & \text{otherwise} \end{cases} \tag{17.9}$$

which is Delbrouck's result.

Note 1: The geometric distribution is a memoryless distribution and a discrete equivalent of the exponential distribution [26]. Because of these features it is highly used as a batch size distribution.

Note 2: In the case of the geometric batch size distribution, replace $\hat{B}_k = 1/1 - \beta_k$ in equation (17.6).

Note 3: The TC and CC probabilities coincide when the arrival process changes from batch Poisson to Poisson, i.e. when $B_r^k = 1$ for $r = 1$ and $B_r^k = 0$ for $r > 1$.

Prior to the application of the BR policy in the BP-EMLM we present, through an example, the difference between TC and CC probabilities. Consider a link of capacity $C = 12$ b.u. which accommodates calls of two services with bandwidth requirements $b_1 = 1$ and $b_2 = 2$ b.u., respectively. In Figure 17.2, we assume that the link has 6 b.u. available for new calls. At the first time-point, a 2nd service call arrives. This call is accepted in the system,

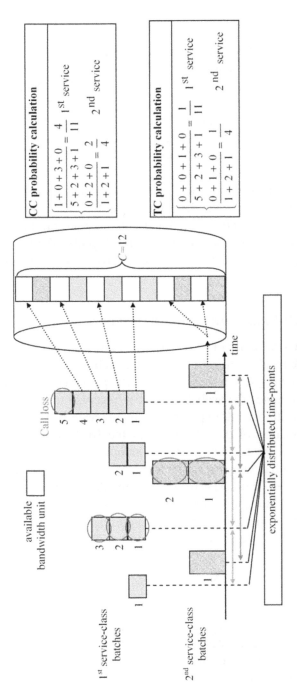

Figure 17.2 Difference between the CC and TC probability (BP-EMLM).

and therefore the available link bandwidth reduces to 4 b.u. At the second time-point, a 1st service batch arrives consisting of 5 calls. One out of these 5 calls is blocked and lost. This type of blocking has to do with CC probability (1 blocking event, see Figure 17.2 "CC probability calculation") and not with TC probability. This is because at the time the 5 calls (batch) arrive, there is available link bandwidth to accept even one of them. At the third time-point, we assume that there is available bandwidth to accept the whole 1st service batch (2 calls) while at the fourth time-point we assume that the whole 2nd service batch (2 calls) is blocked and lost. Now, this type of blocking has to do with both TC probability (1 blocking event, see Figure 17.2 "TC probability calculation") and CC probability (2 blocking events, see Figure 17.2 "CC probability calculation"). At the fourth to sixth time-points, the comments are similar to the abovementioned and thus they are omitted. According to Figure 17.2, the calculation of both CC and TC probabilities gives higher CC probabilities, an expected result since calls arrive following a batch Poisson process.

17.3 The BP-EMLM under the BR Policy

In the BP-EMLM under the BR policy if a new service k call finds that the occupied link bandwidth is j b.u., then this call is accepted in the system if and only if $j + b_k \leq C - t_k$ where t_k is the BR parameter of service k. The BR parameter t_k denotes the available link b.u. reserved to benefit services with higher bandwidth per call requirements than service k.

The application of the BR policy in the BP-EMLM destroys the PFS of the model as the following simple example shows. Consider a link with capacity $C = 4$ b.u. which accommodates $K = 2$ services whose calls require $b_1 = 1$ and $b_2 = 2$ b.u. respectively. Initially, the BR parameters of both services are zero. Let n_1, n_2 be the number of in service calls of the first and the second service, respectively, and $\mathbf{n}$ be the corresponding vector, $\mathbf{n} = (n_1, n_2)$. The system state space Ω associated with the CS policy is defined by

$$\Omega = \left\{ \mathbf{n} : \sum_{k=1}^{2} n_k b_k \leq C \Rightarrow n_1 b_1 + n_2 b_2 \leq 4 \right\}$$

where $\sum_{k=1}^{2} n_k b_k = j$. Figure 17.3a illustrates the system state space Ω of our example (9 system states) together with all the possible transitions.

To guarantee that the BP-EMLM has a PFS (in the case of the CS policy) Kaufman and Rege showed in [19], that there exists local balance across certain levels ("upward" probability flow across a level = "downward" probability flow across the same level). Each level $L_n^{(k)}$ $(k = 1, \ldots, K)$ corresponds to a vector $\mathbf{n} = (n_1, n_2, \ldots, n_k, \ldots, n_K)$ and separates that vector from the vector $\mathbf{n}_k^{+1} = (n_1, n_2, \ldots, n_k+1, \ldots, n_K)$, e.g. according to Figure 17.2a, the level $L_{(1,1)}^{(1)}$ separates the state $\mathbf{n} = (1, 1)$ from $(2, 1)$ while the level $L_{(0,1)}^{(2)}$ separates the state $\mathbf{n} = (0, 1)$ from $(0, 2)$.

Assume now that the BR policy is applied in our example and that the BR parameters are $t_1 = 1$, $t_2 = 0$. Figure 17.3b illustrates the new system state space Ω (8 system states) together with all the possible transitions. By comparing Figure 17.3a with Figure 17.3b it is to see the impact of the BR policy on the state transition diagram of Figure 17.3b: Local balance across the level $L_{(1,1)}^{(1)}$ does not exist ("upward" probability flow across $L_{(1,1)}^{(1)} \neq$ "downward" probability flow across $L_{(1,1)}^{(1)}$); if the system is in the state $\mathbf{n} = (1, 1)$ and a 1st service call arrives then the call will be blocked and lost due to the BR parameter ($t_1 = 1$) while if the system is in the state $\mathbf{n} = (0, 1)$ and a batch of two calls of the 1st service arrives then both calls will be blocked and lost. The fact that local balance across even a single level $L_n^{(k)}$ may not exist destroys the PFS of the BP-EMLM under the BR policy. Because of this, the calculation of the link occupancy distribution, $G(j)$, and therefore of various performance measures, can be approximate only. In the following subsections we apply the BR policy in the BP-EMLM following the Roberts method and the S&G method already proposed for the EMLM/BR.

17.3.1 The Roberts Method

According to the Roberts method, the population of service k calls which require b_k b.u. while $t_k > 0$ is assumed to be negligible inside the RS of service k. The latter consists of the states $j = C - t_k + 1, \ldots, C$. To illustrate the Roberts assumption consider again the example of the link of $C = 4$ b.u. The BR parameters of both services are chosen $t_1 = 1$, $t_2 = 0$ so that $b_1 + t_1 = b_2 + t_2$. This choice allows for equivalence between the TC probabilities of both services. Figure 17.4 presents the one-dimensional Markov chain for our example. The two different types of arrows correspond to the two different services and show all the possible transitions between the states $j = 0, \ldots, 4$. For example, while in state $j = 0$ a 1st service batch may arrive consisting of one, two or three calls and therefore the system "will pass" to states $j = 1$, $j = 2$ or $j = 3$, respectively. Similarly, if the system is in state $j = 0$ and a 2nd service batch arrives consisting of one or two calls the system "will be transferred" to states $j = 2$ or $j = 4$, respectively. As one notices, the

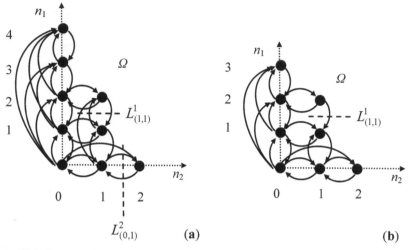

Figure 17.3 State transition diagram in the BP-EMLM, (a) under the CS policy, (b) under the BR policy.

1st service calls are not allowed to enter their RS, which consists of state $j = 4$ only. It is because of this prohibition that the Roberts assumption is reasonable.

To include in equation (17.1) the assumption that the population of service k calls with $t_k > 0$ is negligible inside the RS of service k, we denote the variable:

$$D_k (j - b_k) = \begin{cases} b_k & \text{when } j \le C - t_k \\ 0 & \text{when } j > C - t_k \end{cases} \qquad (17.10)$$

Therefore, equation (17.1) takes the form

$$G(j) = \begin{cases} 1 & \text{for } j = 0 \\ \dfrac{1}{j} \displaystyle\sum_{k=1}^{K} \alpha_k D_k(j - b_k) \sum_{l=1}^{\lfloor j/b_k \rfloor} \hat{B}_{l-1}^k G(j - lb_k) & \text{for } j = 1, \ldots, C \\ 0 & \text{otherwise} \end{cases} \qquad (17.11)$$

Note: If $B_r^k = 1$ for $r = 1$ and $B_r^k = 0$ for $r > 1$, then the EMLM/BR results.

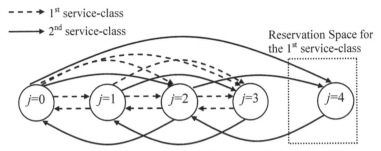

Figure 17.4 One-dimensional Markov chain (Roberts method).

The average number of service k calls given that the system state is j, $E(n_k \mid j)$, is given by

$$E(n_k \mid j) = \begin{cases} \alpha_k \sum_{l=1}^{\lfloor j/b_k \rfloor} \hat{B}_{l-1}^k G(j - lb_k) \Big/ G(j) & \text{when } j \leq C - t_k \\ 0 & \text{when } j > C - t_k \end{cases} \quad (17.12)$$

Based on equations (17.11), (17.12) one can calculate the various performance measures according to equations (17.4) to (17.6) and equation (17.8). To determine the TC probabilities we modify the bounds of equation (17.7) in order to include the BR parameter t_k:

$$P_{b_k} = \sum_{j=C-b_k-t_k+1}^{C} G^{-1} G(j) \quad (17.13)$$

where $G = \sum_{j=0}^{C} G(j)$ is a normalization constant.

17.3.2 The Stasiak–Glabowski Method

In a more realistic situation, calls of a service k with $t_k > 0$ may exist inside their RS. To show how this statement can be true consider again the previous simple example. Assume now that there are two calls of the 1st service in service. Therefore the system is in state $j = 2$. If a 2nd service batch arrives consisting of r calls then only one call can be accepted in the system while the rest $r - 1$ calls will be blocked and lost. In that case the system "will be transferred" in state $j = 4$ (since $b_2 = 2$ b.u.). In that state there can still be the two 1st service calls in service. This can be seen as a transfer of 1st service calls from state $j = 2$ to state $j = 4$ due to a 2nd service call arrival. Therefore it is possible to have 1st service transitions from state $j = 4$ to

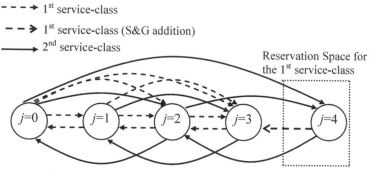

Figure 17.5 One-dimensional Markov chain (S&G method).

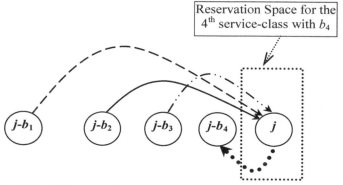

Figure 17.6 Calls of three services of bandwidth b_1, b_2 and b_3 respectively, with $b_1 > b_2 > b_3$, transfer calls of a fourth service to a state j inside its RS.

$j = 3$ as Figure 17.5 shows. Note that the abovementioned transfer of 1st service calls is also illustrated in Figure 17.3b (a 2nd service call arrival in the state $\mathbf{n} = (2, 0)$ transfers the two 1st service calls in state $\mathbf{n} = (2, 1)$).

In a more complicated example there might be many services whose call arrivals help in transferring calls of a certain service inside its RS (see Figure 17.6).

To determine the population of calls of a certain service inside its RS, we denote by $E^*(n_k \mid j)$ the average number of service k calls in state j which belongs to service k RS. In order to calculate the values of $E^*(n_k \mid j)$ for every state j of the RS of service k we follow the analysis of [25]. We denote by $w_{k,i}(j)$ a weight factor that determines the portion of $E^*(n_k \mid j)$ which is transferred to state j (which belongs to the RS of service k) by a service i call, where $i \neq k$. Since more than one service i can contribute in the transfer of calls of a service k to the RS of service k, we use different weights to show that this contribution can be different for each service i (where $i \neq k$). An easy way to decide how the values of $w_{k,i}(j)$ can be calculated is to take into consideration the product of α by b as follows [25]:

$$w_{k,i}(j) = \alpha_i b_i \Big/ \sum_{j=1, j \neq k}^{K} \alpha_j b_j \qquad (17.14)$$

Now the calculation of $E^*(n_k \mid j)$ can be based on

$$E^*(n_k \mid j) = \begin{cases} \alpha_k \sum_{l=1}^{\lfloor j/b_k \rfloor} \hat{B}_{l-1}^k G(j - lb_k) \Big/ G(j) & \text{when } j \leq C - t_k \\[2mm] \sum_{i=1, i \neq k}^{K} E^*(n_k \mid j - b_i) w_{k,i}(j) & \text{when } j > C - t_k \end{cases} \qquad (17.15)$$

where the upper part of equation (17.15) is the same with the upper part of equation (17.12), i.e. the values of $G(j)$'s in the upper part of equation (17.15) can be determined by equation (17.1).

According to equation (17.4), which also holds in the BP-EMLM under the BR policy, if one modifies the value of $E(n_k \mid j)$ then the corresponding value of j should also be modified. Therefore, based on equations (17.15) and (17.4) we continue to calculate the modified occupied link bandwidth, j^*, for every state j:

$$j^* = \sum_{k=1}^{K} b_k E^*(n_k \mid j) \qquad (17.16)$$

Having calculated both $E^*(n_k \mid j)$ and j^* we determine the modified values of $G(j)$ based on:

$$G(j) = \begin{cases} 1 & \text{for } j = 0 \\[2mm] \dfrac{1}{j^*} \sum_{k=1}^{K} \alpha_k b_k \sum_{l=1}^{\lfloor j/b_k \rfloor} \hat{B}_{l-1}^k G(j - lb_k) & \text{for } j = 1, \dots, C \\[2mm] 0 & \text{otherwise} \end{cases} \qquad (17.17)$$

The flow-chart of Figure 17.7 presents the whole procedure of the S&G method. Similar to the Roberts method, the TC probabilities in the S&G method can be calculated by equation (17.13). Based on equations

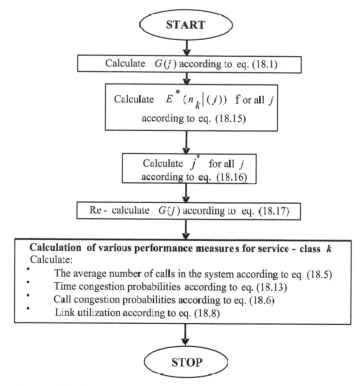

Figure 17.7 The S&G method in the BP-EMLM under the BR policy.

(17.15)–(17.17) one can calculate the average number of service k calls, $\bar{n}_k$, (equation (17.5)), the CC probabilities (equation (17.6)) and the link utilization (equation (17.8)).

17.4 Numerical Results

In this section we compare the analytical results obtained by the Roberts and the S&G methods for three performance measures, namely: the TC probabilities, the CC probabilities and the link utilization. As standards we take simulation results, which are mean values of 7 runs with 95% confidence interval. For comparison we include the analytical results obtained when the CS policy is applied in the BP-EMLM.

The first example is the same as the one presented in [21]. A link of capacity $C = 60$ b.u. accommodates two services whose calls require $b_1 = 1$

Table 17.1 Analytical results of TC, CC probabilities and link utilization for the CS policy.

α_2	TC Prob. (%) (CS – analytical)		TC Prob. (%) (CS – analytical)		Link Utilization (CS – analytical)
	1st service	2nd service	1st service	2nd service	
2.0	2.52	25.56	3.14	47.84	37.14
1.8	2.20	22.99	2.74	45.17	35.84
1.6	1.88	20.32	2.35	42.26	34.38
1.4	1.57	17.54	1.96	39.08	32.72
1.2	1.26	14.69	1.58	35.62	30.84
1.0	0.97	11.79	1.22	31.85	28.70

and $b_2 = 12$ b.u., respectively. As far as the BR parameters are concerned we examine three sets:

(i) $t_1 = 3$ and $t_2 = 0$;
(ii) $t_1 = 7$ and $t_2 = 0$; and
(iii) $t_1 = 11$ and $t_2 = 0$.

The third set equalizes the TC probabilities of both services since $b_1 + t_1 = b_2 + t_2$. The batch size of both services follows the geometric distribution with parameters: $\beta_1 = 0.2$ and $\beta_2 = 0.5$. The call holding time is exponentially distributed with mean $\mu_1^{-1} = \mu_2^{-1} = 1$ without loss of generality. The values of the offered traffic are: $\alpha_1 = 10$ and $\alpha_2 = 2$ (erl).

In the 1st column of Tables 17.1 to 17.10, α_1 remains constant while α_2 decreases by 0.2. Table 17.1 shows, as a reference, the analytical results of the TC probabilities, CC probabilities and link utilization in the case of the CS policy. In Tables 17.2 to 17.4 we present, for both services and each set of BR parameters, the analytical and simulation results for the TC probabilities when the Roberts and the S&G methods are applied. In Tables 17.5 to 17.7 we present, for both services and each set of BR parameters, the analytical and simulation results for the CC probabilities when the Roberts and the S&G methods are applied. Table 17.8 presents, for each set of BR parameters, the analytical results of the link utilization when the Roberts and the S&G methods are applied while Table 17.9 presents the corresponding simulation results.

According to the results of Tables 17.1 to 17.9 we observe that:

Table 17.2 Analytical and simulation results of the TC probabilities for the BR policy $(t_1 = 3)$.

α_2	TC Prob. (%) (Roberts method)		TC Prob. (%) (S&G method)		Simulation (%)	
	1st service	2nd service	1st service	2nd service	1st service	2nd service
2.0	8.27	24.44	8.28	24.45	7.46 ± 0.08	24.82 ± 0.19
1.8	7.23	21.98	7.24	21.99	6.50 ± 0.10	22.32 ± 0.32
1.6	6.20	19.42	6.20	19.43	5.56 ± 0.13	19.61 ± 0.19
1.4	5.18	16.76	5.18	16.77	4.75 ± 0.08	16.80 ± 0.26
1.2	4.18	14.04	4.18	14.04	3.83 ± 0.09	14.27 ± 0.23
1.0	3.22	11.27	3.23	11.27	2.95 ± 0.07	11.35 ± 0.23

Table 17.3 Analytical and simulation results of the TC probabilities for the BR policy $(t_1 = 7)$.

α_2	TC Prob. (%) (Roberts method)		TC Prob. (%) (S&G method)		Simulation (%)	
	1st service	2nd service	1st service	2nd service	1st service	2nd service
2.0	13.88	23.24	13.91	23.27	14.90 ± 0.13	24.05 ± 0.18
1.8	12.24	20.87	12.27	20.90	13.10 ± 0.23	21.55 ± 0.18
1.6	10.58	18.40	10.61	18.43	11.16 ± 0.22	19.16 ± 0.21
1.4	8.93	15.86	8.95	15.87	9.54 ± 0.14	16.45 ± 0.20
1.2	7.30	13.25	7.31	13.27	7.65 ± 0.10	13.94 ± 0.31
1.0	5.71	10.62	5.71	10.63	6.07 ± 0.08	10.98 ± 0.19

- The BR policy is not fair to 1st service calls. This is because the decrease of TC and CC probabilities of the 2nd service (whose calls require more b.u. than the 1st service calls) is less significant compared to the increase of TC and CC probabilities of the 1st service. The impact of this behavior is the decrease of the link utilization (compare the link utilization results of Table 17.1 (CS policy) with those of Table 17.9 (BR policy)).
- The S&G method gives slightly better results than the Roberts method, compared to simulation results. This is because only the 1st service has

Table 17.4 Analytical and simulation results of the equalized TC probabilities for the BR policy ($t_1 = 11$).

α_2	Eq. TC Prob. (%) (Roberts method)	Eq. TC Prob. (%) (S&G method)	Simulation (%)	
			1st service	2nd service
2.0	21.40	21.45	22.32 ± 0.25	22.30 ± 0.23
1.8	19.12	19.16	19.87 ± 0.22	19.89 ± 0.14
1.6	16.78	16.81	17.39 ± 0.16	17.42 ± 0.22
1.4	14.37	14.40	14.80 ± 0.18	14.79 ± 0.29
1.2	11.94	11.96	12.35 ± 0.18	12.38 ± 0.23
1.0	9.51	9.52	9.80 ± 0.12	9.78 ± 0.19

Table 17.5 Analytical and simulation results of the CC probabilities for the BR policy ($t_1 = 3$).

	CC Prob. (%) (Roberts method)		CC Prob. (%) (S&G method)		Simulation (%)	
α_2	1st service	2nd service	1st service	2nd service	1st service	2nd service
2.0	8.74	47.06	8.75	47.07	8.23 ± 0.07	47.06 ± 0.18
1.8	7.65	44.45	7.66	44.45	7.17 ± 0.10	44.41 ± 0.14
1.6	6.56	41.61	6.57	41.62	6.13 ± 0.14	41.77 ± 0.14
1.4	4.43	35.13	4.44	35.13	4.22 ± 0.12	35.16 ± 0.10
1.2	5.49	38.51	5.49	38.51	5.23 ± 0.08	38.39 ± 0.21
1.0	3.42	31.44	3.43	31.45	3.25 ± 0.07	31.61 ± 0.25

a positive BR parameter and therefore only the population of 1st service is assumed to be negligible (in the RS: $C - t_1 + 1, \ldots, C$). Such an assumption does not lead to a high approximation error. Because of the higher complexity of the S&G method together with our extensive study on examples with two services we suggest the usage of the Roberts method when two services are considered.

- Although the third set of BR parameters achieves the TC probabilities equalization, the same set cannot achieve the CC probabilities equal-

Table 17.6 Analytical and simulation results of the CC probabilities for the BR policy ($t_1 = 7$).

α_2	CC Prob. (%) (Roberts method)		CC Prob. (%) (S&G method)		Simulation (%)	
	1st service	2nd service	1st service	2nd service	1st service	2nd service
2.0	14.36	46.22	14.40	46.24	15.57 ± 0.14	46.50 ± 0.05
1.8	12.69	43.66	12.72	43.68	13.68 ± 0.23	43.80 ± 0.12
1.6	10.99	40.88	11.02	40.89	11.71 ± 0.22	41.12 ± 0.13
1.4	9.29	37.84	9.31	37.86	10.02 ± 0.13	37.93 ± 0.18
1.2	7.60	34.54	7.62	34.55	8.05 ± 0.09	34.77 ± 0.06
1.0	5.96	30.95	5.97	30.95	6.41 ± 0.08	31.10 ± 0.21

Table 17.7 Analytical and simulation results of the CC probabilities for the BR policy ($t_1 = 11$).

α_2	CC Prob. (%) (Roberts method)		CC Prob. (%) (S&G method)		Simulation (%)	
	1st service	2nd service	1st service	2nd service	1st service	2nd service
2.0	22.16	44.93	22.20	44.96	22.97 ± 0.27	45.13 ± 0.19
1.8	19.82	42.42	19.86	42.45	20.37 ± 0.22	42.54 ± 0.14
1.6	17.41	39.69	17.44	39.72	17.86 ± 0.16	39.89 ± 0.25
1.4	14.93	36.75	14.96	36.76	15.36 ± 0.18	36.71 ± 0.28
1.2	12.42	33.55	12.44	33.56	12.58 ± 0.26	33.54 ± 0.20
1.0	9.91	30.08	9.92	30.09	10.23 ± 0.11	30.05 ± 0.20

ization (see Table 17.7). According to equation (17.6), which holds when the BR policy is applied in the BP-EMLM, the calculation of CC probabilities depends on the offered traffic load rather than on the BR parameters. For a certain offered traffic load one can find the BR parameters which almost equalize CC probabilities. When the offered traffic load changes then a new set of BR parameters is needed. To show it, we present in Table 17.10 the BR parameters that almost equalize CC probabilities of both services in the case of the Roberts method.

Table 17.8 Analytical results of the link utilization for the BR policy.

α_2	$t_1 = 3$		$t_1 = 7$		$t_1 = 11$	
	Roberts method	S&G method	Roberts method	S&G method	Roberts method	S&G method
2.0	36.82	36.81	36.52	36.50	36.16	36.14
1.8	35.54	35.54	35.25	35.24	34.90	34.88
1.6	34.10	34.10	33.83	33.82	33.48	33.47
1.4	32.47	32.47	32.22	32.22	31.89	31.88
1.2	30.63	30.63	30.40	30.40	30.09	30.08
1.0	28.52	28.52	28.33	28.33	28.04	28.04

Table 17.9 Simulation results of the link utilization for the BR policy.

α_2	$t_1 = 3$	$t_1 = 7$	$t_1 = 11$
2.0	36.82 ± 0.06	36.17 ± 0.04	35.77 ± 0.30
1.8	35.42 ± 0.05	34.95 ± 0.13	34.62 ± 0.14
1.6	34.06 ± 0.09	33.58 ± 0.11	33.33 ± 0.08
1.4	32.38 ± 0.06	32.05 ± 0.08	31.74 ± 0.09
1.2	30.62 ± 0.09	30.20 ± 0.07	29.97 ± 0.05
1.0	28.58 ± 0.06	28.22 ± 0.09	28.00 ± 0.08

Table 17.10 Offered traffic load vs. CC probabilities equalization and the corresponding BR parameters.

	CC Prob. (Roberts method)		BR parameter t_1
α_2	1st service	2nd service	
2.0	42.22	42.33	22
1.8	38.68	39.87	22
1.6	37.05	36.86	23
1.4	32.87	34.05	23
1.2	30.38	30.66	24
1.0	27.07	27.37	25

The second example shows in a convincing way that the S&G method can give better results than the Roberts method when TC probabilities equalization is required.

We consider a link of capacity $C = 100$ b.u. which accommodates three services whose calls require $b_1 = 1$, $b_2 = 4$ and $b_3 = 16$ b.u., respectively. As far as the BR parameters are concerned we examine three different sets:

(i) $t_1 = 5$, $t_2 = 4$ and $t_3 = 0$;
(ii) $t_1 = 10$, $t_2 = 8$ and $t_3 = 0$; and
(iii) $t_1 = 15$, $t_2 = 12$ and $t_3 = 0$.

The third set equalizes the TC probabilities of all services since $b_1 + t_1 = b_2 + t_2 = b_3 + t_3$. The batch size of all services follows the geometric distribution

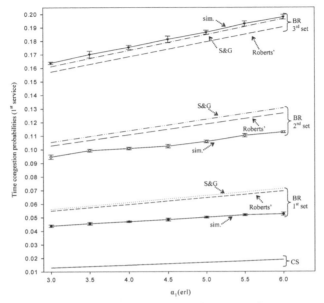

Figure 17.8 TC probabilities of the 1st service.

with parameters: $\beta_1 = 0.75$, $\beta_2 = 0.5$ and $\beta_3 = 0.2$. The call holding time is exponentially distributed with mean $\mu_1^{-1} = \mu_2^{-1} = \mu_3^{-1} = 1$. The values of the offered traffic are $\alpha_1 = 6$, $\alpha_2 = 4$ and $\alpha_3 = 2$ (erl).

In the x-axis of all figures of this example the values of α_2, α_3 remain constant while that of α_1 decreases by 0.5. Figure 17.8 presents the analytical and simulation results of the TC probabilities of the 1st services for both policies; in the case of the BR policy we consider both methods (Roberts and S&G) and all sets of BR parameters. Figures 17.9 and 17.10 present the corresponding results for the 2nd and the 3rd service, respectively. According to Figures 17.8 to 17.10 the S&G method gives always better results than the Roberts method when the 3rd set of BR parameters is used (TC probabilities equalization). In the case of the 1st and the 2nd sets the S&G method gives worse results in Figures 17.8 and 17.9 and better results in Figure 17.10. In order to conclude which method performs better when the 1st and the 2nd sets are used, we present in Figure 17.11 the Mean Relative Approximation

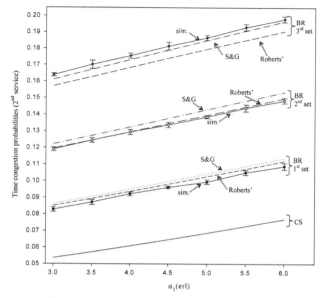

Figure 17.9 TC probabilities of the 2nd service.

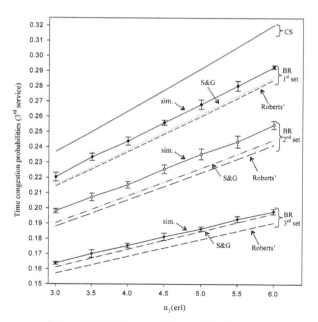

Figure 17.10 TC probabilities of the 3rd service.

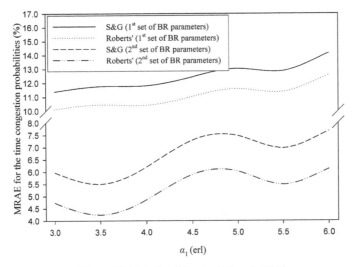

Figure 17.11 MRAE for the TC probabilities.

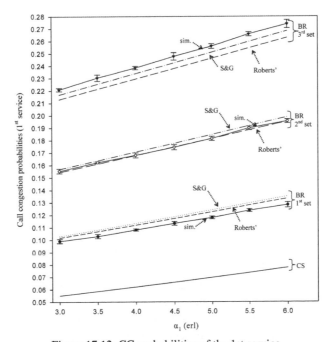

Figure 17.12 CC probabilities of the 1st service.

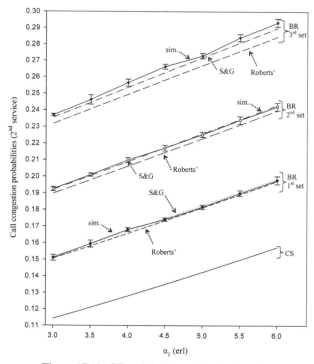

Figure 17.13 CC probabilities of the 2nd service.

Error (MRAE) of TC probabilities. The MRAE calculation is given by

$$\text{MRAE} = \frac{\sum_{k=1}^{K} \text{RAE}_k}{K} \cdot 100\% \qquad (17.18)$$

where RAE_k stands for the Relative Approximation Error for each service k given by

$$\text{RAE}_k = \left| P_{b_k,\text{an}} - P_{b_k,\text{sim}} \right| / P_{b_k,\text{sim}} \qquad (17.19)$$

where $P_{b_k,\text{an}}$, $P_{b_k,\text{sim}}$ are the TC probabilities obtained by equation (17.13) and simulation, respectively.

According to the results of Figure 17.11 the S&G method are slightly worse than the Roberts method and therefore the S&G method is worth to be applied when TC probabilities equalization is required.

Figure 17.12 presents the analytical and simulation results of the CC probabilities of the 1st service for both policies; in the case of the BR policy

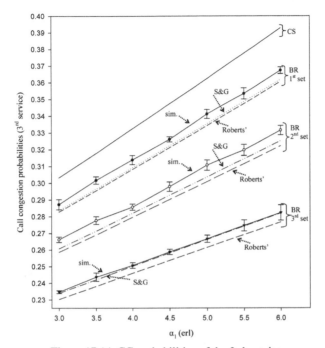

Figure 17.14 CC probabilities of the 3rd service.

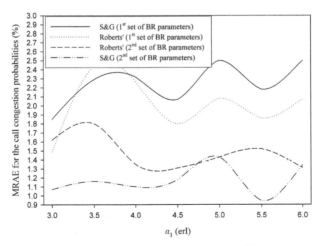

Figure 17.15 MRAE for the CC probabilities.

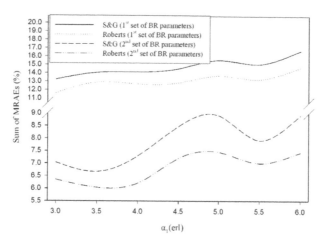

Figure 17.16 MRAE for the TC probabilities and MRAE for the CC probabilities.

we consider both methods (Roberts and S&G) and all sets of BR parameters. Figures 17.13 and 17.14 present the corresponding results for the 2nd and the 3rd service, respectively. According to Figures 17.12 to 17.14 the S&G method gives always better results when the 3rd set of BR parameters is used. In the case of the 1st and the 2nd sets it is uncertain which method performs better and therefore we present in Figure 17.15 the MRAE of CC probabilities. According to Figure 17.15 the S&G method is slightly worse than the Roberts method when the 1st set of BR parameters is used, and slightly better in the case of the 2nd set of BR parameters. In Figure 17.16 we present the sum of the MRAE of the TC probabilities (Figure 17.11) and the CC probabilities (Figure 17.15). The results (for the 1st and the 2nd set of BR parameters) show that the S&G method performs slightly worse than the Roberts one.

17.5 Conclusion

We propose the application of the BR policy in the BP-EMLM with the partial batch blocking discipline and a generally distributed batch size. The importance of BP-EMLM under the BR policy is not only that we can guarantee specific call-level QoS for each service, but also that we can apply the BP-EMLM to model overflow traffic in alternate route systems, where the BR policy is a must in order for the direct traffic in overflow routes to be protected from overflow traffic. To apply the BR policy we investigate two methods, the Roberts and the S&G methods, already proposed in the case of the EMLM/BR. Since the BP-EMLM under the BR policy does not have a PFS, we propose approximate but recursive formulas for the calculation of various performance measures such as the TC and CC probabilities and the link utilization. A comparison between the analytical with simulation results shows that: (a) both methods provide satisfactory results in all circumstances, (b) the Roberts method is preferable when only two service-classes exist, or when the offered traffic-load is low, and (c) the S&G method could be considered only when CBP equalization is required, whereas it is preferable when more than two services exist and the offered traffic-load is high.

References

[1] M. Logothetis and S. Shioda, Medium-term centralized virtual path bandwidth control based on traffic measurements, *IEEE Trans. on Commun.*, vol 43, no. 10, pp. 2630–2640, October 1995.

[2] A. Greenberg and R. Srikant, Computational techniques for accurate performance evaluation of multirate, multihop communication networks, *IEEE/ACM Trans. on Networking*, vol. 5, no. 2, pp. 266–277, April 1997.

[3] M. Logothetis and G. Kokkinakis, Path bandwidth management for large scale telecom networks, *IEICE Trans. Commun.*, vol. E83-B, no. 9, pp. 2087–2099, September 2000.

[4] H. Shengye, Y. Wu, F. Suili and S. Hui, Coordination-based optimisation of path bandwidth allocation for large-scale telecommunication networks, *Comput. Commun.*, vol. 27, no. 1, pp. 70–80, January 2004.

[5] P. Fazekas, S. Imre and M. Telek, Modelling and analysis of broadband cellular networks with multimedia connections, *Telecommun. Systems*, vol. 19, pp. 263–288, March–April 2002.

[6] D. Staehle and A. Mäder, An analytic approximation of the uplink capacity in a UMTS network with heterogeneous traffic, in *Proc. 18th International Teletraffic Congress (ITC)*, Berlin, 31 August–5 September, pp. 81–90, 2003.

[7] A. Mäder and D. Staehle, Analytic modeling of the WCDMA downlink capacity in multi-service environments, in *Proc. ITC Specialist Seminar*, 31 August–2 September, pp. 217–226, 2004.

[8] C. Chigan, R. Nagarajan, Z. Dziong and T. Robertazzi, On the capacitated loss network with heterogeneous traffic and contiguous resource allocation constraints, in *Proceedings of Applied Telecommunications Symposium (ASTC'01) part of Advanced Simulation Technologies Conference*, Seattle, Washington, vol. 33, no. 3, pp. 153–160, April 2001.

[9] J. S. Kaufman, Blocking in a shared resource environment, *IEEE Trans. Commun.*, vol. 29, no. 10, pp. 1474–1481, October 1981.

[10] J. W. Roberts, A service system with heterogeneous user requirements, in *Performance of Data Communications Systems and Their Applications*, G. Pujolle (Ed.), North Holland, Amsterdam, pp. 423–431, 1981.

[11] Z. Dziong and J. W. Roberts, Congestion probabilities in a circuit switched integrated services network, *Performance Evaluation*, vol. 7, no. 4, pp. 267–284, November 1987.

[12] J. S. Kaufman, Blocking with retrials in a completely shared resource environment, *Performance Evaluation*, vol. 15, no. 2, pp. 99–113, June 1992.

[13] I. Moscholios, M. Logothetis and G. Kokkinakis, Connection dependent threshold model: A generalization of the Erlang multiple rate loss model, *Performance Evaluation*, vol. 48, nos. 1–4, pp. 177–200, May 2002.

[14] G. Stamatelos and V. Koukoulidis, Reservation-based bandwidth allocation in a radio ATM network, *IEEE/ACM Trans. Networking*, vol. 5, no. 3, pp. 420–428, June 1997.

[15] I. Moscholios, M. Logothetis and P. Nikolaropoulos, Engset multi-rate state-dependent loss models, *Performance Evaluation*, vol. 59, nos. 2–3, pp. 247–277, February 2005.

[16] M. Mehmet Ali, Call-burst blocking and call admission control in a broadband network with bursty sources, *Performance Evaluation*, vol. 38, no. 1, pp. 1–19, September 1999.

[17] I. Moscholios, M. Logothetis and G. Kokkinakis, Call-burst blocking of ON-OFF traffic sources with retrials under the complete sharing policy, *Performance Evaluation*, vol. 59, no. 4, pp. 279–312, March 2005.

[18] E. A. van Doorn and F. J. M. Panken, Blocking probabilities in a loss system with arrivals in geometrically distributed batches and heterogeneous service requirements, *ACM/IEEE Trans. Networking*, vol. 1, pp. 664–667, December 1993.

[19] J. S. Kaufman and K. M. Rege, Blocking in a shared resource environment with batched Poisson arrival processes, *Performance Evaluation*, vol. 24, pp. 249–263, February 1996.

[20] J. A. Morrison, Blocking probabilities for multiple class batched Poisson arrivals to a shared resource, *Performance Evaluation*, vol. 25, pp. 131–150, April 1996.

[21] G. L. Choundhury, K. K. Leung and W. Whitt, Resource-sharing models with state-dependent arrivals of batches, in *Computations with Markov Chains*, W. J. Stewart (Ed.), Kluwer Academic Publishers, Boston, pp. 255–282, 1995.

[22] H. Akimaru and K. Kawashima, *Teletraffic – Theory and Applications*, 2nd edition, Springer Verlag, Berlin, 1999.

[23] L. E. N. Delbrouck, On the steady state distribution in a service facility with different peakedness factors and capacity requirements, *IEEE Trans. Commun.*, vol. 31, no. 11, pp. 1209–1211, November 1983.

[24] J. W. Roberts, Teletraffic models for the Telecom 1 Integrated Services Network, in *Proc. International Teletraffic Congress, ITC-10*, Montreal, Paper 1.1-2, 1983.

[25] M. Stasiak and M. Glabowski, A simple approximation of the link model with reservation by a one-dimensional Markov chain, *Performance Evaluation*, vol. 41, pp. 195–208, July 2000.

[26] R. Jain, *The Art of Computer Systems Performance Analysis – Techniques for Experimental Design, Measurement, Simulation and Modelling*, Wiley, 1992.

18

End-to-End Quality of Service Prediction Based on Functional Regression

L. Aspirot, P. Belzarena, G. Perera and B. Bazzano

ARTES*, Facultad de Ingeniería, Universidad de la República, 11300 Montevideo, Uruguay; e-mail: artes@fing.edu.uy

Abstract

In this work we propose a methodology to monitor quality of service (QoS) parameters of multimedia flows based on end-to-end active measurements. The system is trained with short multimedia flows and probe packets bursts. With functional nonparametric regression techniques we learn the relation between the QoS parameters of the video sequence and the distribution of interarrival times of the probe packets. We obtain a continuous non intrusive QoS monitoring methodology. As Internet traffic is nonstationary we extend functional nonparametric regression results in order to include this case. We show results from simulations and from real Internet data.

Keywords: End-to-end active measurements, nonstationary, QoS, functional regression.

18.1 Introduction

The increase of the Internet access bandwidth has allowed operators to offer new services to Internet users. New services as video on demand, television

---

* ARTES: Joint Research Group of the Electrical Engineering and Mathematics and Statistics Departments

D. D. Kouvatsos (ed.), Mobility Management and Quality-of-Service for Heterogeneous Networks, 417–440.

and online games were added to the voice and data traditional services. For this reason, the possibility of measuring the network operation parameters has become a very important task in order to verify, monitor, and control the quality of service (QoS) of these applications.

Multimedia services in packet switched networks have some QoS important parameters like delay, jitter and loss rate. Normally the path between two end users is not under the control of only one Internet Service Provider (ISP). In addition, end-to-end QoS parameters cannot be estimated from the data obtained in each isolated router. Therefore, different end-to-end active measurements methodologies have been developed during the last ten years.

However some of the developed methodologies are strongly intrusive. For instance, to estimate end-to-end parameters some techniques are based on generating congestion in a network link.

In this work we analyze the possibility of monitoring some QoS parameters of multimedia flows during long time periods (hours, days). One specific application can be to monitor the degree of a service level agreement (SLA) fulfillment. Another application can be a QoS monitoring system for an operator who offers multimedia services (like video on demand) through an ADSL or cellular access, for example in order to implement access control policies.

A possible measurement technique for these applications is to send a multimedia flow and measure its QoS parameters. However many of these flows may have bandwidth requirements that are not negligible compared with the capacity of the network access. Such a "simplistic" measurement technique could then overload a congested link, thus degrading the QoS perceived by the clients that were using this link. The QoS degradation can be tolerated for short time periods but the previous methodology cannot be used if the operator requires a permanent monitoring.

Other measurement techniques estimate the QoS parameters for any flow sending light probe packets without considering the particular characteristics of the target flow. These light packets do not overload the network. This procedure assumes for example that the packet delay of a specific application can be approximated by the delay measured over the probe packets but QoS parameters depend on the statistical behavior of each service. Therefore, in many cases, this kind of estimation yields inaccurate results. However, probe packets reflect the state of the network and measures over probe packets are correlated with QoS parameters.

In this work we propose a methodology that is an intermediate point between both approaches (to send a multimedia flow during long periods or to send light probe packets during short periods). We aim at developing a monitoring methodology allowing an accurate estimation of the QoS parameters of a multimedia flow without overloading the network for long periods.

Our proposal is to send light probe packets and a multimedia flow during short periods. The goal is to learn the relation between the QoS parameters obtained by the multimedia flow and the statistic of the interarrival times of the probe packets sequence. The rationale behind this is that probe packets characterizes the state of the network. Using a regression model then we learn the relation from these data. Once the model is estimated to predict the QoS of the new multimedia flow is enough to send only the probe packets.

Due to the nonstationary characteristics of the Internet traffic classical results on regression are not suitable for our problem and extensions to this case are proved here.

After discussing some related work in Section 18.2 we present in detail our approach to monitor the QoS parameters of a multimedia flow in Section 18.3. In Section 18.4 we discuss advantages and drawbacks of the proposed solution. The theoretical results needed by our approach are formalized in Section 18.5 whereas the proofs of these results are included in Appendix A. In Section 18.6 the experimental methodology is explained as well as some simulations and results from real data are shown. Finally, in Section 18.7 we discuss the main conclusions and future research directions.

18.2 Related Work

The measurement of different network parameters has been a widely developed area in the last years. In this section we restrict our attention to those works that are related with our proposed solution.

Some work focus on link capacity estimation in a network path or on the estimation of the bottleneck link capacity. In order to estimate the capacity of a link there are many proposed procedures, each one specifically adapted for different conditions of the network and different types of cross traffic (i.e. traffic from other users that share the links with our traffic). In general, all techniques work fine if there is no cross traffic. In a loaded network or in a path with cross traffic in many links, the estimation errors can be large [17]. Link capacity is an important parameter but is not sufficient to estimate the quality of service of a multimedia flow. In order to estimate the QoS, it is a necessary but not a sufficient condition that the link has enough capacity for the bandwidth required by the multimedia flow. The cross traffic introduces delays, jitter or losses to the multimedia traffic. For this reason, we are interested on the whole effect of the link characteristics and the cross traffic over QoS.

Other papers focus on the so called network tomography. They propose different methodologies to estimate network parameters of an interior link from parameters measured with probe packets sent from the endpoints of the network. Literature in this area is extensive and we refer for example to [1,5,7,12,18,19,26]. Although our problem cannot be considered as network tomography the previous references have contributed to our work with some ideas and end-to-end measuring techniques.

Another important topic is available bandwidth (ABW) estimation. The ABW of link i in the time interval $(t, t+\tau)$ is $A_i(t, t+\tau) = C_i(1-u_i(t, t+\tau))$ where C_i is the link capacity and $u_i(t, t + \tau)$ is the average link utilization in this time interval. The ABW of the path is defined as the minimum A_i in that path. There are two main techniques to estimate the ABW. The first one sends a growing volume of probe traffic and analyzes the point where it generates congestion in the path. There are different tools that use this methodology, for example Pathload [13, 14] and PathChirp [24]. The second technique is based on sending packet pairs or a packet train to measure the time dispersion at the end of the path. For example Spruce [25] sends packet pairs with interdeparture time D_{in} and measures at the end of the link the interarrival time D_{out}. From these values and knowing the link capacity, the ABW can be estimated. We refer to [25] for a comparison between Spruce and other tools used to estimate ABW.

ABW estimation is related with our problem because the ABW takes into account the link properties and the cross traffic characteristics. However, the ABW does not give enough information to evaluate the QoS parameters of a multimedia flow. The ABW estimates the average cross traffic in a time interval, while some QoS parameters depend on all the statistics of the cross traffic and not only on its average value. In addition, the ABW and its variability depend on the considered time scale. Further, Jain and Dovrolis [15] have measured the ABW over different Internet paths and they have shown recently that the ABW experiments large variations for each fixed time scale. They also showed that these variations depend not only on the time scale but also on the cross traffic type.

These previous mentioned techniques are not suitable for QoS monitoring for multimedia flows. Moreover, to date an to the best of our knowledge there is not an specific measurement and estimation methodology that address that problem.

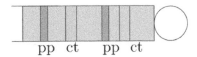

Figure 18.1 Probe packets (pp) and cross traffic (ct) in one queue busy period.

18.3 Problem Formulation and Proposed Solution

In this section we study the case of a path with a single link. The multilink case is discussed in the following section. We assume that the cross traffic of this link, the link capacity, and the buffer size are unknown.

We are interested in the estimation of some QoS parameters (delay, jitter, losses, etc.) of a multimedia flow during long periods. The target parameter that the user wants to estimate is called Y and is a function of the link and traffic characteristics:

$$Y = \Phi(X_t, V_t, C, B)$$

where X_t is the cross traffic stochastic process, V_t is the video or multimedia traffic stochastic process, C is the link capacity, and B is the buffer size. We want to estimate Y without sending video traffic during long time periods. The link capacity C and the buffer size B are not known but it is assumed that are constants during the monitoring process. We want to evaluate the QoS of the process V_t so this can be considered as an input to our problem. Therefore, we can say that $Y = \Phi(X_t)$. In the next section we discuss the previous assumptions about C, B, and V_t.

The previous assumptions, pose two different problems that should be addressed. On the one hand the estimation of the function Φ, and on the other the estimation of the cross traffic process X_t.

We first present the method we use for estimate the cross traffic that is actually a widely used technique which consists in sending probe packets at a known rate and to measure the interarrival times of these packets at the end of the link. We now describe this technique in more detail.

The probe packets are sent with an interdeparture time t_{in}. First we assume that two consecutive probe packets are queued in the same busy period at the link queue, as shown in Figure 18.1.

The interarrival time t_{out}^i between the packets i and $i+1$ is measured at the output of the link. This interarrival time is equal to $X_i/C + K/C$, where X_i is the amount of cross traffic that arrived at the queue between probe packets i and $i + 1$. K is the probe packets size and C is the link capacity. C and K are constant and K should be as small as possible. In addition, when the

pp   video

Figure 18.2 Probe bursts (pp) and probe video (video).

queue is in a busy period, K/C is normally negligible compared with the cross traffic bit rate. Then, the interarrival times are proportional to the cross traffic volume at least of up to a constant.

Suppose now that the packet $i + 1$ is queued after the packet i leaves the queue. In this case, as we infer the cross traffic volume from the values t_{out}^i, we will conclude that there is a cross traffic volume larger than the real one.

Therefore, if two packets arrive at the queue in the same busy period the measuring methodology estimates accurately the accumulated traffic during the time t_{in}^i. If the two packets arrive to the queue in different busy periods the estimation is inaccurate.

Baccelli et al. [3] present a rigorous probabilistic approach to active probing methods for cross traffic estimation. In particular they analyze the system identifiability and show that in the general case different cross traffic types give rise to the same sequence of observed probe delays and that is not always possible to determine the distribution of any desired aspect of the cross traffic using probes.

However we are not looking for an accurate estimator of the cross traffic. We are actually looking for an estimator of Y. Therefore, our interest is only in finding an estimator that allows us to distinguish between the possible types of cross traffic that use the link. We will call X the variable that allows us to distinguish between the different cross traffic processes that are observable in the link. With this variable that identify the cross traffic process we can do a regression and with the observation of the pairs (X, Y) we can estimate the function that relates them.

In what follows we describe the procedure to estimate the function Φ. We assume that

$$Y = \Phi(X) + \varepsilon$$

where X is a variable that characterizes cross traffic and ε is a centered random variable corresponding to the error. We divide the experiment in two phases. In the first phase, called the learning phase, we send a burst of small probe packets of fixed size K separated a fixed time t_{in}. Immediately after the burst, we send a test video stream during a short time period. The procedure is repeated periodically sending a probe burst and a video stream alternatively as shown in Figure 18.2.

With the probe packets we build the variable X that characterizes the cross traffic process observed during each test. We measure at the output of the link the interarrival times t_{out} between consecutive probe packets, which is strongly correlated with the cross traffic process. Specifically we will use the empirical distribution of the interarrival times between consecutive probe packets as X.

Using this variable and measuring the performance metric of interest Y over the test traffic we estimate the function Φ by a function $\widehat{\Phi}$.

In the second phase, called monitoring phase, we send only the probe packets in order to characterize the cross traffic X. Using the function $\widehat{\Phi}$ estimated in the first phase we estimate the performance of the QoS parameter $\widehat{Y}$ by $\widehat{Y} = \widehat{\Phi}(X)$, but avoiding the use of the video stream.

From each probe packet burst and video sequence j we have a pair (X_j, Y_j), and the problem now is how to estimate the function $\Phi : \mathcal{D} \to \mathbb{R}$ from these observations, where $\mathcal{D}$ is the space of the probability distribution functions and $\mathbb{R}$ is the real line.

To this end we will follow a nonparametric functional regression approach.

Another aspect to be considered is that the cross traffic process X_t on the Internet is a dependent and non-stationary process. This topic has been studied by many authors during the last ten years. Zhang et al. [28, 29] show that many processes on the Internet (losses for example) can be well modeled as independent and identical distributed (i.i.d.) random variable within a "change free region", where stationarity can be assumed. They describe the overall network behavior as a series of piecewise-stationary intervals. Karagiannis et al. [16] have found nonstationarity at different time scales analyzing the traffic of a link belonging to a Tier 1 ISP. They found that the traffic can be considered stationary at small time scales with events that change its stationarity at multi-second scale o larger.

The nonstationarity has different causes at different time scales and the "stationary" time scale can be different for different paths and performance metrics. In all cases it is very important to have estimators that can be used with nonstationary traffic. This is one of the main contributions of our work. We develop a measuring methodology that estimates the QoS parameters in a network with nonstationary cross traffic.

In Section 18.5 we extend some results in order to apply them to our problem to avoid the assumption that (X_j, Y_j) are equally distributed.

18.4 Generalizations

In the previous section we have considered the case of one link. In this section we analyze the case of a path with several links. First we must highlight that in many important cases the analysis of a path can be reduced to the analysis of a single link. For example, this is the case when the multimedia service is offered by a server located at the ISP backbone (a video on-demand server for example) and the user has access to the network through a cellular or an ADSL link. In these cases the bottleneck is located at the access link since the backbone links are overprovisioned.

In the case where the packets must wait in more than one queue, these different queues modify the variable X we use to characterize the cross traffic. This means that we estimate an empirical distribution where the influence of all queues are accumulated. Nevertheless even in this case our method will work fine if it is possible to distinguish with the empirical distribution of interarrival times between the cross traffic processes that are observed in the path.

Another assumption of the previous section was that network paths, link capacities and buffer sizes are fixed. For link capacities and buffer sizes this assumption is reasonable. However, the route between two points on the network can change over time. In any case, this problem can be solved by the application that measures the QoS parameters. It is always possible to verify periodically the route between the two points using for example an application like traceroute. If a new route is detected two circumstances can arise. If the system had learned information about the new route, this information can be used for the estimation. If the system had not learned information about the new route it is necessary to trigger a new learning phase. Another possibility is to define a threshold for the distances between the probability distributions X_j. When the application detects that the distance between the present sample and all the learning samples is greater than this threshold a new training phase can be initiated. Finally it is important to note that in the case where the bottleneck is in the access link and the backbone is oversized it is possible that a change in the route does not affect the measures.

Another possible issue is the assumption that the system is trained with a unique kind of video (we assume that V_t is a fixed sequence). This is not so important since the QoS depends on a few characteristics of the video like coding, video bit-rate, frame-rate and motion level. Therefore, it is possible to train the system with different kinds of videos that will be used on that path (using a set of different test video sequences). Later, depending on the specific video that the user or the operator wants to evaluate, the system can use the corresponding training samples.

Finally, it is necessary to note that in our training phase we periodically send a sequence of probe packets and separately a test video sequence. In fact, depending on the specific application, it is possible also to jointly send the video sequence and the probe packets. In this last case we will be able to predict which is the QoS of a video that is being sent now through this path. In the case that they are sent separately we will be able to predict the QoS that a video would receive if it is sent by this path. We used for the previous analysis the case where both sequences are sent separately because for one of the applications that are important for this work (access control) this approach is more suitable. Nevertheless both approaches can be studied with the proposed methodology.

18.5 Theoretical Results

In this section we will explain the mathematical techniques selected to solve the problem stated in the previous sections. We present a brief review of current results and our extensions that are also a main contribution of this work. The proofs of the results are included in Appendix A.

We consider the regression model

$$Y = \Phi(X) + \varepsilon \tag{18.1}$$

where X, Y and ε are random variables. In this work the random variable X is a measure of the state of the network, the response Y is the target QoS parameter (delay, jitter, loss rate) for the multimedia traffic and ε is a centered random variable independent from X that represents an error. It is not assumed an explicit form for the function Φ that relates the state of the network with the QoS parameters, and it is also not assumed any particular probability distribution for the random variables involved in the model. For this reason the model is nonparametric. The random variables Y and ε are real random variables, but the state of the network is represented by the empirical distribution function of interarrival times for a sequence of probe packets, so X is a functional random variable. Then our context will be that of functional nonparametric regression.

There are several results on nonparametric regression for real random variables and for random variables in $\mathbb{R}^d$ since the work of Nadaraya and

Watson [21, 27]. The Nadaraya–Watson estimator of Φ for the real case is

$$\widehat{\Phi}_n(x) = \frac{\sum_{i=1}^{n} Y_i K\left(\frac{\|x - X_i\|}{h_n}\right)}{\sum_{i=1}^{n} K\left(\frac{\|x - X_i\|}{h_n}\right)} = \frac{\sum_{i=1}^{n} Y_i K_n(X_i)}{\sum_{i=1}^{n} K_n(X_i)} \tag{18.2}$$

K is a Kernel, that is a positive function that integrates one and

$$K_n(X_i) = K\left(\frac{\|x - X_i\|}{h_n}\right)$$

h_n is a sequence that tends to zero and is called bandwidth. This estimator is a weighted average of the samples $Y_1, \ldots, Y_n$. The weights given by $K_n(X_i)$ take into account the distances from each point of the sample $X_1, \ldots, X_n$ to the point x.

For functional random variables, i.e. when the regressor X is a random function, Ferraty et al. [8] introduce a Nadaraya–Watson type estimator for Φ, defined by equation (18.2), where the difference with the real case is that $\|\cdot\|$ is a seminorm on a functional space $\mathcal{D}$. In [8] the authors proved the complete convergence of the estimator, the rate of convergence and the uniform complete convergence, when the observations (X, Y) are a sequence of stationary weakly dependent (α-mixing) random variables. One of the main aspects in the functional approach is the "curse of dimensionality". The estimation $\widehat{\Phi}_n(x)$ will be accurate if there are enough samples near x. This problem becomes crucial when the observations come from an infinite dimensional vector space. In order to treat this problem Ferraty, Goia and Vieu introduce in [8] the fractal dimension of the random variable X, that is a measure of samples concentration around x. Convergence results depend on this dimension. In other papers the same problem is addressed with different tools. We refer for example to [6, 9, 11, 20] for different approaches to this problem. These authors prove convergence and in some cases asymptotic distribution of the estimator for stationary and weakly dependent functional random variables.

In this work we follow the approach of Masry [20] that considers a function that determines the concentration of the random variable X in a neighborhood of x.

As our data comes from Internet data traffic and this traffic is typically nonstationary, we extended the previous results about functional nonparametric regression to the nonstationary case. Instead of considering random variables X equally distributed we consider a model introduced by Perera in [23] defined by

$$X_i = \varphi(\xi_i, Z_i) \tag{18.3}$$

where ξ_i takes values in a seminormed vector space with a seminorm $\|\cdot\|$, and Z_i is a real random variable that takes values in a finite set $\{z_1, z_2, \ldots, z_m\}$. The sequence $\varphi(\xi_i, z_k)$ is weakly dependent and equally distributed for every $1 \leqslant k \leqslant m$, but the sequence Z_i may be nonstationary as in [23]. The model represents a mixture of weakly dependent stationary process, but the mixture is nonstationary and dependent. Here ξ represents the usual variations of the traffic, and the variable Z selects between different traffic regimes, and represents different types of network traffic.

In what follows we state the main theoretical result, that is the almost sure convergence of the estimator. The asymptotic distribution of the estimator in the previous nonstationary model is discussed in [2].

Consider

$$\widehat{\Phi}_n(x) = \frac{g_n(x)}{f_n(x)},$$

where

$$g_n(x) = \frac{1}{n\psi(h_n)} \sum_{i=1}^{n} Y_i K_n(X_i),$$

$$f_n(x) = \frac{1}{n\psi(h_n)} \sum_{i=1}^{n} K_n(X_i),$$

and $\psi(h_n)$ is a normalization that depends on the concentration of observations around x and is defined later.

Lemma 1. *Let* $X = (X_n)_{n \geq 1}$, $Y = \Phi(X) + \varepsilon$, *with* ε *a centered real random variable independent from* X *that satisfy*

H1. *There exist two independent processes* $\xi = (\xi_n)_{n \in \mathbb{N}}$, $Z = (Z_n)_{n \in \mathbb{N}}$, *such that* ξ *is stationary, with values in a function space,* Z *is real with values in* $\{z_1, \ldots, z_m\}$ *and exists a function* φ *that takes values in a function space* $\mathcal{D}$ *such that*

$$X_n = \varphi(\xi_n, Z_n).$$

H2. *There exist positive functions* $\psi, \psi_1, \ldots, \psi_m$ *defined en* $\mathbb{R}^+ \times \mathcal{D}$, $c_1, \ldots, c_k$ *defined in* $\mathcal{D}$ *and a subset* $\Delta \subset \{1, \ldots, m\}$ *such that for all* $h > 0$

$$P\left[\|\varphi(\xi_1, z_k) - x\| \leq h\right] = c_k(x)\psi_k(h, x),$$

with

$$\lim_{h \to 0} \frac{\psi_k(h, x)}{\psi(h, x)} = 1 \quad \text{if } k \in \Delta, \quad \text{and}$$

$$\lim_{h \to 0} \frac{\psi_k(h, x)}{\psi(h, x)} = 0 \quad \text{if } k \in \Delta^c,$$

where Δ^c is the complement of subset Δ. In what follows we write $\psi(h_n)$ instead of $\psi(h_n, x)$, in order to simplify notation.

H3. *The functions $u \mapsto \psi_k(u, x)$ are differentiable in $\mathbb{R}^+$, with derivative $\psi'_k(u, x)$ and*

$$\lim_{h \to 0} \frac{h}{\psi_k(h, x)} \int_0^1 K(u)\psi'_k(uh, x)du = d_k(x),$$

where d_k are functions defined in $\mathcal{D}$.

H4. *For all $k \in \{1, \ldots, m\}$ the following limit exists*

$$\lim_{n \to \infty} \frac{1}{n} \sum_{i=1}^{n} P(Z_i = z_k)$$

and we denote it by p_k.

H5. Φ *is a continuous function.*

H6. *K is positive with support in $[0, 1]$.*

H7. *The bandwidth h_n satisfies that $\lim_{n \to \infty} h_n = 0$ and $\lim_{n \to \infty} n\psi(h_n) = \infty$. Then:*

- $\lim_{n \to \infty} E(f_n(x)) = f(x)$, *where $f(x) > 0$ and f is defined for all $u \in \mathcal{D}$ by*

$$f(u) = \sum_{k \in \Delta} p_k d_k(u) c_k(u),$$

- $\lim_{n \to \infty} E(g_n(x)) = \Phi(x)f(x).$

Remark. Hypothesis H2, as in [20] is about concentration of random variables $\varphi(\xi_n, z_k)$, represented by $\psi_k(h)$, in a ball centered at x with radius h. In the real case, for variables in $\mathbb{R}^d$ with a density function, the distribution in a ball centered in 0 and with radius h is proportional to h^d. The components of the mixture that finally determine the normalization ψ are those with indexes in Δ that correspond to the most concentrated variables around x. Hypothesis H3 implies that $\|\varphi(\xi, z_k) - x\|$ has density $c_k(x)\psi'_k$. For variables in $\mathbb{R}^d$ with a density function, d_k is a constant for all $x \in \mathbb{R}^d$. Hypothesis H4 guarantees some kind of stationarity "in mean". This hypothesis is verified for example by periodic random variables. A counterexample can be constructed as in example 2.2 in [23]. Hypotheses H5, H6, H7 are common for kernel estimation.

In order to show almost sure convergence we consider the estimator conditioned to the values of Z, and we work with a random vector in $\mathbb{R}^{2m}$ as follows.

We consider the variable $\tilde{X}^n = (\tilde{X}^{1,n}, \ldots, \tilde{X}^{2m,n})$ with values in $\mathbb{R}^{2m}$ defined for $i \geq 1$ by

- for $l \in \{1, \ldots, m\}$

$$\tilde{X}^{l,n}_i = \frac{1}{\sqrt{\psi(h_n)}} K_n\left(\varphi(\xi_i, z_l)\right)\left(\Phi(\varphi(\xi_i, z_l)) + \varepsilon_i\right)$$
$$- \frac{1}{\sqrt{\psi(h_n)}} E\left[K_n\left(\varphi(\xi_i, z_l)\right)\Phi(\varphi(\xi_i, z_l))\right],$$

- for $l \in \{m+1, \ldots, 2m\}$

$$\tilde{X}^{l,n}_i = \frac{1}{\sqrt{\psi(h_n)}} K_n\left(\varphi(\xi_i, z_l)\right) - \frac{1}{\sqrt{\psi(h_n)}} E\left[K_n\left(\varphi(\xi_i, z_l)\right)\right]$$

Theorem 2. *We assume the hypotheses of Lemma 1 and the following additional hypotheses:*

H8. *For all $B \subset \mathbb{N}$, $n \in \mathbb{N}$, $i \in \{1, \ldots, 2m\}$*

$$E\left[S_n\left(B, \tilde{X}^{i,n}\right)^4\right] \leq C\left(\frac{\text{card}(B_n)}{n}\right)^2.$$

H9. *The bandwidth h_n satisfies*

$$\lim_{n \to \infty} h_n = 0$$

and for some $0 < \beta < 1$

$$\lim_{n \to \infty} \psi(h_n) n^{\beta/2} = \infty.$$

Then $\widehat{\Phi}_n(x)$ converges almost surely to $\Phi(x)$ with $n \to \infty$.

Remark. Hypothesis H8 is about weakly dependence, and it may be obtained for example from α-mixing assumptions. Hypothesis H9 gives necessary bandwidth convergence speed to prove the theorem.

18.6 Simulations and Experimental Results

We simulate cross traffic for a model $X = \varphi(\xi, Z)$ in the following way. We have two Markovian ON-OFF sources and Z is a random variable that takes values in $\{0, 1\}$ selecting periodically between this two sources. Fixing the value of Z we obtain stationary processes $\varphi(\xi, 0)$ and $\varphi(\xi, 1)$.

The first source (source 0) generates Markovian ON-OFF traffic corresponding to $\varphi(\xi, 0)$ with average bit rate varying from 150 to 450 Mb/s and average time T_{on} in the ON state and T_{off} in the OFF state varying from 100 to 300 ms. The second source (source 1) generates Markovian ON-OFF traffic corresponding to $\varphi(\xi, 1)$ with average bit rate varying from 600 to 900 Mb/s and average time T_{on} in the ON state and T_{off} in the OFF state varying from 200 to 500 ms. For each period an independent random variable is sampled to select the average bit rate.

We send this cross traffic to a network link together with the probe traffic that is composed by a number of tests. Each test consists on a probe packet burst with fixed interdeparture time t_{in}^*, and after this burst we send a simulated multimedia traffic (a video traffic trace). For each test j we estimate from the probe packets the empirical distribution function for interarrival times and we measure the average delay Y_j of the multimedia traffic.

The kernel used to estimate Φ is

$$K(x) = \begin{cases} (x^2 - 1)^2 & \text{if } x \in [-1, 1] \\ 0 & \text{if } x \notin [-1, 1] \end{cases}$$

and we use the L^1 norm for the distance between the empirical distribution functions.

The selected kernel (Epachenikov) generally obtains good performance in the real case and has good properties for many distribution functions. The choice of this kernel is due to its generality.

Concerning the time scales in our experiment the probe traffic is sent with fixed time t between consecutive probe packets. The aim is to find some criterion for choosing the best time scale in order to have accurate estimates. We consider different sequences of observations for a finite set of time scales $\{t_1, t_2, \ldots, t_r\}$. In practice, as we send bursts of probe traffic with fixed time t between packets we have observations with time scales in the set $\{t, 2t, \ldots, rt\}$. Consider $n + m$ observations for each time scale

$$\left\{(X_i^{t_j}, Y_i^{t_j}) : 1 \leqslant i \leqslant n + m, \ 1 \leqslant j \leqslant r\right\}.$$

By dividing the sequence for a fixed time scale in two we can estimate the function Φ^{t_j} (for the time scale t_j) by $\widehat{\Phi}_n^{t_j}$ with the first n samples. We then compute the difference

$$\sigma^2_{t_j}(n, m) = \frac{1}{m} \sum_{i=1}^{m} \left(\widehat{\Phi}_n^{t_j}(X_{n+i}^{t_j}) - Y_{n+i}^{t_j}\right)^2$$

that gives a measure of how good is the estimator at time scale t_j. By computing $\sigma^2_{t_j}(n, m)$ for $1 \leqslant j \leqslant r$ we will choose $t_{n,m}^*$ such that

$$\sigma^2_{t_{n,m}^*}(n, m) = \min\{\sigma^2_{t_j}(n, m) : 1 \leqslant j \leqslant r\}$$

The bandwidth is selected with asimilar procedure. For a training sample $\{(X_i, Y_i) : 1 \leqslant i \leqslant n\}$ we consider the estimation of measured value Y_i defined by $\widehat{Y}_i = \widehat{\Phi}_{n,i}(X_i)$ where $\widehat{\Phi}_{n,i}$ is obtained from the training sample without considering (X_i, Y_i). Next we compute

$$\sigma^2(h_n) = \frac{1}{n} \sum_{i=1}^{n} \left(\widehat{\Phi}_{n,i}(X_i) - Y_i\right)^2$$

for a finite set of bandwidths H and select the bandwidth h_n^* such that

$$\sigma^2(h_n^*) = \min\{\sigma^2(h_n) : h_n \in H\}.$$

In the simulations we have 360 values of (X, Y) and we divide the sample in two. The estimation of Φ is obtained from the last 300 samples and the accuracy of the estimation is evaluated over the first 60 samples by comparing $\Phi_n(X_j)$ with the measured average delay Y_j for $j = 1, \ldots, 60$. The relative error in each point j is computed by $|\widehat{\Phi}_n(X_j) - Y_j|/Y_j$. In Figures 18.3 and 18.4 the estimated and the measured value for the average delay and the relative error are plotted, showing a good fitting. The estimates are less accurate for very small values of Y.

Real data were collected with a software tool developed by the Electrical Engineering Department [4]. A user access to the server through a web page

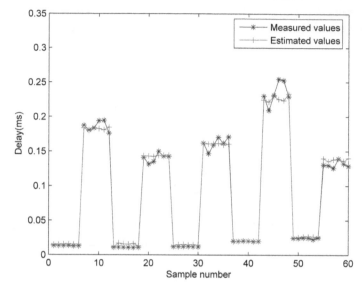

Figure 18.3 Average delay for simulated data.

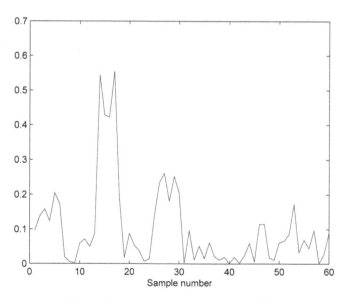

Figure 18.4 Relative error for simulated data.

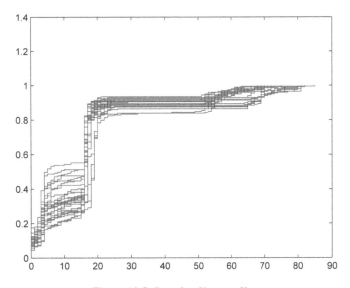

Figure 18.5 Samples $X_1, \ldots, X_{30}$.

and the server performs the experiments described before sending probe packets and short video sequences, then the software sends packets from the user to the server in order to compute the RTT. The data is collected and for future experiments the server only sends probe packets. These data were collected in experiments with the server located at arbolita.fing.edu.uy and with a user connected through an ADSL access. The probe packet sizes are around 50 bytes and were sent at a constant rate of one packet each 15 millisecond. The bit rate of the probe packets was around 30 kbps. The video sequence used has an average bit rate of around 0.5 Mbps.

In Figure 18.5, 30 samples of real values for X are plotted. Each point is estimated taking as training sample all the other 59 samples, using the Epachenikov kernel and computing the bandwidth in same fashion that for simulated data. In Figures 18.6 and 18.7 the estimated and the measured value for the average RTT and the relative error are plotted and we also have accurate results.

18.7 Conclusions

We proposed a methodology to monitor QoS parameters of multimedia flows during long periods. This methodology is based on nonparametric regres-

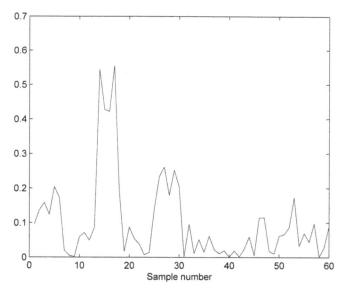

Figure 18.6 RTT for real data.

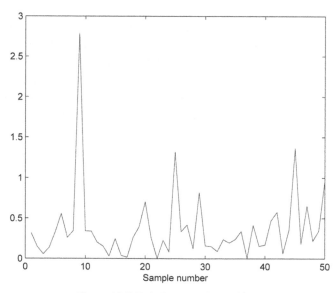

Figure 18.7 Relative error for real data.

sion techniques. We train the system with short multimedia flows and probe packets bursts. With this information we learn the relation between the QoS parameters of the video sequence and the distribution of interarrival times of the probe packets. To predict the QoS parameters for a new video sequence we only want to send the probe packets and this process can be done continuously, obtaining a continuous QoS monitoring.

The main advantage of this method is that the probe packets do not load the network. By all means, in the training phase the video load the network but this situation occurs only during short time periods. Later in the monitoring phase where the probe packets are sent continuously the network is not loaded. We tested our methodology with simulated nonstationary traffic and we obtained good accurate estimations. We also have tested the methodology with a software tool in the Internet and the method also gives in this case a good approximation to the QoS parameters.

In this work we also have extended some results on functional nonparametric regression. These extensions allow us to apply our measuring methodology to nonstationary traffic. This is a very important issue for practical applications because the Internet has nonstationary traffic.

There are many possible improvements. The selection of the Kernel, the time scale for interdeparture times between packets in each burst and the selection of the bandwidth are topics for further analysis.

Appendix

A Proofs

Proof of Lemma 1. We compute

$$E(f_n(x)) = \frac{1}{n\psi(h_n)} \sum_{i=1}^{n} E(K_n(X_i)).$$

As ξ and Z are independent, then

$$E(K_n(X_i)) = E\{E(K_n(X_i) \mid Z_i)\} = \sum_{k=1}^{m} E\{K_n(\varphi(\xi_i, z_k))\} P(Z_i = z_k).$$

As for all k $(\varphi(\xi_i, z_k))_{i \geq 1}$ is stationary sequence we have that

$$E(f_n(x)) = \sum_{k=1}^{m} \left(\frac{1}{\psi(h_n)} E\left(K_n(\varphi(\xi_1, z_k))\right) \frac{1}{n} \sum_{i=1}^{n} P(Z_i = z_k) \right).$$

Hypothesis H2 implies that the function $u \mapsto c_k(x)\psi_k'(u, x)$ is the probability density function for $\|\varphi(\xi_1, z_k) - x\|$ and then

$$E\left[K_n(\varphi(\xi_1, z_k))\right] = h_n c_k(x) \int_0^1 K(u)\psi_k'(uh_n, x)du.$$

Then from Hypotheses H2, H3, H7 we obtain

$$\lim_{n\to\infty} \frac{1}{\psi(h_n)} E\left[K_n(\varphi(\xi_1, z_k))\right] = d_k(x)c_k(x)1_{\{k\in\Delta\}},$$

and from Hypothesis H4

$$\lim_{n\to\infty} E(f_n(x)) = \lim_{n\to\infty} \sum_{k=1}^m \left(\frac{1}{\psi(h_n)} E\left(K_n(\varphi(\xi_1, z_k))\right) \frac{1}{n}\sum_{i=1}^n P(Z_i = z_k)\right)$$

$$= \sum_{k\in\Delta} p_k d_k(x)c_k(x).$$

Analogously,

$$E(g_n(x)) =$$

$$\sum_{k=1}^m \left(\frac{1}{\psi(h_n)} E\left(\Phi\left(\varphi(\xi_1, z_k)\right) K_n(\varphi(\xi_1, z_k))\right) \frac{1}{n}\sum_{i=1}^n P(Z_i = z_k)\right).$$

Then

$$E(g_n(x)) = \Phi(x)E(f_n(x)) + R_n,$$

where

$$R_n =$$

$$\sum_{k=1}^m \left(\frac{1}{\psi(h_n)} E\left[\{\Phi\left(\varphi(\xi_1, z_k)\right) - \Phi(x)\} K_n(\varphi(\xi_1, z_k))\right] \frac{1}{n}\sum_{i=1}^n P(Z_i = z_k)\right)$$

$$\leq \sup_{u:\|x-u\|\leq h_n} |\Phi(u) - \Phi(x)| \, E(f_n(x))$$

and the result $\lim_{n\to\infty} E(g_n(x)) = \Phi(x)f(x)$ is obtained from the continuity of Φ and from

$$\lim_{n\to\infty} E(f_n(x)) = f(x)$$

$\square$

Proof of Theorem 2. As $\widehat{\Phi}_n(x) = g_n(x)/f_n(x)$ it is enough to show that $f_n(x)$ converges almost surely to $f(x)$ and that $g_n(x)$ converges almost surely to $\Phi(x)f(x)$. He have that

$$f_n(x) - f(x) = f_n(x) - E(f_n(x)) + E(f_n(x)) - f(x)$$

$$g_n(x) - \Phi(x)f(x) = g_n(x) - E(g_n(x)) + E(g_n(x)) - \Phi(x)f(x)$$

and from Lemma 1

$$\lim_{n\to\infty} E(f_n(x)) - f(x) = 0,$$

$$\lim_{n\to\infty} E(g_n(x)) - f(x)\Phi(x) = 0,$$

then we must show that $f_n(x) - E(f_n(x))$ y $g_n(x) - E(g_n(x))$ converge almost surely to zero. To prove that $f_n - E(f_n(x))$ converges almost surely to zero is enough to prove complete convergence, i.e. that for all $\varepsilon > 0 \sum P(|f_n(x) - E(f_n(x))| > \varepsilon)$ is a convergent series.

$$f_n(x) - E(f_n(x)) = \frac{1}{n\psi(h_n)} \sum_{i=1}^{n} [K_n(X_i) - E(K_n(X_i))]$$

$$= \frac{1}{\sqrt{n\psi(h_n)}} S_n$$

where

$$S_n = \frac{1}{\sqrt{n\psi(h_n)}} \sum_{i=1}^{n} [K_n(X_i) - E(K_n(X_i))].$$

Applying the Markov inequality

$$P(|f_n(x) - E(f_n(x))| > \varepsilon) \leq \frac{E[f_n(x) - E(f_n(x))]^4}{\varepsilon^4} = \frac{E(S_n)^4}{(n\psi(h_n))^2\varepsilon^4}.$$

To compute $E(S_n^4)$ we condition to the trajectories of Z, i.e. the random variable Z^∞ that take values z^∞ in a space T of sequences.

$$E(S_n^4) = E[E(S_n^4 \mid Z^\infty)]$$

$$= \int_T E(S_n^4 \mid Z^\infty = z^\infty)d\mu$$

where T is the space of sequences that take values in $\{z_1, z_2, \ldots, z_m\}$. It is enough to show that there is a subset of T with probability 1 where

$E(S_n \mid Z^\infty = z^\infty) \leq a_n$ with $\sum a_n/(n\psi(h_n))^2$ a convergent series. For each trajectory

$$E(S_n^4 \mid Z^\infty = z^\infty) = E\left[\frac{1}{\sqrt{n}}\sum_{i=1}^{n}\frac{\{K_n(\varphi(\xi_i, z_{l_i})) - E[K_n(\varphi(\xi_i, z_{l_i}))]\}}{\sqrt{\psi(h_n)}}\right]^4$$

$$= E\left[\frac{1}{\sqrt{n}}\sum_{i=1}^{n}\tilde{X}_i^{l_i,n}\right]^4$$

with $l_i \in \{m+1, \ldots, 2m\}$. Then

$$\sum_{i=1}^{n}\tilde{X}_i^{l_i,n} = \sum_{k=m+1}^{2m}\sum_{i\in A_n^k}\tilde{X}_i^{k,n}$$

where $A_n^k = \{i : Z_i = z_k\} \cap [1, n]$ and

$$\frac{1}{\sqrt{n}}\sum_{i=1}^{n}\tilde{X}_i^{l_i,n} = \sum_{k=m+1}^{2m}\frac{1}{\sqrt{n}}\sum_{i\in A_n^k}\tilde{X}_i^{k,n}$$

$$= \sum_{k=m+1}^{2m}S_n\left(A^k, \tilde{X}^{k,n}\right).$$

Applying Cauchy–Schwartz inequality and hypothesis H8,

$$E\left[\sum_{k=m+1}^{2m}S_n\left(A^k, \tilde{X}^{k,n}\right)\right]^4 \leq m \max_{k\in\{m+1,\ldots,2m\}}E\left[S_n\left(A^k, \tilde{X}^{k,n}\right)^4\right]$$

$$\leq C \max_{k\in\{m+1,\ldots,2m\}}\left(\frac{\operatorname{card}\left(A_n^k\right)}{n}\right)^2$$

Then, as $\operatorname{card}(A_n^k) \leq n$, we obtain $E(S_n^4 \mid Z^\infty = z^\infty) \leq C$. The proof of $E(S_n^4 \mid Z^\infty = z^\infty) \leq C$ is independent of the trajectory. From hypothesis H9 $\sum 1/(n\psi(h_n))^2$ is a convergent series and we obtain complete convergence of $f_n(x) - E[f_n(x)]$ to zero. The proof is analogous for $g_n(x) - E[g_n(x)]$, considering the random variables $\tilde{X}^{k,n}$, $k \in \{1, \ldots, m\}$. $\qquad\square$

Acknowledgments

We would like to thank Paola Bermolen and Federico Larroca for fruitful discussion and the referees for their careful review and helpful comments. This research was partially supported by PDT (Programa de Desarrollo Tecnológico, Préstamo 1293/OC-UR): S/C/OP/ 17/02 and program FCE (Fondo Clemente Estable) 8079.

References

[1] M. Adler, T. Bu, R. Sitaraman and D. Towsley, Tree layout for internal network characterizations in multicast networks, in *Proceedings of 3rd International Workshop on Networked Group Communication.*, London, UK, pp. 189–204, 2001.

[2] L. Aspirot, K. Bertin and G. Perera, Asymptotic normality of the Nadaraya–Watson estimator for non-stationary functional data and applications to telecommunications, *Journal of Nonparametric Statistics*, submitted, 2008.

[3] S. Machiraju, D. Veitch, F. Baccelli, A. Nucci and J. C. Bolot, Theory and practice of cross-traffic estimation, *ACM SIGMETRICS Performance Evaluation Review*, vol. 33, no. 1, pp. 400–402, 2005.

[4] P. Belzarena, V. González Barbone, F. Larroca and P. Casas, Metronet: Software para medición de calidad de servicio en voz y video, in *Proceedings of IV Congreso Iberoamericano de Telemática, CITA, México*, http://iie.fing.edu.uy/publicaciones/2006/BGLC06/, 2006.

[5] R. Cáceres, N. G. Duffield, J. Horowitz and D. Towsley, Multicast-based inference of network-internal loss characteristics, *IEEE Transactions on Information Theory*, vol. 45, pp. 2462–2480, 1999.

[6] H. Cuevas, M. Febrero and R. Fraiman, Linear functional regression: The case of fixed design and functional response, *Canadian Journal of Statistics*, vol. 30, no. 2, pp. 285–300, 2002.

[7] A. B. Downey, Using pathchar to estimate internet link characteristics, in *Proceedings of ACM SIGCOMM*, pp. 241–250, 1999.

[8] F. Ferraty, A. Goia and P. Vieu, Functional nonparametric model for time series: A fractal approach for dimension reduction, *Test*, vol. 11, no. 2, pp. 317–344, 2002.

[9] F. Ferraty and P. Vieu, Nonparametric models for functional data, with application in regression, time-series prediction and curve discrimination, *J. Nonparametr. Stat.*, vol. 16, nos. 1–2, pp. 111–125, 2004.

[10] F. Ferraty and P. Vieu. *Nonparametric Functional Data Analysis: Theory and Practice*, Springer Series in Statistics, Springer, 2006.

[11] F. Ferraty, A. Mas and P. Vieu. Nonparametric regression on functional data: Inference and pratical aspects. *Australian and New Zealand Journal of Statistics*, vol. 49, no. 3, pp. 267–286, 2007.

[12] V. Jacobson. Pathchar – A tool to infer characteristics of Internet path, available from ftp://ee.lbl.gov/pathchar/, 1997.

[13] M. Jain and C. Dovrolis. Pathload: A measurement tool for end-to-end available bandwidth, in *Proceedings of Passive and Active Measurements (PAM) Workshop*, pp. 14–25, 2002.

[14] M. Jain and C. Dovrolis. End-to-end available bandwidth: measurement methodology, dynamics, and relation with TCP throughput, *IEEE/ACM Transactions on Networking*, vol. 11, no. 4, pp. 537–549, August 2003.

[15] M. Jain and C. Dovrolis. End-to-end estimation of the available bandwidth variation range, in *Proceedings of the 2005 ACM SIGMETRICS International Conference on Measurement and Modeling of Computer Systems*, Banff, Canada, pp. 265–276, 2005.

[16] T. Karagiannis, M. Molle, M. Faloutsos and A. Broido, A nonstationary poisson view of internet traffic, in *Proceedings of IEEE INFOCOM 2004*, vol. 3, pp. 1558–1569, 2004.

[17] K. Lai and M. Baker. Nettimer: A tool for measuring bottleneck link bandwidth, in *Proceedings of the USENIX Symposium on Internet Technologies and Systems*, http://www.usenix.org/events/usits01/lai.html, 2001.

[18] E. Lawrence, G. Michailidis and V. N. Nair. Inference of network delay distributions using the EM algorithm, Technical Report, University of Michigan, 2003.

[19] F. Lo Presti, N. G. Duffield, J. Horowitz and D. Towsley, Multicast-based inference of network-internal delay distributions, *ACM/IEEE Transactions on Networking*, vol. 10, pp. 761–775, 2002.

[20] E. Masry, Nonparametric regression estimation for dependent functional data: Asymptotic normality, *Stochastic Process. Appl.*, vol. 115, no. 1, pp. 155–177, 2005.

[21] E. A. Nadaraya, On estimating regression, *Theory of Probability and Its Applications*, vol. 9, no. 1, pp. 141–142, 1964.

[22] P. Oliveira, Nonparametric density and regression estimation for functional data, Prepublication, Departamento de Matemática, Universidade de Coimbra, 2005.

[23] G. Perera, Random fields on $\mathbb{Z}^d$, limit theorems and irregular sets, in *Spatial Statistics: Methodological Aspects and Applications*, M. Moore (Ed.), Lecture Notes in Statistics, vol. 159 Springer, pp. 57–82, 2001.

[24] V. Ribeiro, R. Riedi, R. Baraniuk, J. Navratil and L. Cotrell, PathChirp: Efficient available bandwidth estimation for network paths, in *Proceedings Passive and Active Measurement (PAM) Workshop*, http://www-ece.rice.edu/~vinay/papers/pam03.pdf, 2003.

[25] J. Strauss, D. Katabi and F. Kaashoek, A measurement study of available bandwidth estimation tools, in *Internet Measurement Workshop, Proceedings of the 2003 ACM SIGCOMM Conference on Internet Measurement*, pp. 39–44, 2003.

[26] Y. Tsang, M. Coates and R. Nowak, Network delay tomography, *IEEE Transactions on Signal Processing*, vol. 51, pp. 2125–2136, 2003.

[27] G. S. Watson, Smooth regression analysis, *Sankhya. The Indian Journal of Statistics*, vol. 26, pp. 359–372, 1964.

[28] Y. Zhang, V. Paxson and S. Shenker, The stationarity of internet path properties: Routing, loss, and throughput, ACIRI Technical Report, 2000.

[29] Y. Zhang, N. Duffield, V. Paxson and S. Shenker, On the constancy of internet path properties, in *Proceedings of the ACM SIGCOMM Internet Measurement Workshop*, pp. 197–211, 2001.

19

Ad-Hoc Recursive PCE-Based Inter-Domain Path Computation (ARPC) Methods

Gilles Bertrand and Géraldine Texier

*Telecom Bretagne, RSM Department, 2 Rue de la Châtaigneraie – CS 17607,
35576 Cesson Sévigné Cedex, France;
e-mail: {gilles.bertrand, geraldine.texier}@telecom-bretagne.eu*

Abstract

With the emergence of multimedia applications with stringent requirements,
like IPTV, the need for end-to-end Quality of Service (QoS) is increasing.
In this paper, we investigate the problem of how to route high QoS flows
with end-to-end QoS guarantees in a Path Computation Element (PCE) based
architecture. In this architecture, three main types of path computation methods have been proposed: methods based on aggregated representations of the
network, methods using predetermined domain (AS) sequences or methods
using only the knowledge of the intra-domain topology as well as of the
inter-domain links connected to the domain. The last family of algorithms
is called *ad-hoc methods*. These methods have not been extensively studied
in the context of the PCE architecture yet.

This paper has two main contributions. First, we propose a new *ad-hoc*
PCE-based Recursive Inter-domain Path Computation method (ARPC) that
dynamically determines the AS sequence crossed. This algorithm integrates
complexity-reduction mechanisms. We consider an example application of
this algorithm for simultaneously minimizing the traffic forwarding cost,
guaranteeing a minimum bandwidth and optimizing an additive metric related
to load balancing. We assess the performance of this algorithm with regard to
these objectives. A simulation study demonstrates that the complexity of *ad-*

*D. D. Kouvatsos (ed.), Mobility Management and Quality-of-Service for Hetero-
geneous Networks,* 441–464.

hoc methods is reasonable, in certain topologies. Thus, these methods deserve further study. Second, we propose that PCEs in a domain are aware of the economical cost (price) of the inter-domain links connected to this domain. We demonstrate that this additional knowledge allows the implementation of interesting economical strategies. For that, we implement a detailed example.

Keywords: MPLS, PCE, inter-domain path computation.

19.1 Introduction

The emergence of multimedia applications with stringent requirements, like IPTV or on-line gaming, underlines the need for end-to-end Quality of Service (QoS) in networks. There are two main principles for providing QoS in a telecommunication network. The first involves allocating resources per flow or per class of service. The second principle involves optimizing the performance of the network by Traffic Engineering (TE).

Multi-Protocol Label Switching with Traffic Engineering (MPLS-TE) is an architecture that allows each MPLS domain to perform constrained source routing in the head-end Label Source Router (LSR) of a Label Switched Path (LSP). Path computations are based on one or several metrics associated with the links of the domain. These user-oriented metrics can be related to the level of QoS or to the protection of the path. Alternatively, network-oriented metrics can be related to traffic engineering objectives.

A domain is a set of network equipments administrated by the same entity and with homogeneous configuration and policy. More precisely, in the scope of this paper, a domain is an autonomous system (AS). ASes have a very limited collaboration due to scalability (information aggregation) and confidentiality constraints. Each AS advertises inter-domain routes toward a subset of its neighboring ASes according to its *export policy*: the route advertisements depend on inter-AS relationships. ASes can be bound by three main types of agreements: customer-provider, peering and provider-customer. Customer ASes have to pay when they forward traffic to a provider AS. Therefore, it is economically interesting to select appropriate egress nodes and inter-domain routes in order to minimize the cost induced by inter-domain traffic forwarding [1].

The Border Gateway Protocol (BGP-4) is the most used inter-domain protocol in the Internet. It provides a reduced path diversity due to information aggregation. Therefore, inter-domain routing is far from being optimal in the current Internet [2]. Moreover, Internet Service Providers (ISPs) only have a limited control on path selection. This makes load balancing and maintenance operation planning more complicated [3].

In practice, BGP only advertises reachability information between domains. This approach has been proved to be scalable but hinders the deployment of inter-domain TE mechanisms. Reference [4] provides an overview of open issues in inter-domain routing. Several extensions and modifications of BGP for solving previous issues and for providing end-to-end QoS have been suggested in recent papers (e.g. [5, 6, 7]). However, BGP replacement is not envisioned because of its worldwide deployment [4].

The interest for computing constrained paths crossing several domains is rising, because it would allow end-to-end flow protection, optimal path computations and inter-domain traffic engineering. A control plane enabling inter-domain path negotiation is standardized by the Internet Engineering Task Force (IETF) in the Path Computation Element (PCE) working group [8]. In the PCE architecture, LSRs can delegate path computations to a specialized network node called PCE. PCEs in different ASes are able to cooperate in order to compute constrained inter-domain LSPs. This way, constrained inter-domain paths can be computed even if neither topology nor traffic information is advertised out of the domains.

PCE standards do not impose a path computation algorithm. A Backward Recursive PCE-based Computation (BRPC) method is under standardization [9]. BRPC assumes that the sequence of ASes crossed by the path is predetermined. Due to this assumption, good paths crossing different AS sequences are ignored. In addition, because two paths crossing the same AS sequence have a larger probability to share a common risk, the simultaneous computation of a primary and of a backup path has a lower success probability [10]. Moreover, the paths computed are less likely to be globally optimal[1] due to the constraint added by a predetermined AS sequence [11].

This paper has two main contributions. First, we propose a new PCE-based *ad-hoc* Recursive Inter-domain Path Computation method (ARPC) that dynamically determines the AS sequence crossed. This algorithm integrates complexity-reduction mechanisms. We consider an example of application of ARPC for simultaneously minimizing the traffic forwarding cost, guaranteeing a minimum bandwidth and optimizing an additive metric related to load balancing. We assess the performance of ARPC with regard to these objectives. A simulation study demonstrates that the complexity of ARPC

---

[1] In this paper, the paths that would be computed by an omniscient element are referred to as optimal.

is reasonable, in certain topologies. This indicates that ad-hoc methods deserve further study. Second, we propose that the PCEs in a domain know the economical cost (price) of the inter-domain links connected to this domain. We demonstrate that this additional knowledge makes the implementation of interesting economical strategies possible. For example, we integrate price information in the algorithm proposed, in order to guarantee that each AS advertises its most profitable paths.

The rest of the paper is structured as follows. In Section 19.2, we introduce the architecture considered and related work. In Section 19.3, we describe the metrics considered and our routing algorithm. In Section 19.4, we present an evaluation of the proposed mechanisms based on a simulation model.

19.2 Inter-Domain Path Computation

19.2.1 Forwarding Policies

Unlike intra-domain routing, inter-domain routing in the Internet is mainly policy driven. There are two main types of inter-AS relationships in the Internet: customer-provider and peer-to-peer. These relationships define *export policies* describing which prefixes are advertised toward neighboring ASes. Each AS *A* advertises [10]:

- to its providers: its own IP prefixes and those learned from its customers, but not those learned from its peers or from other providers,
- to its customers: all the reachable IP prefixes it knows,
- to its peers: its own IP prefixes and those learned from its customers, but not those learned from its providers or other peers.

Thus, the relationship between two ASes *A* and *B* respects the following *forwarding policies* [11]:

- If *B* is a customer of *A*, then, *B* can forward packets from its provider *A* to its customers but not to its peers or other providers.
- If *B* is a provider of *A*, then, *B* can forward packets from its customer *A* to any of its neighboring domain (provider, customer or peer).
- If *B* is a peer of *A*, then, *B* can forward packets from its peer *A* to its customers but not to its providers or other peers.

Export policies and forwarding policies have to be taken into account when computing an inter-domain path. This makes the computation of constrained inter-domain paths a much more difficult problem than the computation of paths at intra-domain level.

19.2.2 The PCE Architecture

The MPLS architecture was mainly designed in order to provide improved packet forwarding performance compared to IP packet forwarding, thanks to label switching. It was later extended into MPLS-TE for traffic engineering. In the MPLS-TE architecture, given a source-destination couple in the MPLS domain, the head-end LSR (source) is able to compute a path toward the destination, by using a Constrained-Shortest Path First (C-SPF) algorithm. CSPF is a link-state routing algorithm that requires full knowledge of the topology of the domain, as well as of the resource availability. This information is retrieved by the head-end LSR thanks to traffic engineering (TE) enabled protocols like OSPF-TE or ISIS-TE. For confidentiality and scalability reasons, it is not possible to generalize this method for inter-domain traffic engineering: network operators do not want to advertise information on their topology outside of the domain and inter-domain protocols can only advertise aggregated representations of the network, due to the number of domains (ASes) in the Internet. As a result, each domain has a *partial visibility* of the network.

The Path Computation Element architecture provides a solution to the partial visibility problem. This architecture is standardized by the PCE working group of the IETF [8, 12]. The PCE architecture provides a method for computing inter-domain paths fulfilling a set of constraints in an MPLS network. In the PCE architecture, LSRs can act as Path Computation Clients (PCC) and send path computation requests to a PCE. The PCE performs path computation operations and replies back to the PCC. During path computation operations, a PCE can forward a request to other PCEs. Consequently, PCEs in several domains can be involved in the computation of inter-domain paths. Thus, it solves the partial visibility problem. A simple PCE network is depicted in Figure 19.1. In this example, there is a single PCE per AS and ASes are not divided into sub-domains (areas).

PCE standards describe a request-response protocol (the PCE protocol) [13]. This protocol is not appropriate for the periodic advertisement of aggregated representations of the network [10].

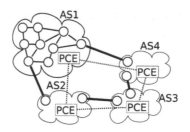

Figure 19.1 Architecture considered.

19.2.3 The BRPC Method

The Backward Recursive PCE-based Computation (BRPC) method is the only PCE-based path computation algorithm under standardization [9][2]. It assumes that the sequence of domains crossed by the path computed is pre-determined. The path computation request is sent by the PCC to a PCE in its domain and then, forwarded until it reaches a PCE in the destination domain. Then, the algorithm builds a Virtual Shortest Path Tree (VSPT). This structure is built recursively, from the destination. Each domain (AS(k)) concatenates its topology with the VSPT and computes a set of constrained shortest paths from its entry border routers to the path destination, in the resulting topology. The VSPT is updated in order to include the shortest paths computed and forwarded to a PCE in AS($k - 1$). Finally, PCE1 receives the VSPT and computes one or several constrained shortest paths satisfying the constraints. Note that the constrained shortest path computations are performed according to the same link metric in all domains.

A simple example is depicted in Figure 19.2. In this example, a computation request for a path from 1 to 9 is sent by a LSR denoted as 1 to a PCE (PCE1) in AS1. This request is forwarded by PCE1 toward a PCE (PCE3) in the destination domain AS3. PCE3 computes constrained shortest paths from the entry border routers (7 and 8) to the destination 9 according to the metric considered. It forwards the paths computed to a PCE (PCE2) in AS2, in a VSPT. AS2 concatenates the VSPT with its topology and computes constrained shortest paths from its entry border routers (3 and 4) to the destination 9, in the resulting topology. The shortest paths computed are composed of two segments: 3→7 + 7→9 and 4→8 + 8→9, for example. The VSPT is updated with the paths computed and forwarded to a PCE (PCE1) in AS1.

---

[2] Note that the per-domain method [14] does not require the presence of a PCE but could use a PCE to compute intra-domain routes.

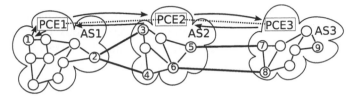

Figure 19.2 BRPC method: an example.

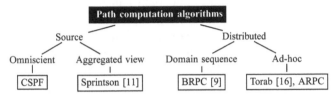

Figure 19.3 PCE-based inter-domain path computation algorithms.

Then, PCE1 computes the final paths from 1 to 9, which are made up of three segments, for example $1\rightarrow3 + 3\rightarrow7 +7\rightarrow9$ or $1\rightarrow4 + 4\rightarrow8 + 8\rightarrow9$. With this procedure, optimal[3] end-to-end paths across a predetermined sequence of ASes can be computed.

19.2.4 Related Work and Our Contributions

PCE standards do not impose a path computation algorithm. A few algorithms have been proposed in the literature (Figure 19.3). Among them, only one is currently under standardization: the BRPC method [9]. The BRPC method assumes that the sequence of ASes crossed by the path is predetermined. This assumption leads to several limitations listed in [10, 11]. The main concerns expressed by Sprintson et al. in [11] are related to the optimality of the paths computed, because of the additional constraint introduced by a predetermined AS sequence.

In [11], Sprintson et al. propose an alternative method based on advertising an aggregated representation of the domains. Their method solves the problems introduced by predetermined AS sequences. In summary, their algorithm provides optimal solutions for the multi-domain disjoint path problem both in the general setting, as well as subject to the export policy constraints. It is based on the Suurballe and Tarjan method [15]. The ag-

---

[3] Here, optimal means equivalent to the shortest paths fulfilling the constraints that would be computed in a flat topology without domains.

gregated network representation proposed is designed to enable optimal path computations and is based on line graph representations of the domains, in order to take forwarding policies into account. The size of the aggregated representation of a domain is $O(B^4)$ for a domain containing B border routers.

In [11], the authors mention that in the destination domain, the destination t of the request considered must be included into the aggregated representation, so that the end-to-end path can be computed by a PCE in the source domain. In our opinion, this may introduce two problems. First, in a stub domain, the number of potential destinations of an inter-domain request can be high, thus a large number of nodes may have to be included in the aggregated representation of this domain. Therefore, the size of the aggregated representation of each stub AS may be much larger than $O(B^4)$, which may introduce scalability concerns. Second, a domain has to be able to identify the potential destinations of inter-domain requests, when it computes its aggregated representation.

The complexity of the forwarding policies makes that each AS has its view of the network that can be different from the view from another AS. Thus, it is rather difficult to find a concise aggregated representation of the network. Consequently, in this paper, we study alternative, distributed methods called *ad-hoc methods*, which rely on local choices in each domain [16]. Each domain is aware of the relationships it has with its neighbors. Therefore, *ad-hoc* methods take forwarding policies into account more easily. *Ad-hoc* methods require only a local view of the network: they need to be aware of the intra-domain topology and of intra-domain traffic engineering information, as well as of the inter-domain links connected to the domain considered and of their remaining bandwidth. *Ad-hoc* methods have not been extensively studied in the context of the PCE architecture yet.

Unlike the methods relying on the diffusion of aggregated representations of the network, *ad-hoc* methods rely on simple response-request exchanges that could be implemented with a protocol such as the one described in [13] with only slight modifications. The main advantage of *ad-hoc* methods is that they rely on dynamic network exploration for finding paths satisfying the constraints. This allows fast auto-adaptation to network and topology changes. The main drawback of these methods seems to be the potential latency and overhead introduced by network exploration. Therefore, it is important to explore the paths efficiently, even if, according to the PCE working group's charter, the PCE architecture considers small set of domains. Thus,

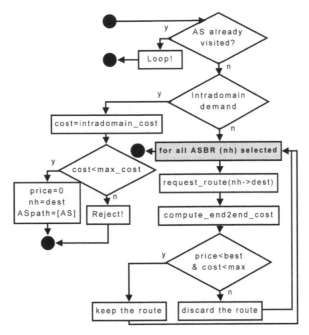

Figure 19.4 Ad-hoc path computation algorithm (simplified).

this work focuses on the evaluation of the complexity of *ad-hoc* algorithms that includes simple complexity-reduction mechanisms.

To the best of our knowledge, this paper is the first to study complexity-reduction mechanisms for *ad-hoc* algorithms in the PCE architecture. *Ad-hoc* algorithms have been proposed for peer-to-peer networks and *ad-hoc* networks, for example. The use of *ad-hoc* algorithms in the PCE architecture was first suggested in [16], where Torab et al. provide a short example of *ad-hoc* cooperation. They summarize the principle of *ad-hoc* algorithms adapted to the PCE architecture. This principle is illustrated in Figure 19.4 and is explained in Section 19.3.3. Unlike in this paper, Torab et al. [16] do not study any mechanisms for limiting the complexity of *ad-hoc* path computation algorithms.

The advantages and drawbacks of *ad-hoc* PCE-based path computation methods can be summarized as follows:

+ *Ad-hoc* methods rely on distributed path computations requiring only local visibility. Thus, *ad-hoc* methods do not require large amount of information to be advertised throughout the network.

+   *Ad-hoc* methods consider a larger path diversity than methods imposing a predetermined AS sequence.
−   But, the complexity of the network exploration performed by *ad-hoc* methods has to be controlled.

A summary of existing PCE-based path computation methods is presented in Figure 19.3. In this figure, PCE-based path computation methods are classified according to the location of the path computations as well as the information used for these computations. The Constrained Shortest Path First (CSPF) method refers to the computation of constrained shortest paths by an omniscient PCE in the source domain, this method is inapplicable in the networks considered, due to forwarding policies and scalability concerns. The algorithm presented in [11] is classified as source routing, because the path computation is performed by a PCE in the source domain (but the aggregated representation computations are distributed throughout the network).

Our contributions can be summarized as follows:

*   We propose two simple complexity-reduction mechanisms for *ad-hoc* PCE-based path computation algorithms and demonstrate their efficiency.
*   We propose that PCEs are aware of the price of the inter-domain links connected to their domain. We show that this knowledge can be used in order to minimize the traffic forwarding cost paid by the domains and to increase their profits.
*   We describe a new *ad-hoc* recursive PCE-based path computation (ARPC) algorithm respecting forwarding policies. This algorithm minimizes, first, the traffic forwarding price paid by each domain crossed, second, an additive metric m associated with the path. In addition, it guarantees a minimum bandwidth and a maximum value M of the additive metric m considered.

19.3 Metrics and Path Computation Algorithm

19.3.1 Assumptions

We consider that the PCE-based routing algorithm is used only for a part of the traffic, namely the traffic with high QoS requirements or traffic with stringent protection requirements (e.g. traffic related to mission critical services). For example, in a DiffServ able network, the PCE-based mechanism could be used for Expedited Forwarding (EF) traffic. The PCE has to find a path satisfying some constraints (e.g. QoS, protection, price). In the example considered in this paper, we assume that the purposes of the path computation are to minimize traffic forwarding costs, to guarantee a minimum bandwidth and to balance network load. Stated differently, the algorithm optimizes, first, a price metric and, second, an additive link metric (cost), while respecting a bottleneck metric constraint (bandwidth).

A link that connects two nodes in the same domain is referred to as an *intra-domain* link, while a link that connects different domains is referred to as an *inter-domain* link. We propose that PCEs are not only aware of the cost of intra-domain links in their domain, but also of the price of the inter-domain links connected to their domain. The number $N_{er,A}$ of inter-domain links connected to a domain A is usually much lower than the number $N_{ra,A}$ of intra-domain links. Thus, storing $N_{er,A}$ values of a price metric should not introduce any scalability problem. PCEs use price information in order to minimize the inter-domain traffic forwarding cost. In a BGP based routing configuration, each domain would also minimize this cost by allocating the highest local preference to the least expensive inter-domain paths. Thus, the learning of inter-domain link prices by PCEs is quite similar to the configuration of local preferences in BGP routers. However, in this work, we do not study how PCEs learn the prices.

Globally, we assume that each PCE knows the following characteristics of the ASes it manages:

- the intra-domain topology (intra-domain links, their remaining bandwidth, and the value of the link metric considered), as is assumed for the BRPC method,
- the inter-AS links connected to the AS considered, their type (e.g. customer to provider, peering), as well as associated price information.

Part of this information is static (e.g. peering agreements) and part is dynamic (e.g. internal topology). The topology can be learned by means of ISIS or OSPF, if the AS is made of a single Interior Gateway Protocol (IGP) area. In all other cases, the PCE must be able to request intra-domain path computations to one or several intra-domain PCEs. For the sake of simplicity, we consider in this paper that each AS is made of a single area. The case of an Inter-AS TE LSP spanning multiple ASes where some of those ASes are themselves made of multiple IGP areas can be easily derived from this case by applying the path computation procedure described in this paper, recursively.

19.3.2 Metrics Considered

In this paper, two metrics are considered for the recursive computation of inter-domain paths: the value of a *cost metric* on all links and the *traffic forwarding price* on the first inter-domain link of the path. The *cost metric* $m_\mathfrak{p}$ characterizes the quality of a path $\mathfrak{p}$ regarding user-oriented or operator-oriented objectives. The *price* $p_\mathfrak{p}$ is related to the economical interest of the path $\mathfrak{p}$. As a result, the optimization of user-oriented and operator-oriented metrics is considered.

The price of intra-domain paths is considered to be zero. The end-to-end price $p_\mathfrak{p}$ of an inter-domain path $\mathfrak{p}$ is the price $p_{(e,nh)}$ incurred by the first inter-domain link: (e, nh). This link connects the egress node e of the first AS crossed to the next-hop nh, where we call *next-hop* the first AS border router (ASBR) crossed in the next AS. We consider a demand d routed on a path whose first inter-domain link is (e, nh). The price paid by an AS for serving d is proportional to the bandwidth b_d requested by this demand:

$$p^d_{(e,nh)} = p_{(e,nh)} \cdot b_d \qquad (19.1)$$

Each domain is expected to carry the traffic on the paths that are the most profitable. In other words, if an AS A knows several paths that are suitable for a demand, then, A *advertises its most profitable paths*.

We consider additive positively-valued link costs $m_{(i,j)}$ in order to penalize long paths. As a result, the end-to-end cost of an inter-domain path is the sum of the costs for each section of the path: ingress(s)-egress(e), egress-next_hop(nh) and next_hop-destination(t). Note that all domains have to use the same cost metric for computing intra-domain paths.

We seek to balance the load over network links. Thus, the least loaded links get the smallest weights and are most likely to be used. We define the weight $m_{(i,j)}$ of each link (i, j) as 10^4 divided by its remaining capacity $c_{r,(i,j)}$ in Megabit per second:

$$m_{(i,j)} = \frac{10^4}{c_{r,(i,j)}} \qquad (19.2)$$

19.3.3 An Ad-Hoc Recursive PCE-Based Path Computation (ARPC) Method

19.3.3.1 Intra-Domain Routing Algorithm

A standard on-line algorithm can be used for the routing of any intra-domain demand d with source s, termination t and bandwidth b_d. This algorithm computes segments of inter-domain paths. If the source and the termination of the demands are the same single node, then, the cost of the path is arbitrarily set to zero. In all other cases, a variant of Dijkstra's algorithm is run on a graph of the network where all links with remaining capacity $c_{r,(i,j)} < b_d$ are pruned [17]. Alternative methods can be used depending on the constraints considered. For instance, reference [18] reviews methods for finding a path subject to many additive constraints.

19.3.3.2 Inter-domain Routing Algorithm

The inter-domain routing algorithm running in the PCEs computes constrained inter-domain paths recursively. In the example considered in this paper, the aim of this algorithm is to minimize the price paid for a demand, while fulfilling the two following QoS constraints. First, the bandwidth allocated on each link must be equal to the demanded bandwidth. Second, a given threshold related to a specified additive cost metric must not be exceeded by the end-to-end path and the value of that cost metric should be minimized. The cost metric considered in this paper is provided in equation (19.2). The overall algorithm, is depicted (with a few simplifications) in Figure 19.4. The termination of the algorithm can be guaranteed by defining the maximum number of times a single request can be forwarded. As a path is selected only if its price is better than the one of the previous best path, the algorithm converges toward the paths with the best prices.

The head-end LSR s of a path sends a demand d to a PCE (PCE α) in its AS A. The request contains the following information:

- the address of the source s in A of the demand d,
- the address of the destination t of the demand d,
- the bandwidth b_d requested,
- the maximum cost M_d allowed,
- the list of the ASes already visited (this list is empty for the first request), and
- the type of the metric m to be used for the path computations.

When a PCE (PCE α) in AS A receives a request, it performs the following operations, summarized in Figure 19.4:

- It checks if its AS number appears in the list of the AS already visited. If yes, it returns a loop advertisement (cost $= \infty$) and terminates. If no, it performs the following operations.
- It checks if the destination of the demand belongs to its AS (AS A).

– If yes (the demand is intra-domain), the PCE computes a constrained intra-domain path or retrieves IGP information.

* If it finds an intra-domain path with a cost which is lower or equal to the maximum cost allowed, then, it returns it this path the requesting PCC (s). The price of this path is zero, its next-hop address is the destination address t and the AS-path contains only the AS number of A.
* If such a path is not found, the algorithm returns an infinite cost and the PCC is advised that no path could be found.

– If no (the demand is inter-domain), the request is forwarded to a subset of the domains with which A has an agreement, depending on the forwarding policies of A.

The request forwarded by a PCE α in an AS A to a PCE β in an AS B is slightly modified, so that the recursive procedure can continue:

• the address of the source is set to the address of the next-hop in B,
• the maximum cost allowed is decreased by the cost of the inter-domain link used and
• A is included in the list of the AS already visited.

However, the address of the destination t, the bandwidth b_d requested as well as the type of metric m used for the path computations remain the same.

When a PCE finds a path, it returns:

• the path,
• the cost characterizing its QoS, and
• the address of the next-hop to be used.

When PCE α receives an answer from PCE β, it computes the end-to-end cost corresponding to the path received. The computation of this end-to-end quantity requires that information related to three segments of the path is combined: ingress – egress, egress – next-hop and next-hop – destination. If the path $\mathfrak{p}$ returned by PCE β has a finite price $p_\mathfrak{p}$ (no AS loop), then, PCE α verifies if the path $\mathfrak{p}$ is strictly cheaper than the cheapest path $\wp^*$ known yet ($p_\mathfrak{p} < p_{\wp^*}$). If no, the path $\wp$ is discarded. If yes, the new path replaces the best path. If both have the same price, smaller than the maximum price allowed, then, the path with the lowest cost is kept. Thus, the traffic load is balanced between inter-domain links.

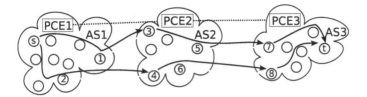

Figure 19.5 Computation of two disjoint paths: an example.

19.3.3.3 Computation of Disjoint Paths

The simultaneous computation of disjoint paths (primary and backup paths) is not studied in this paper. However, the algorithm presented could be extended for disjoint paths computations. For example, during each recursion of the algorithm, each inter-domain request could be forwarded through two different next-hops, simultaneously. A simple example is depicted in Figure 19.5: s sends a request to PCE1 for two link-disjoint paths. PCE1 selects two next-hops in AS2 (nodes 3 and 4), and computes two link-disjoint paths $s \rightarrow 3$ and $s \rightarrow 4$. Then, PCE1 forwards the request through nodes 3 and 4 to a PCE of the next domain (AS2). The PCE in AS2 (PCE2) selects two next-hops (7 and 8) and computes two link-disjoint intra-domain paths from the entry ASBRs 3 and 4 to the next-hops (e.g. $3 \rightarrow 7$ and $4 \rightarrow 8$). Then, PCE2 forwards the request through nodes 7 and 8 to PCE3. PCE3 computes the final segment of the paths: from node 7 to t and from node 8 to t and sends a response including the two segments to PCE2. The concatenation of the segments computed in each AS provides two link-disjoint paths: $s, 3, 7, t$ and $s, 4, 8, t$.

19.3.3.4 Complexity

Each path computation request is likely to trigger the computation of a constrained intra-domain path. Thus, complexity-reduction mechanisms are required in order to avoid flooding the network with path computation requests and overloading the PCEs. Several simple strategies can be applied for decreasing the complexity of our algorithm. The termination of the algorithm can be guaranteed by defining the maximum number of times a single request can be forwarded. The additional methods implemented in the model for reducing the complexity of the network exploration are presented in Section 19.4.1 and listed below:

- A maximum-cost constraint is used. If the maximum cost associated with a potential request is negative, then, the related request is not sent, which decreases the complexity and the overhead.
- In addition, a "take first good" mode is used, which means that the first suitable path found is returned back to the requester. In this mode, the neighboring PCEs have to be visited in a variable order so that the load can be shared between the inter-domain links more fairly.

We have implemented a greedy approach where requests are forwarded on the inter-domain links with lowest price first, in a variable order. This method offers good chances to find a suitable path at an affordable price, quickly. However, it does not work in some cases: for example, if the peers (price=0) have no suitable path to advertise, then, they will be explored needlessly. We have implemented a model, presented in the next section, in order to assess the viability of our path computation algorithm. This model allows us to evaluate the efficiency of the complexity-reduction mechanisms and their impact on the quality of the paths computed.

19.4 Performance Evaluation

19.4.1 Simulation Model

We have modeled our path computation algorithm in Matlab. The model takes a topology and a traffic load as input. A topology is described as:

- a link capacity matrix, and
- a matrix providing the price per unit of bandwidth associated with each inter-domain link.

The main topologies considered in this article are fictitious. The main topology considered is depicted in Figure 19.6. Forwarding policies are taken into account: links are oriented according to the point of view of the AS considered (AS1). We consider demands having their source in AS1 and their destination in any AS. We associate a price with each directed inter-domain link. This models the fact that an AS A pays another AS B for forwarding its traffic ($A \rightarrow B$) and does not accept to forward the traffic from B ($B \nrightarrow A$). Two types of directed links are modeled, one is described as expensive (15\$ per unit of bandwidth) and the other as normal (10\$ per unit of bandwidth), where \$ refers to an arbitrary unit of price. We consider that bidirectional links are free of charge (price = 0).

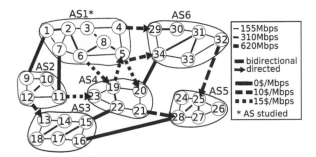

Figure 19.6 Topology considered in the simulations (view from AS1).

In the model, demands are generated between random pairs of nodes (s, t) with s in AS1 and t in any AS. The bandwidth requested for the demands follows a lognormal distribution in order to have a large variability of the demands, like in real networks. The demands are routed using the *ad-hoc* recursive PCE-based algorithm presented in Section 19.4. The complexity of the network exploration is a potential limitation of *ad-hoc* PCE-based path computation algorithms. Therefore, we consider the topology in Figure 19.6, which represents a pessimistic case for the algorithm because AS degrees are relatively high, which makes network exploration more complicated. As a result, the complexity of the path computation procedure (e.g. the number of intra-domain computations required) is expected to be higher than usual. For the same reason, the success rate of the path computation procedure is expected to be slightly greater than usual, thanks to path diversity. Policy constraints are implemented in the model but not considered in the simulations of Section 19.4.2 so as to increase the path diversity.

Outputs of on-line routing models usually depend on the initial state of the network. In our model, the initial state of the network is computed from a randomly generated traffic matrix T with a known intra-domain and inter-domain load expectation. This matrix represents a traffic aggregate corresponding to the average amount of traffic exchanged between a source and a destination during a certain period of time.

The impact of inter-domain traffic on intra-domain links is modeled by adding a fraction of the total inter-domain traffic generated by the other ASes to the demands originating or terminating in the AS border routers. This increase reflects the fact that the inter-domain paths usually cross several ASes and thus require bandwidth on links belonging to several ASes. The inter-domain traffic load is shared fairly between the inter-domain links because the simulations represent a deviation of the network from an ideal state, obtained by offline traffic engineering, for example.

Intra-domain routing is performed with an offline *max-flow* routing algorithm. The objectives of this linear program are: to route the demands, to balance the load and to maximize the admitted traffic. It is solved by the COmputational INfrastructure for Operations Research (COIN-OR) Linear Programming solver [19].

In the simulations, the maximum-cost parameter is set to 400. Several replications of the simulations are realized in order to evaluate the performance metrics considered with satisfying confidence intervals. An example of path computed in the model is provided below.

```
Run 19/625000 -- demand 1/1:   src=6, dest=30,
 bw=250kbps, max_cost=400
Route: price=0.4, cost=345.9, nh=11, AS_path=1 2 4 6,
 node_path=6 7 11 23 19 20 34 31 30
```

Note that, in the algorithm implemented, the node sequence is not forwarded to the PCCs, due to confidentiality and scalability reasons. However, our implementation of the model memorizes this sequence, in order to compute statistics and to check the validity of the results.

19.4.2 Numerical Results

The aim of our PCE-based inter-domain routing algorithm is, first, to optimize the price paid for inter-domain transit by the network operators considered and then, to optimize the cost of the path computed, while respecting two QoS constraints: a minimum bandwidth and a maximum value for the cost metric. We evaluate the performance of ARPC regarding these objectives, with the metrics described in Section 19.3.2 and using the model described in Section 19.4.1.

BRPC uses predetermined AS sequences, but the mean by which the AS sequences are determined is not standardized. Thus, it is hard to compare the quality of the paths computed by either our method or BRPC. Consequently, we compare the paths computed by our algorithm to the optimal paths in the network. The optimal paths are computed by an algorithm without complexity-reduction mechanism, that explores the whole network. The results of the simulations are presented in Figures 19.7 and 19.8. In these figures, the x axis represents the average intra-domain link load in the whole network. The load of a link is defined as its total used capacity divided by its

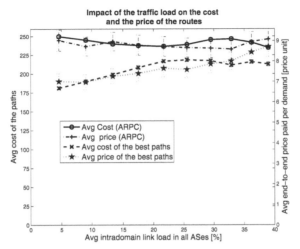

Figure 19.7 Comparison of the paths computed with ARPC to the best (cheapest) paths.

nominal capacity. The average load of intra-domain and of inter-domain links is approximately equal for $x \leq 40\%$.

Figure 19.7 shows the average cost (left y axis) and the average price (right y axis) of the paths computed. This curve demonstrates that our algorithm provides paths with a price and a cost close to the optimal values: in average about 15% higher. The cost is up to 40% higher than the optimum if the traffic load is very low: in this situation the value of the cost metric is not a problem because there is no congestion. The path diversity decreases when the load increases. Thus, the distance to the cost and to the price of the optimal path decreases as the traffic load increases. The deviation from the optimal cost allows us to divide the complexity of the network exploration by a factor of more than 2.5 in average. Moreover, this deviation can be adjusted by configuring the complexity-reduction mechanisms implemented. In average, the cost of the paths returned is approximately two times smaller than the maximum-cost constraint ($M = 400$). When the average intra-domain link load is larger than 35%, the maximum-cost constraint is reached on most paths. Thus, suitable paths become difficult to find (the success rate decreases). As a result, the cost and the price of both the paths computed by ARPC and the optimal paths begins to decrease. The average price of the paths computed is relatively stable for $x \leq 40\%$. However the price of the optimal paths rises with the traffic load, because the best links become congested and alternative paths must be used. Additional simulations on the

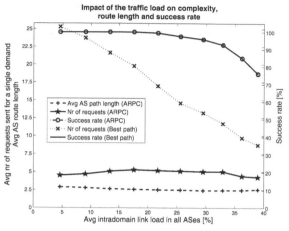

Figure 19.8 Impact of traffic load on path computations with ARPC.

same topology with lower AS connectivity produce similar results, except that the deviation from the optimal, as well as the number of requests required for the computation of a single path are smaller.

We have assessed the performance of simple strategies for reducing the number of calls to distant PCEs and, thus, the number of intra-domain constrained routing computations. Figure 19.8 presents the complexity of the path computation procedure as the average number of inter-PCE requests sent per demand (left y axis). In this Figure, we compare the number of requests sent for the computation of the optimal path (dotted curve) and for the computation of a "good" path with ARPC (stars). We notice that the complexity of ARPC is quite stable when the traffic load increases. The complexity of ARPC is much lower than the complexity of the optimal path computations (in average 2.8 times lower).

The success rates (Figure 19.8, right y axis) of the optimal path and the "good" path computation algorithms are the same in the simulations: the two curves are superimposed. This means that the network exploration heuristic is efficient: when paths satisfying the request exist, ARPC finds one. Moreover, ARPC is able to find a suitable path in most cases (in more than 95% of the cases when the average intra-domain link load is below 30%). Simulations on topologies with lower AS degrees, all other parameters being identical, confirm that the success rate is high (e.g. > 95% for $x < 17\%$). When the average intra-domain link load is larger than 35% the success rate decreases strongly, both for ARPC and the optimal path computations, and reaches about 70% when the load is 40%. This means that the maximum-cost limit is reached on almost all paths.

Figure 19.8 depicts the length of the paths computed, in term of the number of ASes crossed (dashed curve, left y axis). Note that the length of the AS-paths is quite stable. The effect of the maximum-cost constraint is more and more important when the traffic load increases. The paths crossing the lowest number of links are less likely to reach the maximum value of the additive cost metric considered. Thus, the length (number of links crossed) of the paths computed decreases when the load increases. The length of the paths computed decreases faster when the load is higher than 35%, because the maximum-cost constraint is reached on more and more paths. In addition, in the topology considered (Figure 19.6), the average number of requests required for finding a suitable path is small: only about 2 requests per AS-hop. In a topology with lower AS connectivity this number decreases to only 1.5 request per AS-hop, in average.

19.4.3 Discussion

Simulation studies reveal that *ad-hoc* algorithms have promising performance on the topologies studied. They demonstrate that the complexity of the exploration phase can be easily and significantly decreased without deviating too far from the optimal. This suggests that *ad-hoc* PCE-based path computation methods deserve further study and may represent viable alternatives to the BRPC method, in certain topologies. The tradeoff between quality of the paths computed and complexity of the path computation procedure can be adjusted easily, by configuring the complexity-reduction mechanisms (e.g. the maximum cost).

Simulation results depend on the topology simulated. Simulations on larger topologies are required in order to determine in which topologies *ad-hoc* methods are the most useful. This work is a first study for evaluating the potential of PCE-based *ad-hoc* methods. Thus, we have considered fictitious topologies. As the results are promising, we plan to perform more detailed simulations on real topologies. Note although, that the PCE working group's charter specifies that the PCE architecture is adapted for small sets of domains. Temporal aspects of *ad-hoc* PCE-based algorithms are not represented in our model. They require further study in order to evaluate the delay between path-computation requests and answers.

The proposition of adding price information related to every inter-domain link connected to a domain reveals interesting. We demonstrate that this information can easily be integrated into ARPC. The knowledge of price information allows our algorithm to minimize the inter-domain transit cost, which is an important feature for network operators. With ARPC the paths advertised are the most profitable for all the domains crossed. The integration of price information also paves the way for the implementation of more complicated mechanisms such as dynamic transit price negotiation.

19.5 Conclusion

In this paper, we address the problem of constrained inter-domain path computation in the PCE architecture. We consider two metrics: we call them price and cost. The price is defined for each link connecting two different ASes (inter-domain links). The cost is defined for each link as a function of the remaining bandwidth. We propose that each PCE is aware of the price of the inter-domain links connected to its domain. This proposition is new and its implications require further study. We show how price information can be integrated into *ad-hoc* path computation methods for allowing the domains to advertise the most profitable paths.

We have studied *ad-hoc* path computation methods, which rely on network exploration for finding the best paths for a request. Unlike the BRPC method, which is the only PCE-based method that is currently under standardization, these algorithms do not suppose that the sequence of ASes crossed is predetermined. We have described an *ad-hoc* PCE-based path computation method (ARPC) including complexity-reduction mechanisms. This work is, to the best of our knowledge, the first to study *ad-hoc* path computation algorithms with complexity-reduction algorithms for the PCE architecture.

The main limitation of *ad-hoc* PCE-based methods seems to be the complexity of the exploration of the network for finding suitable paths. Thus, this work is focused on evaluating simple complexity-reduction mechanisms for *ad-hoc* PCE-based algorithms. We consider an algorithm (ARPC) that minimizes, first, the traffic forwarding price paid and, second, the value of an additive metric. In addition, ARPC guarantees a minimum bandwidth, as well as a maximum value of the metric considered. We demonstrate by simulation that the complexity of this algorithm remains reasonable thanks to simple complexity-reduction methods. More precisely, simulation studies show that the paths computed have a price and a cost close to optimal, while the number of requests and, thus, the number of intra-domain constrained routing computations, remain reasonable, in the topologies considered. Therefore, *ad-hoc* PCE-based path computation methods seem to deserve further study.

As perspectives for the model, we intend to implement and evaluate additional complexity-reduction mechanisms. For example, sending the requests to the PCEs that are the most likely to provide a suitable path. For this, we may exploit information available through BGP, as well as information on former requests, in order to grade the distant PCEs according to their ability to provide the best paths. We will also model and evaluate several temporal schemes for sending PCE requests, in order to test the delay introduced by the path computation procedure. We are also working on an *ad-hoc* PCE-based diverse path computation algorithm for the simultaneous computation of primary and backup paths.

References

[1] T. Bressoud, R. Rastogi and M. Smith, Optimal configuration for BGP route selection, in *IEEE INFOCOM*, San Francisco, CA, 1–3 April, vol. 2, pp. 916–926, 2003.

[2] M. P. Howarth, M. Boucadair, P. Flegkas, N. Wang, G. Pavlou, P. Morand, T. Coadic, D. Griffin, A. Asgari and P. Georgatsos, End-to-end quality of service provisioning through inter-provider traffic engineering, *Computer Communications*, vol. 29, no. 6, pp. 683–702, March 2006.

[3] R. Teixeira and J. Rexford, Managing routing disruptions in internet service provider networks, *IEEE Communications Magazine*, vol. 44, pp. 160–165, 2006.

[4] M. Yannuzzi, X. Masip-Bruin and O. Bonaventure, Open issues in interdomain routing: A survey, *IEEE Network*, vol. 19, pp. 49–56, 2005.

[5] D. Griffin, J. Spencer, J. Griem, M. Boucadair, P. Morand, M. Howarth, N. Wang, G. Pavlou, A. Asgari and P. Georgatsos, Interdomain routing through QoS-class planes, *IEEE Communications Magazine*, vol. 45, pp. 88–95, 2007.

[6] W. Xu and J. Rexford, MIRO: Multi-path Interdomain ROuting, in *ACM SIGCOMM*, Pisa, Italy, 11–15 September, pp. 171–182, 2006.

[7] C. Pelsser and O. Bonaventure, Path selection techniques to establish constrained interdomain MPLS LSPs, in *Proceedings of Networking 2006*, Coimbra, Portugal, 15–19 May, pp. 209–220, 2006.

[8] A. Farrel, J.-P. Vasseur and J. Ash, A Path Computation Element (PCE)-based architecture, RFC 4655, August 2006. [Online]. Available: www.ietf.org/rfc/rfc4655.txt

[9] J. P. Vasseur, R. Zhang, N. Bitar and J. L. Le Roux, A Backward Recursive PCE-based Computation (BRPC) procedure to compute shortest constrained inter-domain traffic engineering label switched paths, draft-ietf-pce-brpc-09.txt, work in progress, 2008.

[10] M. Yannuzzi, X. Masip-Bruin, S. Sanchez, J. Domingo-Pascual, A. Orda and A. Sprintson, On the challenges of establishing disjoint QoS IP/MPLS paths across multiple domains, *IEEE Communications Magazine*, vol. 44, pp. 60–66, December 2006.

[11] A. Sprintson, M. Yannuzzi, A. Orda and X. Masip-Bruin, Reliable routing with QoS guarantees for multi-domain IP/MPLS networks, in *INFOCOM*, Anchorage, Alaska, 6–12 May, pp. 1820–1828, 2007.

[12] J. L. Le Roux, Path Computation Element Communication Protocol (PCECP) specific requirements for inter-area MPLS and GMPLS traffic engineering, RFC 4927, June 2007.

[13] J. P. Vasseur and J. L. Roux, Path Computation Element (PCE) communication Protocol (PCEP), draft-ietf-pce-pcep-19.txt, November 2008.

[14] J. Vasseur, A. Ayyangar and R. Zhang, A per-domain path computation method for establishing inter-domain Traffic Engineering (TE) Label Switched Paths (LSPs), RFC 5152, 2008.

[15] J. W. Suurballe and R. E. Tarjan, A quick method for finding shortest pairs of disjoint paths, *Networks*, vol. 14, no. 2, pp. 325–336, 1984.

[16] P. Torab, B. Jabbari, Q. Xu, S. Gong, X. Yang, T. Lehman, C. Tracy and J. Sobieski, On cooperative inter-domain path computation, in *ISCC '06: Proceedings of the 11th IEEE Symposium on Computers and Communications*, Pula-Cagliari, Sardinia, Italy, 26–29 June, pp. 511–518, 2006.

[17] S. Chen and K. Nahrstedt, An overview of quality-of-service routing for the next generation high-speed networks: Problems and solutions, *IEEE Network Magazine, Special Issue on Transmission and Distribution of Digital Video*, vol. 12, pp. 64–79, 1998.

[18] G. Xue, A. Sen, W. Zhang, J. Tang and K. Thulasiraman, Finding a path subject to many additive QoS constraints, *IEEE/ACM Transactions on Networking*, vol. 15, no. 1, pp. 201–211, 2007.

[19] COmputational INfrastructure for Operations Research Linear Programming solver. [Online]. Available: http://www.coin-or.org/Clp

Author Index

Subject Index